2nd International Conference on Aerospace Technology, Communications and Energy Systems (ATCES 2018)

AA006565

IOP Conference Series: Materials Science and Engineering Volume 449

Shanghai, China
15 – 17 September 2018

ISBN: 978-1-5108-7662-0
ISSN: 1757-8981

Printed from e-media with permission by:

Curran Associates, Inc.
57 Morehouse Lane
Red Hook, NY 12571

CURRAN ASSOCIATES INC.
proceedings
.com

Printed by Curran Associates, Inc. (2019)

For permission requests, please contact the Institute of Physics at the address below.

Institute of Physics
Dirac House, Temple Back
Bristol BS1 6BE UK

Phone: 44 1 17 929 7481
Fax: 44 1 17 920 0979

techtracking@iop.org

Additional copies of this publication are available from:

Curran Associates, Inc.
57 Morehouse Lane
Red Hook, NY 12571 USA
Phone: 845-758-0400
Fax: 845-758-2633
Email: curran@proceedings.com
Web: www.proceedings.com

TABLE OF CONTENTS

IOP Conf. Series: Materials Science and Engineering **449** (2018) 011001 doi:10.1088/1757-899X/449/1/011001

PREFACE

ATCES 2018 was held in Shanghai, China, September 15-17, 2018. ATCES 2018 was organized by Asia Pacific Institute of Science and Engineering (APISE). The conference provides a useful and wide platform both for display the latest research and for exchange of research results and thoughts in broad Aerospace Technology, Communications and Energy Systems. The participants of the conference were from almost every part of the world, with background of either academia or industry, even well-known enterprise. The success and prosperity of the conference is reflected high level of the papers received.

The proceedings are a compilation of the accepted papers and represent an interesting outcome of the conference. This book covers 2 chapters: 1. Aerospace Technology; 2. Communications and Energy Systems.

We would like to acknowledge all of those who supported ATCES 2018. Each individual and institutional help were very important for the success of this conference. Especially we would like to thank the organizing committee for their valuable advices in the organization and helpful peer review of the papers.

We sincerely hope that ATCES 2018 was a forum for excellent discussions that will put forward new ideas and promote collaborative researches. We are sure that the proceedings will serve as an important research source of references and the knowledge, which will lead to not only scientific and engineering progress but also other new products and processes.

Prof. Yury A. Nozhnitskiy, Central Institute of Aviation Motors (Moscow), Russia

IOP Conf. Series: Materials Science and Engineering **449** (2018) 011002 doi:10.1088/1757-899X/449/1/011002

Committees

Honorable Chair

Prof. Nabil Ayoub, American University of Madaba AUM in Jordan, Jordan

Prof. Wagdi G. Habashi, McGill University, Canada

Prof. Alexander M. Mebel, Florida International University, USA

Conference Committee Co-Chair

Prof. Evgeniy V. Shakhmatov, Samara University, Russia

Prof. Yury A. Nozhnitskiy, Central Institute of Aviation Motors (Moscow), Russia

Prof. Andrey B. Prokofiev, Samara University, Russia

Program Committee

Prof. Mikhail Kolotnikov, Research Institute of Mechanics , Moscow State University, Russia

Prof. S. A. Channiwala, S. V. National Institute of Technology,Surat, India

Dr. Yaser Qudaih, American University of Madaba (AUM) - Jordan

Financial Chair

Dr. Jinping Lei, The Hong Kong University of Science and Technology, Hong Kong

General Conference Secretary

Assoc. Prof. Dmitry A. Uglanov, Samara University, Russia

Technical Committee Member

Prof. Vladimir A. Zrelov, Samara University, Russia

Prof. OKUYAMA Keiichi, Kyushu Institute of Technology, Japan

Prof. Yehia Abdel-Aziz, Head of Space Research Lab, NRIAG, Egypt.

Prof. Aleksandr A. Igolkin, Samara University, Russia

Prof. Dmitry K. Novikov, Samara University, Russia

Prof. Tieying Wu, Nanjing University of Aeronautics and Astronautics, China

Prof. Adel A. Elbaset, Minia University, EGYPT

Assoc. Prof. Viktor Ya. Sverbilov, Samara University, Russia

Assoc. Prof. Aleksandr S.Vinogradov, Samara University, Russia

Dr. Mohammed Mahdi Ali, Philadelphia University, Jordan

Dr. Tarek Hassan Mohamed, Aswan University, Egypt

Dr. Thongchart Kerdphol, Kyushu Institute of Technology, Japan

IOP Conf. Series: Materials Science and Engineering **449** (2018) 011002 doi:10.1088/1757-899X/449/1/011002

IOP Conf. Series: Materials Science and Engineering **449** (2018) 011003 doi:10.1088/1757-899X/449/1/011003

Peer review statement

All papers published in this volume of *IOP Conference Series: Materials Science and Engineering* have been peer reviewed through processes administered by the proceedings Editors. Reviews were conducted by expert referees to the professional and scientific standards expected of a proceedings journal published by IOP Publishing.

IOP Conf. Series: Materials Science and Engineering **449** (2018) 012001 doi:10.1088/1757-899X/449/1/012001

An Un-differenced Carrier Phase Time Transfer Method Between Stations

WANG Xiang-lei*, SHI Feng and DU Yan

No.22, Beiqing road, haidian district, Beijing, China

*E-mail:chxywxl@163.com

Abstract. GNSS carrier phase time transfer use PPP technology generally, has become the satellite two-way time comparison (TWTFT) widely used high precision time comparison method, due to its simple equipment, high precision, the global time become the important means of backup TWTFT of lab.An un-differenced carrier phase time transfer method between stations is proposed. Using the relations between un-differenced observation equations which established through common view satellites, the results of the satellite clock offset and inter-station time transfer can be calculated jointly. It is not limited by the accuracy and real-time performance of the satellite clock. As long as there is enough total station between apparent between satellite can gain standing time transfer as result, the precision clock difference from satellite product precision and real limit, time pass accuracy can reach the nanosecond level, good real-time, more suitable for real-time and high precision time transfer. Experimental results show that the time transfer accuracy of the proposed algorithm can reach sub-nanosecond magnitude, and it can be applied to high precision real-time time transfer.

1. Introduction

GNSS carrier phase time transfer use PPP technology generally, has become the satellite two-way time comparison (TWTFT) widely used high precision time comparison method, due to its simple equipment, high precision, the global time has become the important means of backup TWTFT of lab[1,2]. The PPP is single station operation, unable to effectively eliminate the time passed relevant error between standing over short distances, and the PPP need to use the IGS provides poor precision orbit and clock products, the lack of real time, its application mainly stay in post-processing model[3]. Satellite orbit changes due to gently can forecast, the IGS provides super quick forecast precision of orbit is 5 cm, basic can meet the demand of real-time Dagoberto (2010)[4], however, because of the satellite clock change is complicated and difficult to accurately forecast, make the low precision ultra fast clock difference product is only about 3 ns[5,6,7], far below the precision clock difference products, can achieve the effect of the result of the time passed a number of nanoseconds, and will extend linearly increase with time, the error can't satisfy the need of real time transmission[8,9,10].

Based on the deficiency of the PPP in the application, An un-differenced carrier phase time transfer method between stations is put forwarded, this method is compared with the traditional PPP algorithm for time transfer, do not need to use satellite precision clock difference product, without being limited by the clock difference interpolation and sampling rate; Depending on the time relay algorithm, compared with the traditional although precision clock difference don't need the product, but a slightly different processing methods, the traditional common depending on the time relay algorithm is through the station intercropping differential elimination the satellite clock error parameters, this method is not eliminated by stand poor intercropping would satellite clock error, but the satellite clock error as unknown parameters online, via satellite were apparent to establish the connection between the two station observation equation and calculating the satellite clock difference between parameters and standing time passing results. Finally using WTZR, GRAZ, GOPE and BOR1 4 IGS stations such as actual data are analyzed, and the analysis results show that the algorithm of time transfer accuracy can reach the nanosecond level, can be applied in real-time and high precision time transfer.

2. The Traditional Model of PPP

The error equation based on GNSS carrier phase time transmission is as follows[4,7]:

$$v_{k,\Phi}^{j} = \delta t_k - \delta t^j + \delta\rho_{k,trop} / c + \lambda \cdot N_k^j / c + \rho_k^j / c + \varepsilon_{k,\Phi}^j - \lambda \cdot \Phi_k^j / c \tag{1}$$

$$v_{k,P}^{j} = \delta t_k - \delta t^j + \delta\rho_{k,trop} / c + \rho_k^j / c + \varepsilon_{k,P}^j - P_k^j / c \tag{2}$$

Where, k is station number, j is satellite number, δt_k , δt^j are relative clock difference between the receiver and the satellite clock is relative to the reference time, $\delta\rho_{k,trop}$ is Tropospheric delay, N_k^j is fuzziness of the ionospheric combination, Φ_k^j , P_k^j are respectively represents the phase and pseudo - distance combined observations, $v_{k,\Phi}^j(i) , v_{k,P}^j(i)$ are measurement error, $\varepsilon_{k,\Phi}^j , \varepsilon_{k,P}^j$ are the unmodeled error.

Through PPP calculation, the results of the clock difference can be obtained for the reference time of each station, and then the single station results of different stations will be reduced, and the reference time can be eliminated to realize the time transfer between stations.

3. A Difference Model Based on Satellite Clock Error

In order to overcome the PPP algorithm for IGS satellite precision clock difference products rely on, put forward the difference between a station GNSS carrier phase time transfer methods, this method via satellite were apparent to establish the connection between the different station observation equation, transfer time between satellite clock error and station real-time estimation results as unknown parameters.

The GNSS measurement is the time delay between the station and the satellite, and the equation is singular when satellite clock difference parameters and receiver clock parameters are got from formula (1) and formula (2), and the solution is infinite, the time transfer results which the station and satellite clock relative to the reference can be got when one reference clock is fixed. Where l station clock error is δt_l, and put it as the reference clock, $\delta \bar{t}^j$ is the relative clock error which satellite j relative to the reference clock error.

$$\delta \bar{t}^j = \delta t^j - \delta t_l \tag{3}$$

After selection reference station clock, the observation equation of station no longer contains the receiver clock difference parameters, only satellite clock error, the troposphere, and the fuzzy degree of parameters, satellite clock difference is between satellite clock and benchmark clock. The error observation equation of the station is.

$$v_{l,\Phi}^{j} = -\delta \bar{t}^j + \delta\rho_{l,trop} / c + \lambda \cdot N_l^j / c + \rho_l^j / c + \varepsilon_{l,\Phi}^j - \lambda \cdot \Phi_l^j / c \tag{4}$$

$$v_{l,P}^{j} = -\delta \bar{t}^j + \delta\rho_{l,trop} / c + \rho_l^j / c + \varepsilon_{l,P}^j - P_l^j / c \tag{5}$$

$\delta \bar{t}_k$ is the relative clock error which is the station k relative to reference clock.

$$\delta \bar{t}_k = \delta t_k - \delta t_l \tag{6}$$

Where $\delta \bar{t}_k$ is the time transfer result between station k and station l, the error equation of station k can be got in the followed formula.

IOP Conf. Series: Materials Science and Engineering **449** (2018) 012001 doi:10.1088/1757-899X/449/1/012001

$$v_{k,\Phi}^{j} = \delta\bar{t}_k - \delta\bar{t}^{j} + \delta\rho_{k,trop} / c + \lambda \cdot N_k^j / c + \rho_k^j / c + \varepsilon_{k,\Phi}^j - \lambda \cdot \Phi_k^j / c \quad (7)$$

$$v_{k,P}^{j} = \delta\bar{t}_k - \delta\bar{t}^{j} + \delta\rho_{k,trop} / c + \rho_k^j / c + \varepsilon_{k,P}^j - P_k^j / c \quad (8)$$

It is exactly the same form relative to the tradition form, except that the meaning of the difference between the satellite clock and the receiver clock in the error equation is changed, which is relative to the relative clock of the reference clock.

Standing between the differential GNSS carrier phase time transfer algorithm is to use common view satellite the relationship established between each station observation error equation, the satellite clock error as unknown parameters with standing time transfer between real-time estimation results. This method is free from the dependence on the precision and the sampling rate of the clock. This method only need to use the information on the precise orbit can be sent as a result, direct access to stand between the satellite orbit changes due to gently can forecast, the IGS provides super quick forecast precision of orbit is 5 cm, in the real time transmission, can use the ultra fast forecast track for calculating.

4. Test result anlysis

In order to verify the effectiveness of the proposed method, here taking WTZR, GRAZ, GOPE and plant area network composed of four IGS reference station observation data analysis, select station WTZR clock as reference clock, at the same time calculating result time passed between the station and satellite clock relatively poor, observation data sampling interval is 30 s. Here IGS provides the post - precision clock differential product as the reference value.

Figure. 1, Figure. 2 and Figure. 3 respectively show the time transfer results of GRAZ-WTZR, GOPE-WTZR and BOR-WTZR1 by the method in this paper, and the comparison diagram of the results of IGS network solution. Figure. 4 shows the sequence of difference between GRAZ-WTZR, GOPE-WTZR and BOR-WTZR1 link time transfer results minus IGS net solution results.

Figure. 1, Figure. 2 and Figure. 3 show that the time transfer sequence of the obtained stations is consistent with the change trend of IGS network solution, and the results are basically the same, which is in good agreement with the IGS results. It can be seen from figure 4 and table 1 that the time transmission accuracy of different time links can reach 0.16ns to 0.42ns, which is basically equivalent to the traditional GNSS carrier phase time delivery precision.

The Table 1 give the time transfer precision result of GRAZ-WTZR, GOPE-WTZR, BOR1-WTZR.

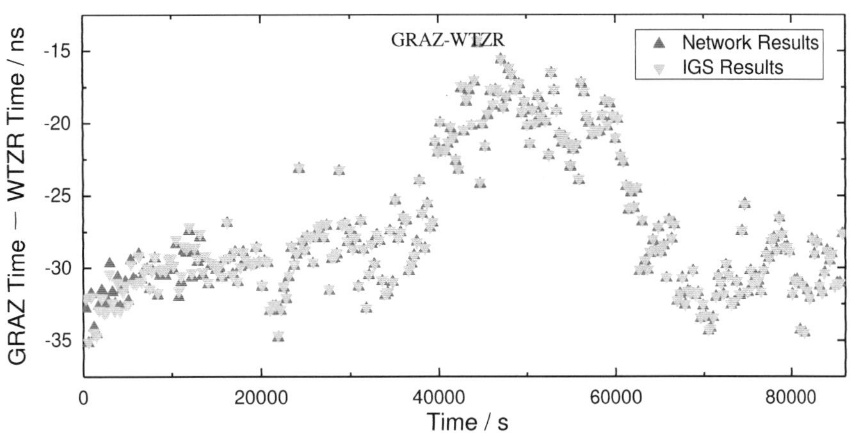

Figure 1. The test result and IGS result on link GRAZ-WTZR

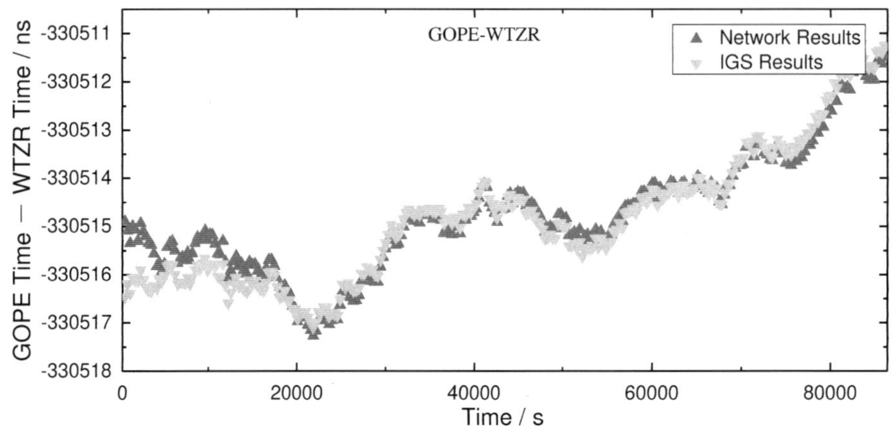

Figure 2. The test result and IGS result on link GOPE-WTZR

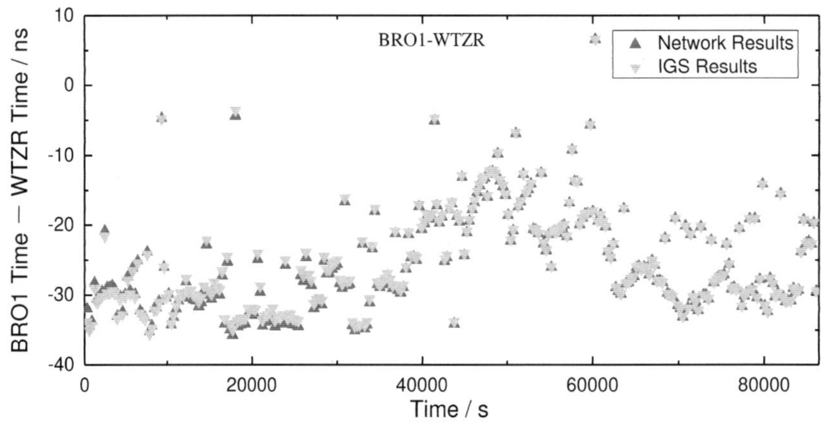

Figure 3. The test result and IGS result on link BOR1-WTZR

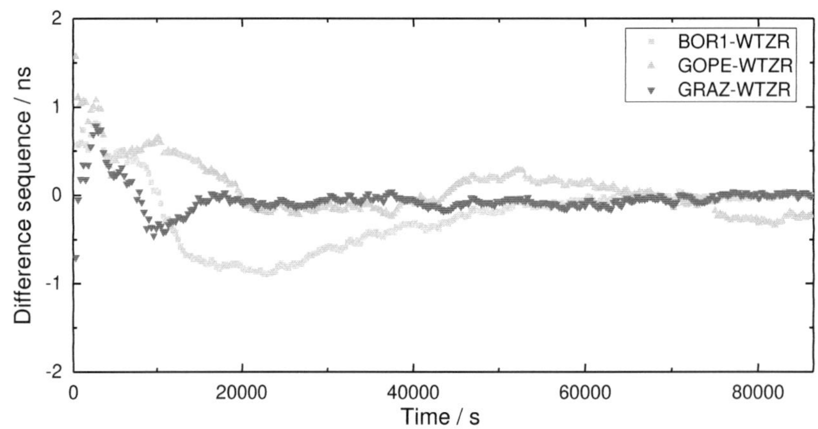

Figure 4. Time passes the result of the difference sequence

Table 1. The time transfer precision on difference link

	Mean	STD	RMS
GRAZ - WTZR	-0.05	0.15	0.16
GOPE - WTZR	0.07	0.30	0.31
BOR1 - WTZR	-0.21	0.37	0.42

5. Conclusion

The proposed standing between the differential GNSS carrier phase time transfer methods through total depending on the satellite station, the difference between the observed equation between the satellite clock error as unknown parameters and standing time estimate deliver results. As long as there is enough total station between apparent between satellite can gain standing time transfer as a result, the precision clock difference from satellite product precision and real time limit, time pass accuracy can reach the nanosecond level, good real-time, more suitable for real-time and high precision time transfer.

Reference

[1] Dach R, Schildknecht T, Hugentobler U, et al. "Continuous Geodetic Time Transfer Analysis Methods," IEEE Transactions on Ultrasonics, Ferroelectrics, and Frequency Control, vol. accepted for publication, 2006.
[2] Dach R, Schildknecht T, Springer T, et al. Continuous time transfer using GPS carrier phase.[J]. IEEE Transactions on Ultrasonics Ferroelectrics & Frequency Control, 2002, 49(11):1480 - 1490.
[3] Defraigne P, Bruyninx C, Guyennon N. PPP and Phase-only GPS Time and Frequency transfer[C]. Frequency Control Symposium, 2007 Joint with the 21st European Frequency and Time Forum. IEEE International, May 29 2007-June 1 2007, 904-908.
[4] Defraigne P, Aerts W, Pottiaux E. Monitoring of UTC(k)' s using PPP and IGS real-time products[J]. Gps Solutions, 2014, 19(1):165-172.
[5] Defraigne P, Sleewaegen J M. Code-Phase Clock Bias and frequency offset in PPP clock solutions.[J]. IEEE Transactions on Ultrasonics Ferroelectrics & Frequency Control, 2016, 63(7):986.
[6] Defraigne P, Baire Q, Guyennon N. GLONASS and GPS PPP for Time and Frequency Transfer[C]. Frequency Control Symposium, 2007 Joint with the 21st European Frequency and Time Forum. IEEE International, May 29 2007-June 1 2007, 909-913.
[7] Defraigne P, Banerjee P, Lewandows. Time transfer through GPS[J].Indian Journal of Radio &Space physics, 2008,36:303-312.
[8] Defraigne P, Sleewaegen J M. Correction for code-phase clock bias in PPP[C]. Frequency Control Symposium & the European Frequency and Time Forum. IEEE, 2015:662-666.
[9] Defraigne P, Guyennon N ,Bruyninx C. GPS Time and frequency transfer PPP and phase-only analysis[J]. International Journal of Navigation and Observation, Volume 2008, Article ID 175468, 7 pages.
[10] Delporte J, Mercier F, Laurichesse D, et al. Fixing integer ambiguities for GPS carrier phase time transfer[C]. Proceedings of the joint European Frequency and Time Forum and IEEEF requency Control Symposium, Geneva, Switzerland, 2007, 29 May-1 June, pp.927-932.

IOP Conf. Series: Materials Science and Engineering **449** (2018) 012002 doi:10.1088/1757-899X/449/1/012002

Imitators of Dynamic Systems with Landing

Danilov Alexander[1] and Garkina Irina[2]

[1,2]Street Titov 28, Penza, Russia, 440028
Penza State University of Architecture and Construction

E-mail: [1]dampguas@mail.ru; [2]i.a.naum@mail.ru

Abstract. The dependencies of the simulation characteristics of simulators (for the preparation of the operators of transport ergatic systems) on the step of integrating the equations of motion of the Runge -Kutta control object of the second order of accuracy (updating the initial conditions at each integration step) are analyzed. Clarification of the time-dependent parameters of the control object (aerodynamic coefficients, mass, etc.) is made at each step. Approximate methods for estimating the effect of delay in subsystems (the latent period of the operator's motility, the play in the control channels, etc.) are given. An analytical evaluation is made of the dependence of the required integration step on the parameters of the control object. The admissible values of the integration steps are given for various criteria for estimating the integration accuracy. The methods underwent practical approbation when setting up a number of simulators of the dynamics of objects of ergatic systems.

1. Introduction

Simulation characteristics of training complexes for the training of operators of transport systems are largely determined by the latent period of the operator's formation of control actions, the presence of backlashes in the control channels, etc. The presence of delay complicates the choice of numerical methods for simulation of the control object of the ergatic system.

Indeed, consider a second-order system:

$$\left.\begin{aligned} x'(t) &= a_{11}x(t)+a_{12}y(t)+b_1(p_1x(t-\tau_1)+p_2y(t-\tau_2)) \\ y'(t) &= a_{21}x(t)+a_{22}y(t)+b_2(p_1x(t-\tau_1)+p_2y(t-\tau_2)) \end{aligned}\right\} (t \ge 0)$$
$$x(t)=S_1(t),\ \ t\le 0;\ \ x(+0)=S_1(0),$$
$$y(t)=S_2(t),\ \ t\le 0;\ \ y(+0)=S_2(0), \qquad (\tau_1, \tau_2 \ge 0). \tag{1}$$

For very small $\tau_1,\ \tau_2$ such that they can be neglected (sufficient integration accuracy is ensured even at step $h > \max(\tau_1,\ \tau_2)$), we arrive at the problem without delay:

$$\left.\begin{aligned} x'(t) &= (a_{11}+b_1p_1)\,x(t)+(a_{12}+b_1p_2)\,y(t) \\ y'(t) &= (a_{21}+b_2p_1)\,x(t)+(a_{22}+b_2p_2)\,y(t) \end{aligned}\right\} (t \ge 0);$$
$$x(+0)=S_1(0);\ \ y(+0)=S_2(0). \tag{2}$$

At large delays, the integration step must be commensurate with the delay; at $\tau_1=\tau_2=\tau$ one can take $\tau = mh$, where $m \in \{1,2,...\}$.

In the vector-matrix form we have:

$$\mathbf{x}'(t)=\mathbf{A}\mathbf{x}(t)+\mathbf{QT}\,\mathbf{x}(t)\quad (t\geq 0),$$
$$\mathbf{x}(t)=\mathbf{S}(t),\ \ t\leq 0,\ \ \mathbf{x}(+0)=\mathbf{S}(0);$$
$$\mathbf{x}'(\mathrm{t})=\begin{bmatrix} x(\mathrm{t}) \\ y(\mathrm{t}) \end{bmatrix},\ \ \mathbf{A}=\begin{bmatrix} a_{11} & a_{12} \\ a_{21} & a_{22} \end{bmatrix},\ \ \mathbf{Q}=\begin{bmatrix} b_1 p_1 & b_1 p_2 \\ b_2 p_1 & b_2 p_2 \end{bmatrix},\ \ \mathbf{T}=\begin{bmatrix} T_\tau & 0 \\ 0 & T_\tau \end{bmatrix}=\mathbf{T}_\tau\mathbf{E}; \tag{3}$$

T_τ - shift operator: $\mathbf{T}_\tau z(t)=z(t-\tau)$, $\mathbf{E}=\begin{bmatrix} 1 & 0 \\ 0 & 1 \end{bmatrix}$, $\mathbf{S}(t)=\begin{bmatrix} S_1(t) \\ S_2(t) \end{bmatrix}$.

For $0\leq t\leq \tau=mh$ the problem reduces to determining the solution

$$\mathbf{x}'(\mathrm{t})=\mathbf{A}\mathbf{x}(\mathrm{t})+\mathbf{QTS}(\mathrm{t}),\ \ \mathrm{x}(+0)=\mathbf{S}(0); \tag{4}$$

it is possible to reduce the step (instead of h take step h^*, where $h=nh^*$, $n\in\{1,2,...\}$; $\tau=mh=mnh^*$, $0\leq t\leq \tau=mnh^*$).

2. The integration by the Runge - Kutta method of the second order of accuracy

Here we have:

$$\mathbf{x}_{\mathrm{j+1}}=\mathbf{x}_{\mathrm{j}}+\mathrm{h}^*\mathbf{A}\mathbf{x}_{\mathrm{j}}+\frac{(\mathrm{h}^*)^2}{2}\mathbf{A}^2\mathbf{x}_{\mathrm{j}}+h^*\mathbf{QS}_{\mathrm{j-mn+\frac{1}{2}}} \tag{5}$$
$$(j=0,1,...,mn-1),\ \ \mathbf{x}_{\mathrm{j}}=\mathbf{x}(jh^*),\ \ \mathbf{S}_j=\mathbf{S}(jh^*).$$

At $t\geq\tau=mh$ the initial problem with delay is solved approximately with step h. True:

$$\mathbf{x}_{k+1}=\mathbf{x}_k+h\mathbf{A}\mathbf{x}_k+\frac{h^2}{2}\mathbf{A}^2\mathbf{x}_k+hQ\overline{x}_{k-m}+\frac{h^2}{2}\mathbf{Q}^2\mathbf{x}_{k-2m}\quad (k=0,1,...), \tag{6}$$
$$\mathbf{x}_k=\mathbf{x}(\tau+kh),\ \ \mathbf{S}_k=\mathbf{S}(\tau+kh).$$

In view of the continuous dependence of the correctly posed problem on delay for small values of the delay, the dependence between step h, frequency ω, attenuation decrement $-\sigma/2$, and a specified accuracy of ε (when fixing the remaining parameters of the system) remains close to the dependence that exists for systems without delay. In other cases, in order to obtain the required simulation characteristics of the training complex, the choice of the step when integrating the equations of motion of the object must be carried out taking into account the retardation values in various control loops.

3. Simulation of the longitudinal motion of a transport aircraft

We confine ourselves to the short-period component of the longitudinal motion of the transport aircraft [1,2]. The problem reduces to solving equation

$$\dot{\mathbf{x}}=\mathbf{A}\mathbf{x},\ \ \mathbf{x}(0)=\mathbf{x}_0, \tag{7}$$

where $\mathbf{x}=x(t)$ is a two-dimensional vector function, $\mathbf{A}$ is a square matrix. The task is defined by six parameters: the components of matrix $\mathbf{A}$ and vector $\mathbf{x}_0$.

In the case of imaginary roots $-\frac{\tilde{\sigma}}{2}\pm i\omega=\frac{\sigma}{2}\pm i\omega\ (\tilde{\sigma}\geq 0,\ \omega>0)$ of the characteristic equation, the solution of the problem has the form:

$$\mathbf{x}=e^{-\frac{\tilde{\sigma}}{2}t}(\mathbf{p}\cos\omega t+\mathbf{q}\sin\omega t), \tag{8}$$

where $\mathbf{p}$, $\mathbf{q}$ are linearly independent vectors.

With an approximate solution by a numerical method with step h for the error of calculation by the Runge-Kutta method of the second order of accuracy we have

$$\mathbf{r}_k = \mathbf{x}_k - \mathbf{x}(t_k) = \frac{-1}{24} h^2 f''(\xi_k),\ \mathbf{f}(t) = \mathbf{A}\mathbf{x},\ 0 < \xi_k < t_k. \tag{9}$$

Since $\mathbf{f}(t) = \dot{\mathbf{x}}$, then

$$\mathbf{r}_k = \frac{-1}{24} h^2 \ddot{\mathbf{x}}(\xi_k),\ 0 < \xi_k < t_k. \tag{10}$$

We have:

$$\dot{\mathbf{x}} = e^{-\frac{\tilde{\sigma}}{2}t}\left(\mathbf{p}\left(-\frac{\tilde{\sigma}}{2}\cos\omega t - \omega\sin\omega t\right) + \mathbf{q}\left(\omega\cos\omega t - \frac{\tilde{\sigma}}{2}\sin\omega t\right)\right),$$

$$\ddot{\mathbf{x}} = e^{-\frac{\tilde{\sigma}}{2}t}\left(\mathbf{p}\left(\left(\frac{\tilde{\sigma}^2}{4} - \omega^2\right)\cos\omega t + \tilde{\sigma}\omega\sin\omega t\right) + \mathbf{q}\left(-\tilde{\sigma}\omega\cos\omega t + \left(\frac{\tilde{\sigma}^2}{4} - \omega^2\right)\sin\omega t\right)\right), \tag{11}$$

$$\dddot{\mathbf{x}} = e^{-\frac{\tilde{\sigma}}{2}t}\mathbf{p}\left(\left(\frac{3}{2}\tilde{\sigma}\omega^2 - \frac{\tilde{\sigma}^3}{8}\right)\cos\omega t + \left(\omega^3 - \frac{3}{4}\tilde{\sigma}^2\omega\right)\sin\omega t\right) +$$

$$+ e^{-\frac{\tilde{\sigma}}{2}t}\mathbf{q}\left(\left(\frac{3}{4}\tilde{\sigma}^2\omega - \omega^3\right)\cos\omega t + \left(\frac{3}{2}\tilde{\sigma}\omega^2 - \frac{\tilde{\sigma}^3}{8}\right)\sin\omega t\right).$$

Obviously, $\mathbf{p} = \mathbf{x}_0$, $\omega\mathbf{q} - \frac{\tilde{\sigma}}{2}\mathbf{p} = \mathbf{A}\mathbf{x}_0$. Therefore, as the parameters determining the system, we can take the following six: $\overline{\sigma}$, ω and the components of the vectors $\mathbf{p}$, $\mathbf{q}$.

Taking into account

$$|\dddot{\mathbf{x}}(t)| \le e^{-\frac{\tilde{\sigma}}{2}t}\left(\frac{\tilde{\sigma}}{2} + \omega\right)^3 (|\mathbf{p}| + |\mathbf{q}|) \tag{12}$$

we obtain

$$|\mathbf{r}_k| \le \frac{h^2}{24} e^{-\frac{\tilde{\sigma}}{2}\xi_k}\left(\frac{\tilde{\sigma}}{2} + \omega\right)^3 (|\mathbf{p}| + |\mathbf{q}|) \le \frac{h^2}{24}\left(\frac{\tilde{\sigma}}{2} + \omega\right)^3 (|\mathbf{p}| + |\mathbf{q}|). \tag{13}$$

From $|\mathbf{r}_k| < \varepsilon$ follows $h^2 < \dfrac{24\varepsilon}{(|\mathbf{p}| + |\mathbf{q}|)\left(\frac{\tilde{\sigma}}{2} + \omega\right)^3}$.

As we see, for fixed components of vectors $\mathbf{p}$ and $\mathbf{q}$ we have:

$$h^2 < \frac{B}{\left(\frac{\tilde{\sigma}}{2} + \omega\right)^3},\quad B = \frac{24\varepsilon}{|\mathbf{p}| + |\mathbf{q}|} \tag{14}$$

and step h is determined by the damping decrement $\frac{\tilde{\sigma}}{2}$ and the frequency ω (the dependence for $B = 1$ is given in figure 1).

For the control objects under consideration, the band of operating frequencies lies in the interval $(2,10)s^{-1}$; and the dimensionless attenuation coefficient ξ varies between 0.4 and 0.9.

With increasing ξ, the quantity

$$-\frac{\sigma}{2} + \omega = -\frac{\xi\omega}{\sqrt{1-\xi^2}} + \omega = \omega_0\left(\sqrt{1-\xi^2} - \xi\right) \tag{15}$$

decreases ($\frac{\partial\left(-\frac{\sigma}{2}+\omega\right)}{\partial\xi} = -\omega_0\left(1+\frac{\xi}{\sqrt{1-\xi^2}}\right) < 0$); the admissible step h increases with increasing ξ.

It is no accident when tuning simulators there is an aspiration to increase the damping coefficient for sufficiently large real values of h (leads to a distortion of the characteristics of the aircraft: an increase of ξ causes a corresponding decrease in the natural frequency ω). Point (ξ, ω) in areas of equal assessments of flight characteristics is shifted ([3-6] the evaluation of the piloting characteristics of the simulator is improved, and the correspondence of the aircraft and the model is deteriorating).

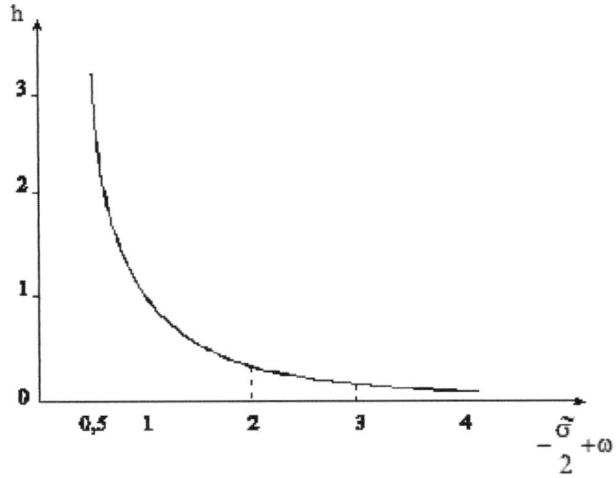

Figure 1. Dependence of the integration step h on $(-\frac{\tilde{\sigma}}{2}+\omega)$

4. Some estimates for the integration step

For a system without damping ($\xi = 0$) we have:

$$h^2 < \frac{B}{\omega^3}. \tag{16}$$

For the ratio of the maximum permissible steps for systems with natural frequencies ω and $\Omega = k\omega$ (the upper bound) is valid:

$$\frac{h_\Omega}{h_\omega} = \sqrt{\frac{\omega^3}{\Omega^3}} = \frac{1}{\sqrt{k^3}}. \tag{17}$$

In a strip of working frequencies from 2 rad/sec. to 10 rad/sec. admissible step of integration change more than by 10 times.

We obtain an analogous estimate for step h, starting from condition

$$|\mathbf{R}_k| < \delta;$$

$$\mathbf{R}_k = \dot{\mathbf{r}}_k = \mathbf{A}\mathbf{r}_k = \mathbf{A}\left(-\frac{h^2}{24}\ddot{\mathbf{x}}(\xi_k)\right) = -\frac{h^2}{24}\mathbf{x}^{IV}(\xi_k), \quad 0 < \xi_k < t_k. \tag{18}$$

We have

$$\mathbf{x}^{IV} = e^{-\frac{\tilde{\sigma}}{2}t}\mathbf{p}\left(\left(\frac{\tilde{\sigma}^4}{16} - \frac{3}{2}\tilde{\sigma}^2\omega^2 + \omega^4\right)\cos\omega t + \left(\frac{\tilde{\sigma}^3\omega}{2} - 2\tilde{\sigma}\omega^3\right)\sin\omega t\right) +$$

$$+ e^{-\frac{\tilde{\sigma}}{2}t}\mathbf{q}\left(\left(2\tilde{\sigma}\omega^3 - \frac{\tilde{\sigma}^3\omega}{2}\right)\cos\omega t + \left(\frac{\tilde{\sigma}^4}{16} - \frac{3}{2}\tilde{\sigma}^2\omega^2 + \omega^4\right)\sin\omega t\right); \quad (19)$$

$$\left|\mathbf{x}^{IV}(t)\right| \le e^{\frac{\tilde{\sigma}}{2}t}\left(\frac{\overline{\sigma}}{2} + \omega\right)^4 (|\mathbf{p}| + |\mathbf{q}|);$$

$$|\mathbf{R}_k| \le \frac{h^2}{24} e^{-\frac{\tilde{\sigma}}{2}\xi_k}\left(\frac{\tilde{\sigma}}{2} + \omega\right)^4 (|\mathbf{p}| + |\mathbf{q}|) \le \frac{h^2}{24}\left(\frac{\tilde{\sigma}}{2} + \omega\right)^4 (|\mathbf{p}| + |\mathbf{q}|) < \delta .$$

From whence

$$h^2 < \frac{24\delta}{(|\mathbf{p}| + |\mathbf{q}|)\left(\frac{\tilde{\sigma}}{2} + \omega\right)^4} \quad \text{or} \quad h^2 < \frac{C}{\left(\frac{\tilde{\sigma}}{2} + \omega\right)^4}, \quad C = \frac{24\delta}{|\mathbf{p}| + |\mathbf{q}|}. \quad (20)$$

If $\delta = \varepsilon$, then $C = B$.

The maximum values of steps h_ε and h_δ, obtained under conditions $|\mathbf{r}_k| < \varepsilon$ and $|\mathbf{R}_k| = |\dot{\mathbf{r}}_k| < \delta$, respectively, are shown in figure 2 (taken B = C = 1).

The obtained estimates were used in the development of simulation models of the object of control of the transport ergatic system with the integration of the equations of motion in real time with the updating of the initial conditions and variable parameters at each integration step should be avoided whenever possible. If required they should be used only for brief notes that do not fit conveniently into the text.

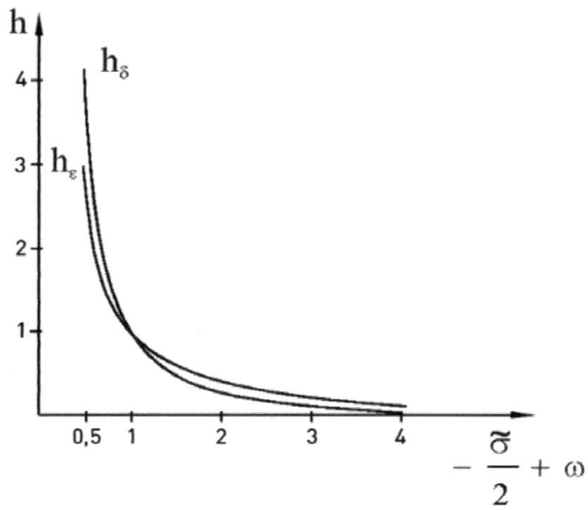

Figure 2. The dependence of the integration step h on ε and δ

5. Conclusion

An analytical evaluation is made of the dependence of the integration parameters on the characteristics of the control object are given.

The admissible values of the integration steps are indicated for various criteria for estimating the accuracy of integration.

The methods were tested in practice when tuning the flight simulators of a transport aircraft.
figure should have a brief caption describing it and, if necessary, a key to interpret the various lines and symbols on the figure.

References

[1] Irina Garkina, Alexander Danilov 2017 *Decomposition of the spatial movement of ergatic system* J. Ponte vol 73 Issue 4 p 94-99
[2] Irina Garkina, Alexander Danilov 2017 *Ergatic systems of control in space: assessment of the required duration of the computing cycle* J. Ponte vol 73 Issue 5 p 170-174
[3] Irina Garkina, Alexander Danilov 2017 *Formation of control style by the object in the training complexes: engineering and psychological bases* J. Ponte vol 73 Issue 1 p 127-132
[4] Alexander Danilov 2017 *Evaluate of operator control actions from technical parameters of the objec*t J. Ponte vol 73 Issue 4 p 100-104
[5] Alexander Danilov 2017 *Formalized assessment of dynamic characteristics of complex systems* J. Ponte vol 73 Issue 3 p 289-294
[6] E. Budylina, A.Danilov, I. Garkina 2015 *Control of multiobjective complex systems* J. Contemporary Engineering Sciences vol 8 no 10 p 441-445

A Design of Space-borne AIS Scene Simulation Based on Density Distribution of Global Vessels

CHEN Lihu*, **Liu Xianfeng, Yu Sunquan and LI Shiyou**

College of Aerospace Science and Engineering, National University of Defense Technology, Changsha, 410073, China

*Clh2055@163.com

Abstract. At present, the demand for application of space-borne Automatic Identification System (AIS) is increasing. Since more and more satellites are equipped with AIS payload, it is especially important to test the effectiveness of its performance. Owing to difficult test on AIS payload and high cost, this paper proposes a method of space-borne AIS scene simulation platform based on density distribution of global vessels, focusing on basic principle and design method of simulation scene. It contains detection range computing, the number of vessels within range and the detection probability of AIS receiver. Moreover, corresponding software and hardware have been designed. 'Ocean-2B' satellite-based AIS payload is used to test the performance of the scene simulation system, which indicates that the method can simulate real scene of receiving signals in orbit in a favourable manner.

1. Introduction

The AIS uses the very high frequency (VHF) band and Self-Organized Time Division Multiple Access (SOTDMA) to receive dynamic and static information of vessels, which can monitor and identify various vessels. SOTDMA is a core of this system, which is capable of navigation, positioning, collision avoiding and information exchange (the position, the route and the velocity) of vessels. However, the space-borne AIS is an information gathering system loaded with AIS payload which has high sensitivity in the satellite. It can easily receive a great deal of vessel information within satellite coverage to perform vessel-tracking and logistics controlling [1, 2].

The big difference between space-borne AIS and ground AIS is that the space-borne AIS has a wide coverage so that it can receive many signals from several SOTDMA networks. The LEO (Low Earth Orbit) whose orbit height is 600km can cover almost 12 million square kilometers of the area where contains a large number of vessels. For instance, if the covered area is located in the Mediterranean, it can include more than eight thousand vessels [3]. The coverage radius of SOTDMA network formed by various vessels at sea about 40~100 kilometers, because every SOTDMA sub-network is distributed independently. And there is no time synchronization among these sub networks. Thus, multi-network signal collision will inevitably occur for space-borne AIS, which causes drastic reduction of signal detection probability of space-borne AIS and even complete blockage of the channel [4, 5].

2. Coverage of the satellite

The location of satellite nadir point needs to be obtained before analyzing coverage of the satellite at certain time. The satellite nadir point is the intersection of the line which connects geocentric point, the satellite and the earth's surface. With the satellite moving, the track of satellite nadir point is gradually formed. Generally, the earth can be regarded as globular, assuming orbit height is h, eccentricity ratio e ,orbit inclination i, right ascension of ascending node Ω, argument of perigee ω, true anomaly f. Considering the earth's rotation, geocentric latitude φ and geocentric longitude λ of the satellite at certain moment t can be computed as [6],

$$\varphi = \arcsin(\sin i \sin u) \tag{1}$$

$$\begin{cases} \lambda = \arctan(\cos i \tan u) + \Omega - \overline{S}(t) & \text{ascending} \\ \lambda = 180^\circ + \arctan(\cos i \tan u) + \Omega - \overline{S}(t) & \text{declining} \end{cases} \tag{2}$$

Where $u = \omega + f$ and $\overline{S}(t)$ is Greenwich mean sidereal time (GMST). Hypothetically, the AIS satellite is deployed omni-directional antenna to receive AIS signals. The width of the AIS satellite can be computed after antenna beam angle and orbit height are set[7]. It is shown in figure 1.

For instance, there is a target A. **S** represents the satellite. S' is the nadir point at the moment. **P** points to the projective direction at local level where the line connects the target and the satellite. θ_1 is the elevation angle of the target to the satellite. θ_3 represents the covered angle. The relationship among these angles can be known as

$$\theta_3 = 90^\circ - \theta_1 - \theta_2 \tag{3}$$

In the triangle ΔAOS , we can obtain

$$\frac{R_E}{\sin \theta_2} = \frac{R_E + h}{\sin(\theta_1 + 90^\circ)} \tag{4}$$

The width of the AIS antenna coverage on the ground is

$$l = R_E * \theta_3 \tag{5}$$

Thus, the upper and lower limits of the latitude and longitude can be calculated according to the width of the AIS antenna coverage to the ground and the latitude and longitude of satellite nadir point. In addition, the antenna coverage can cover the outer quadrilateral area of the circle for approximation when the width of the antenna is set on a small value. It is shown in figure 2.

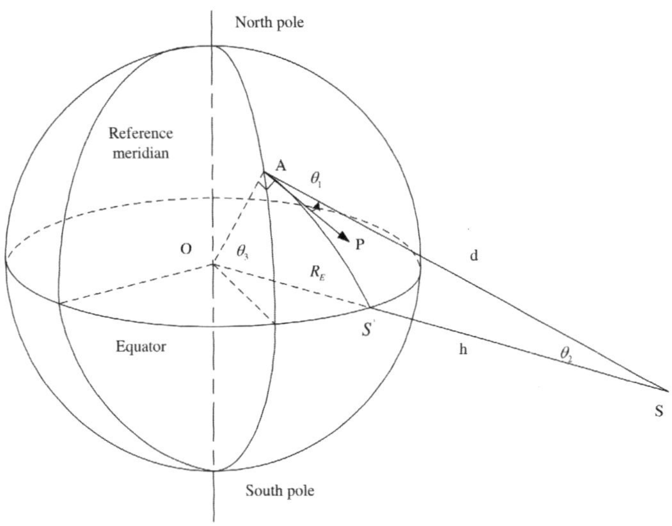

Figure 1. coverage of one satellite

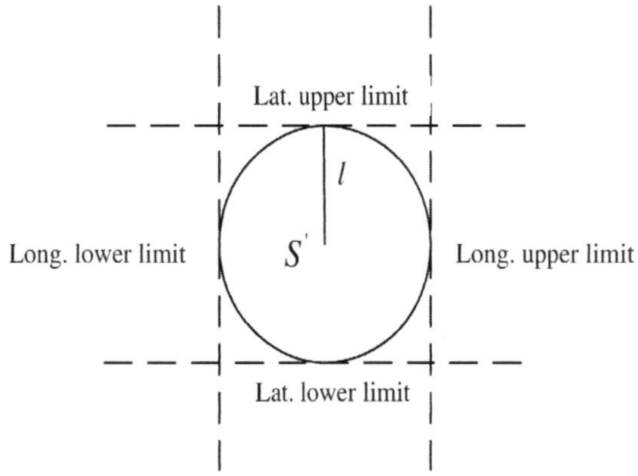

Figure 2. Satellite-to-the-ground coverage approximate diagram

3. Computing the number of vessels within one AIS satellite coverage

This paper makes out a global vessel distribution chart at a certain time based on the data collected by 'TT3' of NUDT at certain time and that by certain business company within corresponding time [8, 9], shown in figure 3. In which, the red figure represents the number of vessels within certain latitude and longitude range.

The earth's surface can be divided by some longitude interval $\Delta\lambda$ and latitude interval $\Delta\varphi$. Actually, longitude interval can be equal to latitude interval. The AIS receiving signals' scene would be simulated more accurately on condition that the number of divided grids is large. But it also needs large computation resource. In principle, $\Delta\lambda \geq l$, $\Delta\varphi \geq l$. l is the width of antenna coverage. For the convenience of grid division, let $\Delta\lambda = \Delta\varphi$ c Thus, the number of grids n_{cell} in the global wide area is

$$n_{cell} = (\frac{180}{\Delta\varphi} + 1) * (\frac{360}{\Delta\lambda} + 1) \tag{6}$$

From Figure 4, we can see that there will be a certain number of vessels within part of sub-grids. And in every sub-grid, hypothetically, the distribution of the vessels is also even.

Generally, the satellite coverage area can be divided into nine parts which are shown in figure 5.

The yellow part represents approximation of satellite coverage area. φ_i and λ_j are the latitude and longitude respectively. At the moment, the relevant latitude and longitude form 9 blocks whose area is indicated by $S_N^1 \sim S_N^9$ .And $s_n^1 \sim s_n^9$ is the area of divided satellite coverage. The number of vessels represented by $n_1 \sim n_9$ in every block is equal or greater than zero and it is an integer. Assuming the global vessels are divided evenly in every block, the number of the vessels is

$$\begin{aligned} N_1 &= \frac{s_n^1}{S_N^1} * n_1 \\ N_2 &= \frac{s_n^2}{S_N^2} * n_2 \\ &\vdots \\ N_9 &= \frac{s_n^9}{S_N^9} * n_9 \end{aligned} \tag{7}$$

Thus, at the time t_n, the satellite covers $N(t_n)$ vessels.

$$N(t_n) = N_1 + N_2 + ... + N_9 \tag{8}$$

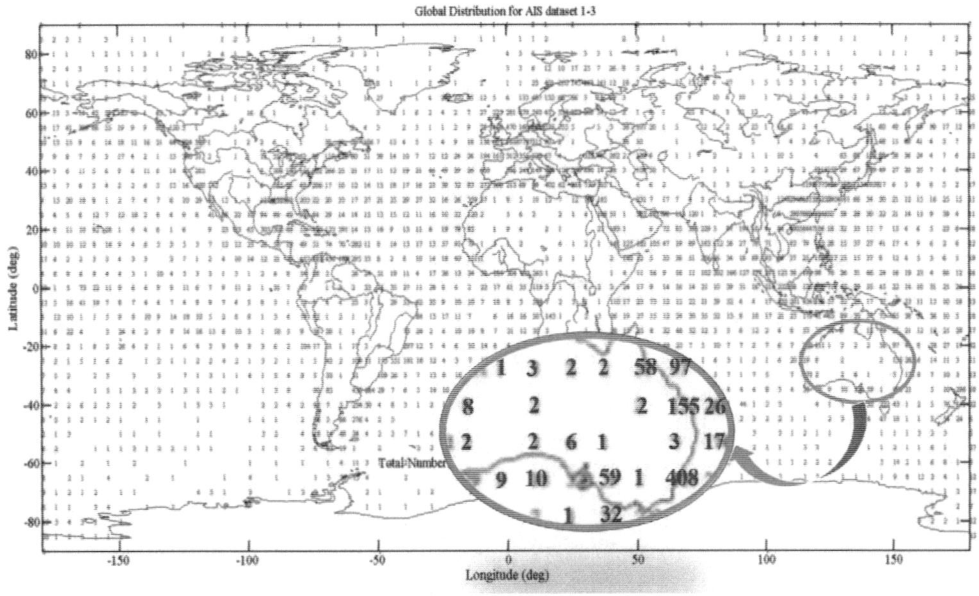

Figure 3. The global vessels distribution

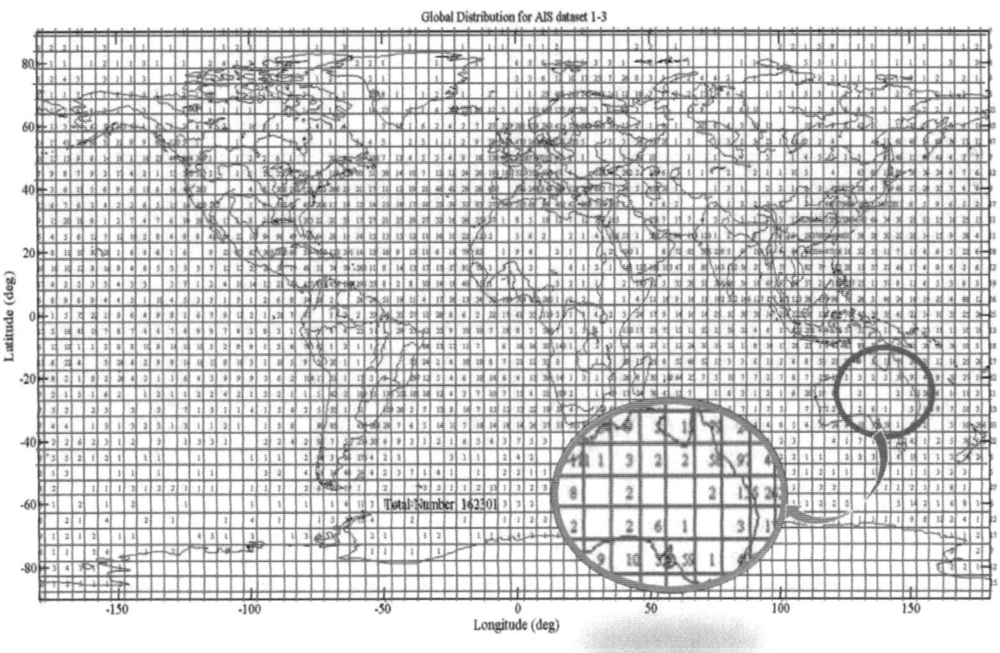

Figure 4. Gridding of distribution of the global vessels

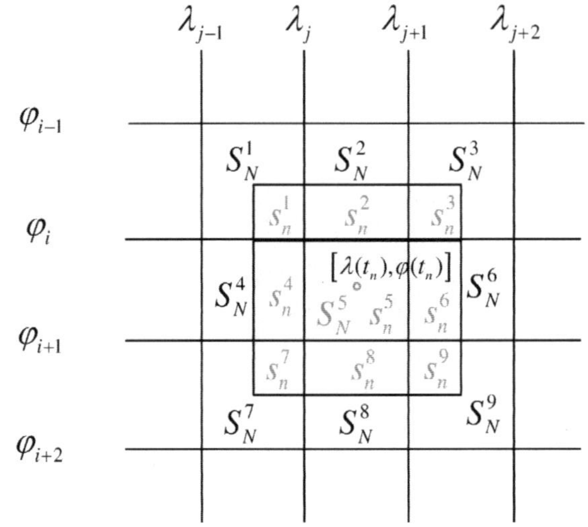

Figure 5. Division of coverage area

4. Simulation of transmitting signals

Besides the influence caused by the number of vessels within coverage range at a certain moment, factors like transmitting power, Doppler frequency shift, channel, transmitting cycle are also affecting the capability of receiving signals.

Transmitting power is established by AIS criterion [1]. And the velocity of the satellite to vessels is relatively high. Thus, considering vessels are relatively immobile, the Doppler frequency shift is

$$f_d = \frac{v_{sat}}{c} * f_c * \cos\theta \tag{9}$$

where v_{sat} is relative velocity. c is the light velocity. f_c is the carrier frequency. θ is the angle between satellite motion and signal direction. In addition, the frequency of AIS channels has increased to four which are 161.975Mhz, 162.025Mhz, 156.775Mhzand156.825Mhz[10]. And the transmitting cycle is set by AIS criterion[11].

For evaluating the performance of AIS receiver, detection probability of space-borne AIS is considered as an important indicator. In order to compute it, the MMSI code of every independent vessel must be gained from AIS message. The MMSI code is an identity of vessel. At the moment t_n, vessels covered by satellite randomly send out AIS message. The number of vessels whose information can be obtained by the AIS payload is counted with distinctive MMSI code.

The discretization of satellite motion can be beneficial for counting vessels shown in Figure 6. During the motion of the satellite, when the upper and lower limits of the latitude and longitude reach the following condition, the simulator accumulates the MMSI code once

$$\lambda(t_{n+1}) - \lambda(t_n) = 2\Delta\lambda \tag{10}$$

$$\Delta\lambda = \frac{l}{R_e} * \frac{180}{\pi} \tag{11}$$

$\Delta\lambda$ transformed by the width of antenna is the difference of longitude. And the direction of satellite motion cannot affect the above-mentioned condition. Finally, gather message collected by AIS receiver within certain period of testing time based on MMSI code. More specifically, assuming that the space-borne AIS-receiver obtains N_m^{re} MMSI codes, and there are N_m^{trans} vessels-covered by satellite during the simulated time, the detection probability of the receiver is

$$P = \frac{N_m^{re}}{N_m^{trans}} \tag{12}$$

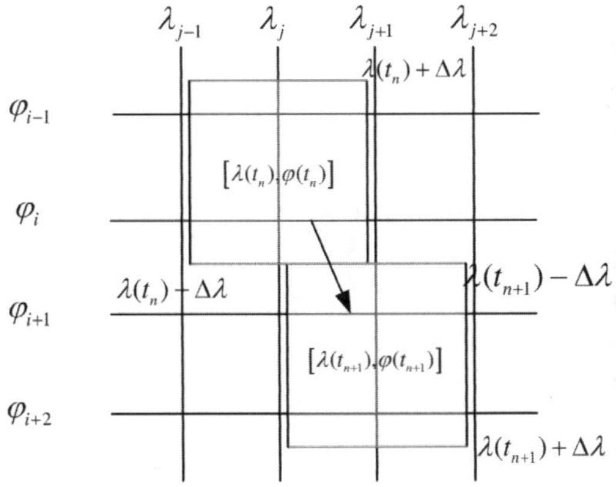

Figure 6. The motion of satellite coverage

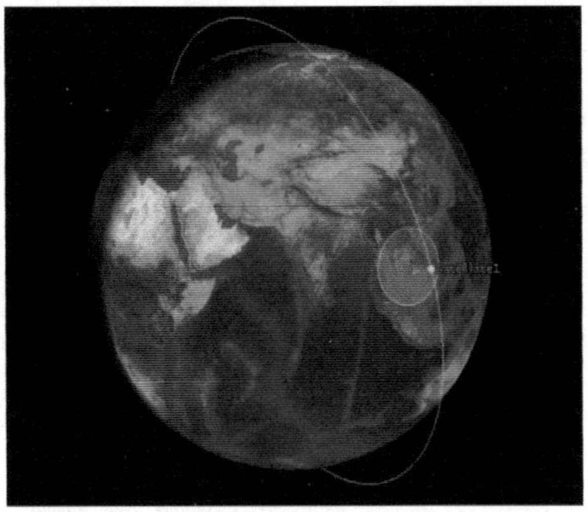

Figure 7. 'Ocean 2B'satellite motion

5. Simulation verification

Table 1. 'Ocean 2B' satellite parameters

Orbit semi-major axis	7341.732km
Inclination	99.34015°
Eccentricity	0.00117
Argument of perigee	60°
Right ascension of ascending node	20°

Table 2. Different sea area and the whole detection probabilities(%)

	Pacific	Atlantic	Indian Ocean	Arctic Ocean	Mediterranean	Global Average
Single satellite loading single receiver	83.40	42.64	45.75	59.58	0.64	49.55
Single satellite loading two receivers	97.24	67.09	70.57	83.23	1.27	67.53

The AIS signal generator and corresponding software have been designed according to part II and III. The AIS payload of 'Ocean 2B' Satellite whose major orbit parameters are shown in Table 1 has been considered as the testing object. And ten seconds is regarded as the interval to compute the position of the satellite nadir point. The satellite motion can also be displayed in STK shown in Figure 7.

Moreover, there is a table about global vessels distribution to count the number of vessels. At every moment, the AIS signal generator sends AIS message according to vessels kept in satellite's coverage and preinstalled parameters which are carrier frequency, sending signal cycle and signal power, respectively. Especially, every vessel sends AIS message complying with SO-TDMA (Self-Organized Time Division Multiple Access). And when the simulation operates, the monitoring window is shown in Figure 8.

The window contains latitude and longitude, the upper and lower limits of the latitude and longitude within coverage range and the number of MMSI code of the vessels which have transmitted message. And the rest of the window displays the monitoring information of AIS signal simulator.

More specifically, the four channels represent four carrier frequency points of AIS signal. The signal power and sending cycle of the signal have been set in advance. The simulating time lasts 24 hours. The diameter of the FOV (field of view) is 2000km. The sensibility of AIS receiver is -111dBm. Finally, through the simulating test, the 'Ocean 2B' detection probability shown in Table 2 can be obtained through the number of vessels from several areas and the independent MMSI code.

From the table 2, we can know that serious signal collision happens in the Mediterranean since it has the highest density of vessels, which leads to the fact that the AIS receiver cannot obtain AIS message correctly.

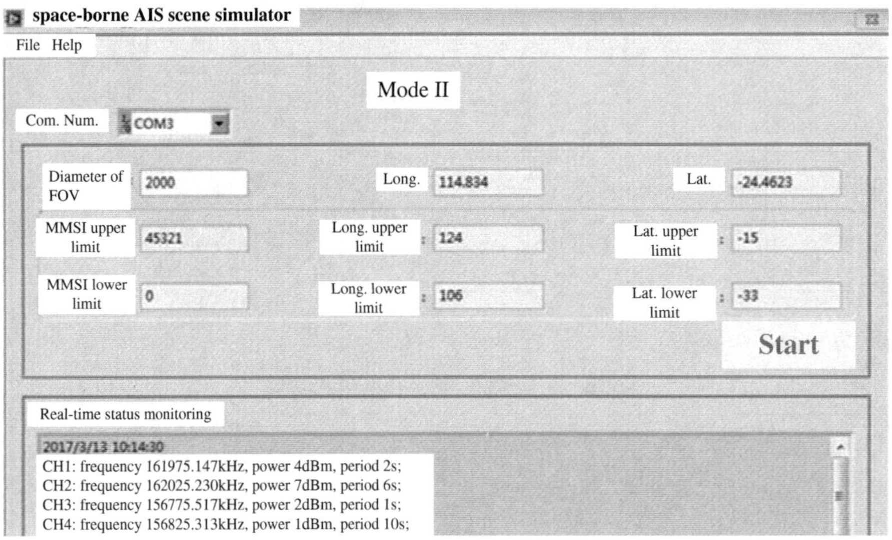

Figure 8. Monitoring window of simulation

6. Conclusions

This paper introduces the principle and method of designing simulating scene based on global vessel distribution. Meanwhile, corresponding software and hardware have been designed to test its performance combined with the 'Ocean 2B'satellite-based AIS payload. The outcome indicates that this AIS scene simulating technology can guarantee the testing performance of AIS receiver on the ground.

Acknowledgments

This research has been supported by the program under grant 2016GK2001, ZDYYJCYJ20140701 and ZK16-03-58.

References

[1] International Telecommunication Union. Technical characteristics for an automatic identification system using time division multiple access in the VHF maritime mobile frequency band [R]. ITU-R.M.1371-2, 2007

[2] F. teHennepe, R Rinaldo, A. Ginesi, C. Tobehn, M. Wieser, Ø. Helleren. Feasibility of an European constellation for space-based detection of AIS signals[J]. *European symposium on satellite-AIS*, December 2010

[3] F. teHennepe, R Rinaldo, A. Ginesi, C.Space-Based Detection of AIS Signals[C]. *2010 5th Advanced Satellite Multimedia Systems Conference and the 11th Signal Processing for Space Communications Workshop*.2010

[4] Davies M E, James C J. Source separation using single channel ICA[J].*Signal Processing*, 2007, 87(9): 1819-1832
[5] HoeherP A, Badri-Hoeher S, Xu W, et al. Single-antenna co-channel interference cancellation for TDMA cellular radio systems[J]. *Wireless Communications, IEEE*. 2005, 12(2): 30-37.
[6] Xi Xiaoning, Wang Wei, Gao Yudong. *Fundamentals of Near-earth Spacecraft Orbit*[M].2003
[7] Liu Xianfeng.A design and optimism of monitoring sea-sky targets and data collection system based on micro-satellite[D]. National University of Defense Technology. 2016
[8] Cheng, Y., Chen, L.H. and Chen, X.Q. Beam Scanning Method based on the Helical Antenna for Space-based AIS. *Journal of Navigation*, 2015(68), 52–70
[9] Shiyou Li, Lihu Chen, Xiaoqian Chen, Yong Zhao, Lei Yang.Statistical analysis of the detection probability of the "TianTuo-3" space-based AIS[J]. *Acta Astronautical*. 2017
[10] Eriksen T, Høye G, Narheim B, et al. Maritime traffic monitoring using a space-based AIS receiver [J]. *Acta Astronautical*. 2006, 58(10): 537-549
[11] International Telecommunication Union. Satellite detection of automatic identification system messages[R]. *ITU-R.M.2084*, 2006

Theoretical Study of A Three-Swing Blades Rotary Engine

H.K. Ma[1*], N.T. Liu[1], M.H. Hsu[2], Y.W. Liu[2] and C.Y. Lin[2]

[1]Research Center of Climate Change & Sustainable Development, National Taiwan University, Taiwan
[2]Mechanical Engineering Dept., National Taiwan University, Taiwan

[*]Email: skma@ntu.edu.tw

Abstract. The innovative eccentric swing blade rotary engine is internally divided into a compression section, a combustion chamber, and a power section. With the three swing blades on each rotor, the swing blade comprises a cylindrical roller mounted to a front end and a curved back, and each of the roller is provided, at an inner side, with a support device that is capable of sustaining a counteracting force applied by a cylinder wall to the roller. After completing the combustion of the mixed gas in the combustion chamber, the generated high-pressure gas is injected into power section and push the swing blades rotation, resulting in power. In this study, the theoretical model has been established to calculate the trajectory/the tip speed of the swing blade, pressure variation, compression torque and to predict the performance of the rotary engine. For a prototype 120 cc rotary engine, it can provide an estimated power output between 6.26 to 50.16-hp at about 1000 to 8000 rpm.

1. Introduction

Invented in 1933 by Felix Wankel, the Wankel rotary engine was used in many famous automotive manufactures like Toyota, Daimler Benz and General Motor etc. Previously, Mazda, former Toyo Kogyo [1], had launched many commercial vehicles form RX-3 to RX-8 which provides up to 247-hp with its RENESIS engine. Although having advantages like lower vibration, smaller engine size and weight, the Wankel engine suffers from higher fuel consumption and maintenance issues. Through the optimization of combustion, the rotary engines are now being used on small UAVs of some countries, providing an estimated power output between 20 and 90-hp [2]. One Research has also proved the viability through real manufacturing, making a 35.6kg Wankel engine prototype providing 55-hp at about 8000 rpm [3]. Recently, a new form of powertrain that is lightweight and small size will have numerous applications in motor cycles, unmanned aerial vehicles and the automotive industry. The Mazda2/3/CX-3, Chevrolet Volt and BMW i3 have announced new range extenders driving an electric generator which charges a battery to supply the vehicle's electric motor with electricity. Due to the compact size and the high power to weight ratio of a Wankel engine, it has been used for electric vehicles as range extenders to provide supplementary power [4]. In this study, an innovative eccentric swing blade rotary engine with three swing blades has been developed and internally divided into a compression section, a combustion chamber, and a power section [5-6]. Also, the theoretical model has been established to calculate the trajectory/the tip speed of the swing blade, pressure variation, compression torque and to predict the performance of the rotary engine.

2. Actuation principle of an eccentric swing blade rotary engine

The eccentric swing blade rotary engine shown in Figure 1 is comprised of a compression section and a power section each provided with a rotor device. Each rotor device is made up of a rotor body and three swing blades. Each swing blade is pivotally connected to the rotor body and may swing about an axis defined by a pivot pin.

Figure 2 shows the actuation of the eccentric swing blade rotary engine. The swing blades continuously moves and propels the compressed air/fuel mixture into the combustion chamber. When the compressed gas is ignited, the high pressure combustion gas will push the blade and generate power.

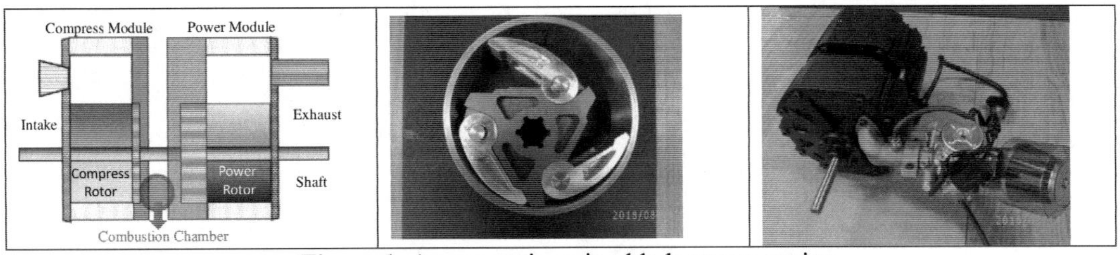

Figure 1. An eccentric swing blade rotary engine

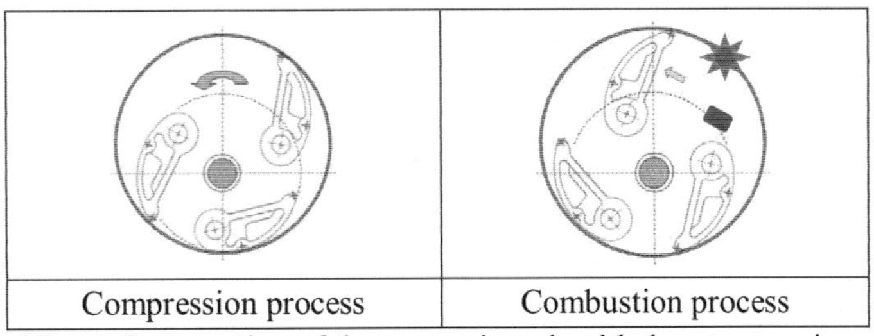

Figure 2. Actuation of the eccentric swing blade rotary engine

3. Analysis of the trajectory of an eccentric swing blade

Table 1. The parameters of 120 cc swing blade rotary engine

Parameter	Definition	Unit (mm)
R	Radius of the cylinder	62.5
r	Radius of the rotor	32
e	Eccentricity of the rotor	12.5
l	Length of the blade	57
H	Height of the cylinder	35

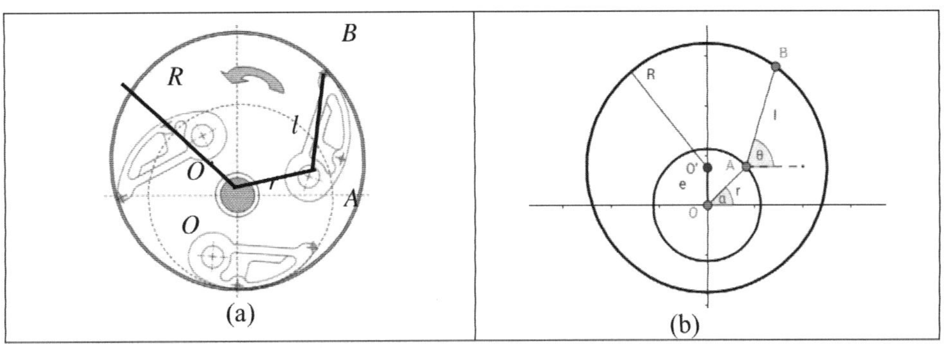

Figure 3. Symbol definition of an eccentric swing blade rotary engine

To establish the theory of eccentric swing blade rotary engine, we construct the model of the engine and make the analysis of the trajectory of the blade. Table 1 shows the parameters of the 120cc swing blade rotary engine. And, Figure 3 shows the symbol definition of the model. The motion of the swing blade is an "arc" along the cylinder wall.

3.1 The tip trajectory equation and the maximum open of swing blade

The tip of the blade B(x,y) moves along the cylinder, forming an eccentric circle as the trajectory Eqn. (1).

$$-2el\sin(\theta) - 2er\sin(\alpha) + 2lr\cos(\alpha - \theta) = R^2 - r^2 - l^2 - e^2 \tag{1}$$

The relation of θ and α can be described by Eqn. (1). In the design of 120 cc engine, the modification changes the ratio of the radius of rotor to the radius of cylinder to be 10:8 to adjust the maximum open angle (θ - α). To avoid the stagnation of moving, the open angle (θ - α) ranges from 80° to 130°.

3.2 Pressure variation in the compression section and power section

The estimated maximum volume (TDC) is roughly 120 cc and the minimum volume at BDC is 24 cc by considering the volume of blade (23 cc). The volume of combustion chamber is 10 cc. The designed volume in power section leads the volume in compression section by 30 deg.

The inlet gas pressure is atmospheric pressure $P_1 = 1.003$ bar and the volume of the working chamber when the compression starts $V_1 = 120$ C.C. Assuming the process is an isentropic compression process, the pressure variation of the compression section can be obtained from the relation $PV^k = C$. Figure 4 shows the pressure variation in both power and compression section, the maximum pressure 4.671 bar is obtained at α = 145° (BDC), which is the pressure in combustion chamber. The pressure at α = 145°~220° is due to the residual gas remained in the compression section. The exhaust and intake begins at α = 220° and ends at α = 300° (TDC) where the compression starts. After heat adding in the combustion chamber, the maximum pressure is 21.68 bar at α = 145°. Isentropic expansion is happened during α = 145°~ 300° . The exhaust port opens at α = 300° (TDC) and closes at next combustion starts.

3.3 The Thermodynamic properties in power cycle

The proportional ratio of CVC and CPC affects the pressure and temperature after combustion. This proportional ratio relates to the open timing of the exhaust port at the outlet of combustion chamber to the power section. Maximum heat can be released shown in Figure 5 during the combustion of octane fuel for the case CVC/CPC=0.6/0.4.

3.4 Torque

$$Torque = \Delta P * h * (\frac{1}{2}l^2 + rl(cos(\theta - \alpha)) \tag{2}$$

ΔP is the pressure difference between two sides of blade and h is the height of compression section. When we only consider the effect of single blade, the variation of torque output is shown in Figure 4. We can notice that the value of torque output is negative in some periods, which means the blade is pushed by the gas and output power. The major reason of this phenomenon is that, in these periods, the compression section ends the connection to the combustion chamber but there is still a small space containing residual high pressure gas between the two adjacent blades. As a result, the high pressure air will push the blade and causes the torque output to have a negative value.

3.5 The influence of combustion chamber on compression section and power section

In this section, Table 2 Summary of power balance with combustion chamber. Also, the influence of combustion chamber and the possible reasons causing the discontinuous points (sudden increase or decrease) are discussed. There are several discontinuous points can be found in Figure 4 (yellow circles). To realize this phenomenon, we compare the pressure variation in front of and behind the blade with the torque output. The red and yellow curves shows the variation of the pressure in front of and behind the blade respectively. We can realize these are caused by starting and ending the connection between combustion chamber and the compression section. Since the variation of torque output is associated with the variation of pressure, we can confirm that it is the sudden open and closure of combustion chamber make the torque output discontinuous. And then, Figure 5 shows the variation of torque output with respect to (α) for a single blade in power section. Finally, Figure 6 combines the total torque input in compression section and the total torque output in power section. It allows us to derive the total power output by integrating the net torque output with respect to (α), and the effect of a single blade contributing to the total torque output can also be observed in Figure 6.

Table 2. Summary of power balance with combustion chamber

Power	**With combustion chamber**
Power consumption in compression section	9.1887J/round
Power generation in power section	154.6925J/round
Net power output	145.5038J/round

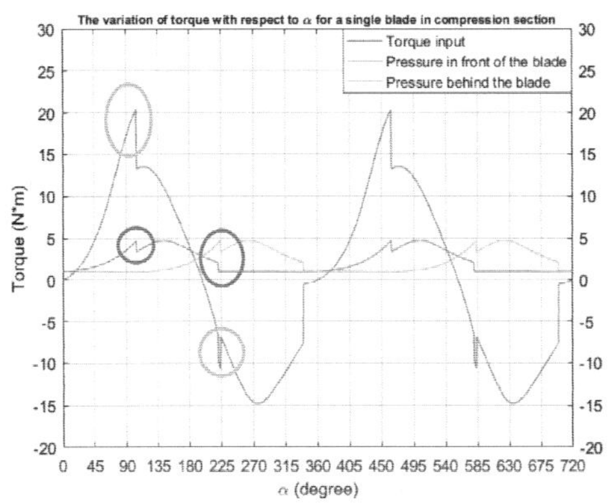

Figure 4. The variation of torque with respect to (α) for a single blade and the variation of the pressure in front of and behind the blade compression section

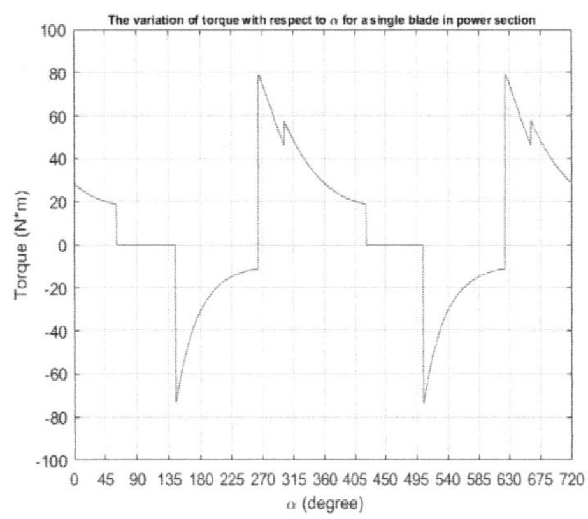

Figure 5. The variation of torque with respect to (α) for a single blade in power section

3.6 Performance of the rotary engine

The performance of swing blade rotary engine can be evaluated by the following formulas in Table 3.

Table 3. The performance of 120 cc swing blade rotary engine

Definition	Formula	Calculated Result	Note
Thermal efficiency	$\eta_{th} = \frac{W_{net}}{Q_{in}}$	63.84%	W_{net} is calculated from PVdiagram
Indicated mean effective Pressure (IMEP)	$P_{imep} = \frac{W_{net}}{V_d}$	2.96~3.75 bar	V_d is the displacement volume (V_{TDC} - V_{BDC})
Indicated horse power	$N_i = \frac{P_{imep} \times V_d \times \omega}{44700}$	6.26~50.16 hp (rpm 1000~8000)	ω : (rpm)
Shaft horse power	$N_s = \frac{W_{torque} \times \omega}{44700}$	2.80~22.38 hp (rpm 1000~8000)	ω : (rpm)
Mechanical efficiency	$\eta_{me} = \frac{N_s}{N_i}$	15.97~44.63%	

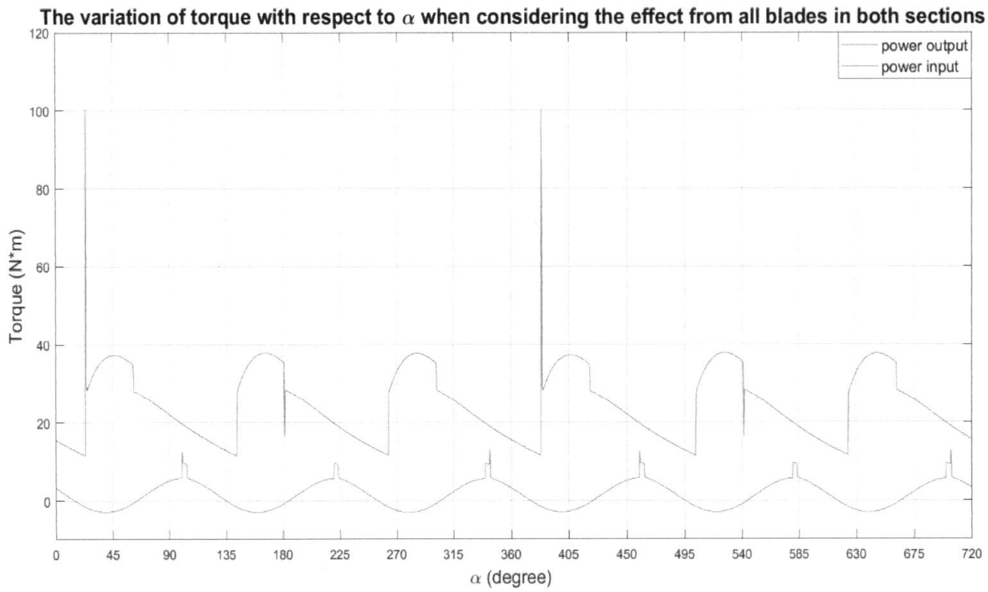

Figure 6. The variation of torque with respect to (α) when combining the compression section and the power section, and the area between two curves is the net power output

4. Conclusions

The innovative eccentric swing blade rotary engine with three swing blades has been developed and internally divided into a compression section, a combustion chamber, and a power section. At the same time, the theoretical model has been established to calculate the trajectory/the tip speed of the swing blade, pressure variation, compression torque and to predict the performance of the rotary engine. This study has proved the viability through the cold model test and the predicted results of theoretical model. For a prototype 120 cc rotary engine, it can provide an estimated power output between 6.26 to 50.16-hp at about 1000 to 8000 rpm. The innovative eccentric swing blade rotary engine is only one-fifth of the multi-cylinder piston engine by volume, weighs only one-sixth, which allows the vehicle to operate with higher power intensity, light in weight and low CO_2 emission.

Appendix

A	Axis of the blade
A_{tot}	Area of the working chamber
B	Tip of the blade
D	The distance from the rotor axis to the tip of blade
e	Eccentricity of the rotor
l	Length of the blade
R	Radius of the cylinder
r	Radius of the rotor
α	The angle between the blade axis and horizontal line
θ	The angle between the blade tip and horizontal line 。
ΔP	The pressure difference between two sides of blade
h	The height of the compression section

IOP Conf. Series: Materials Science and Engineering **449** (2018) 012004 doi:10.1088/1757-899X/449/1/012004

Acknowledgments
The study group thanks the Ministry of Science and Technology for funding support (MOST 107-3113-E-002-008-).

References

[1] K. Yamamoto, "Rotary Engine", Toyo Kegyo Co., Ltd., 1981.
[2] C.B.M. Kweon, "A Review of Heavy-Fueled Rotary Engine Combustion Technologies (OMB No. 0704-0188)". U.S. Army Research Laboratory, (2011), 15.
[3] K.M. Pillai, A.J.S. Mithran, V.K.W. Grips, K.S. Kumar, U.K. Sinha, M.N. Varadarajan, J.J. Isaac and Y.V.S. Murthy, "Design and development of an indigenous 55 HP Wankel Engine", *Proceedings of the International Conference on Aerospace Science and Technology,* (2008), 4.
[4] P. Harrop, "Hybrid vehicle range extenders: goodbye pistons". IDTechEx, *Nature America*, 304(1), (2015), 25.
[5] N.T. Liu, US patent no. 6,082,324 , 9,458,719 and Taiwan patent no. I-468581 , I-479075
[6] H.K. Ma, etc, Taiwan patent no.106218131.

Anupadin1.0

Harshit Jain[1], Arun Kumar[2], Omkar Lawate[3] and Bhavishs[4]

M.V.J. College of Engineering, Bangalore-560067

[1]jainharshit8@gmail.com, [2]aru497@gmail.com, [3]omkarslawate505@gmail.com, [4]bhavishs8055@gmail.com

Abstract. The Unmanned Aerial Vehicle is commonly an Aircraft of feasible sizes with its pilot or controller aboard. Over the years we have seen the usage of UAVs in various multicultural activities that include:

1. Military & Defence
2. Conservation
3. Irrigation
4. Mapping
5. Patrolling, etc.

In this System, we Introduce and propose a new application for the use of drones in the field of Search and Rescue of human body insights of regions affected by Natural and Manmade Calamities and is stuck under debris, where it is impossible for human being to conduct a wide scale search mission. This system uses eight highly efficient and lightweight UAVs that is directed towards eight different geographical axes. Each drone initially detects the zones with high possibilities of human existence and then the drones' uses SWARMING techniques and gives the data about the possibility of human existence in and around the particular geographical positions. This in return increases the drone traffic in that particular geographical positioning leading to a better effective search in the most affected zones. The drone carries a location tracker, a Ground Penetrating Radar, Pixhawk and a specially assembled system that generates Microwave Radiations. This uses a technique of detection of a human body through the generation of Microwave Radiations. This Microwave system operates at a frequency of 1150 MHz and 450 MHz and can easily detect the human being through the heartbeat and the breathing even if the subject is trapped in between a 10 ft. deep construction rumble.

1. Introduction

Natural calamities are scarily becoming a matter of daily news. Each incident results in great loss of human lives and millions of dollars' worth losses to infrastructure. To help overcome the losses of lives to quite some extent, a new drone system known as ANUPADIN 1.0 is devised. Drone is a flying robot or unmanned aerial vehicle which can be remotely controlled or operate being entirely autonomous through the use of different software's and inbuilt technologies. This system is created to locate alive and trapped human beings under the earthquake debris. ANUPADIN means 'rescuer' and 1.0 indicates the first version of this device. This paper is based on the theoretical study of disasters and various technologies available to save human lives, exact feasibility will be confirmed only after implementation. The system, ANUPADIN 1.0 is a set of fully autonomous drones which are spread across 8 different geographical axes. These drones detect the presence of a human body, state of the human body whether alive or dead and the location of a human body by using inbuilt devices such as ground penetrating radar, pixhawk flight controller, Swarm Intelligence, microwave life detection system.

This essential piece of information is passed to the Rapid Action Forces and can help reduce the time taken for detection by the teams and also their workload, thereby increasing the number of people rescued. Also, there is a dire need for a system which can operate in human-inaccessible areas in times of a disaster. The drones offer high degrees of accuracy and operate quickly with maximum efficiency. As the technology in this arena advances, the system will only improve upon its present capabilities.

The drones are entirely autonomous and do not require any skilled human control in the course of their operation.

2. Literature Survey

Earthquakes measuring high on the Richter scale are turning out to be common around the world, causing major destruction to infrastructure and affecting lives of millions of people at one go, almost all of them happening just because of lack of time and reach. Detection of alive and trapped humans in these difficult to access areas is a big factor in reducing the number of people losing their lives in these earthquakes, stuck under the rubble.

Searching for alive and breathing humans beneath the rubble is a time-consuming job and requires the use of technology and skill. Existing methods like detection by a dog through its sniff is turning out to be outdated as they won't be able to access all degrees of rubble. Rapid action forces are responsible for removing trapped humans but at the time of the disaster, this is a time-consuming process which causes the humans to eventually die.

Autonomous drones are being employed in various fields due to its increased mobility in tough places coupled with the speed which drastically decreases the time consumed for a task. To add to it, the technologies that make it are reliable and efficient.

Our solution to the problem described above uses a group of drones equipped with various types of equipment's to detect life. These drones uses techniques to scan the area, share vital information with each other and confirm this information with the local team of rapid action forces along with the exact co-ordinates it has read which cuts down the time taken by rapid action forces in finding the humans and thus, they can, more effectively save the humans.

3. Proposed System

3.1 Principle of working

ANUPADIN 1.0 is a search and rescue drone employed to find the existence of life over a particular search zone. The drone uses effective ground penetrating search radars to identify the presence of life and its state under the surface of the target. ANUPADIN 1.0 will be a team of 8 drones loaded with a server and surveillance screen. The different drones in the unit are directed towards 8 different geographical axes and these inspect and survey the zones. A detailed surveillance report is then sent to the surveillance screen. The process of inspection uses ground penetrating radar to find the location of the victim. Through swarming, the information is passed to the other drones which then increases the traffic in that target zone. A microwave detection system checks the state of the victim's body and if found alive, sends the location of the detected body to the surveillance screen which will be shared with the Rapid Action Forces.

3.2 Ground Penetrating Radar [1]

A GPR system contains the Control Unit, Antenna, and the Power Supply.

The control unit consists of the electronics emitting the radar energy sent by the antenna to the ground. The unit comes with a built-in computer which stores the data for later examination and analysis. The electrical pulse received by the antenna is amplified and transmitted to the ground at a particular frequency. This antenna frequency is a major factor in determining the depth of penetration, more the frequency, lesser is its penetrating power. Also, a high-frequency signal 'identifies' smaller targets.

GPR sends a small pulse of energy towards the material and records the time taken by the signal to return and its strength. This process of sending pulses in a series is called a scan as shown in Figure 1. The reflections occur when there are any differences in the electrical conduction properties or di-electric permittivity in the material. The strength of the signal received is determined by considering the dielectric constants and conductivities of both the materials. While some portion of the energy is reflected back, the remaining portion does travel through the material until it dissipates or until the GPR control unit's time window is closed. The rate of attenuation varies widely and this depends on the properties of the material to which the pulse is striking or passing through.

To enhance the portability and convenience so that it can be easily carried and can access tough locations easily, the device, ANUPADIN 1.0 has a portable and rugged design and is compact and extremely lightweight. The drone operates on low frequencies for increasing the penetration depth. The built-in electronics ensure low battery consumption so that the drones can fly for a longer time. The system is capable of operating in temperatures ranging from -20 to 50 degrees centigrade.

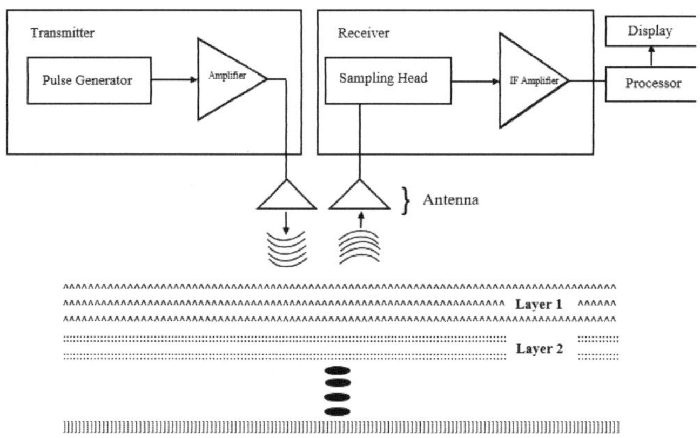

Figure 1. GPR block diagram [2]

3.3 Swarm Intelligence Technique

Swarming is a unique Artificial Intelligence Technique through which the drones can be linked through their mechanical outputs. The drones communicate with each other through the process of swarming. This technique is used for locating the surface effectively and for identifying the axis at which each of the drones is placed. They communicate the surface positioning and pair up successfully.

As soon as there are fewer targets in a particular axis, the paired up drones are informed to reduce the traffic prevalent in that particular zone and these divert to places where the incidence of the target is more. Swarming is a smart search and rescue method which consumes much lesser time than conventional methods.

Simultaneous Localization and Mapping (SLAM) is an advanced mapping technique used in mapping the environment without GPS. The map is generated by processing multiple images that are captured using RGB and depth imaging.

All the images are stitched together to form a single 3-D structure. In ANUPADIN, the SLAM follows a loop pattern to ensure effective search and minimal error. SLAM makes it easier for ANUPADIN to detect the other drones and identify their positions thus reducing the factor of the collision.

IOP Conf. Series: Materials Science and Engineering **449** (2018) 012005 doi:10.1088/1757-899X/449/1/012005

3.4 Microwave Detection System

Microwave Technology is the major life-detecting system employed in the unit. The use of low-frequency microwaves (l band or s band) makes it possible for it to penetrate even through the highly thick rubble. It senses the heartbeat and breaths of the people underneath the rubble. When the microwave beam hits the human body, it is reflected back with additional modulations created by the heartbeat and the lungs as shown in Figure 2. This modulated beam is received and checked for any clutter signals from other objects. These clutter waves are cancelled to maintain sensitivity by an automatic clutter cancellation system.

This system contains three major components. First is the microwave circuit which generates, amplifies and distributes signals to different components. The second is a dual antenna system which consists of two antennas energized sequentially and lastly, a microwave controlled clutter cancellation system to cancel the clutter signals.

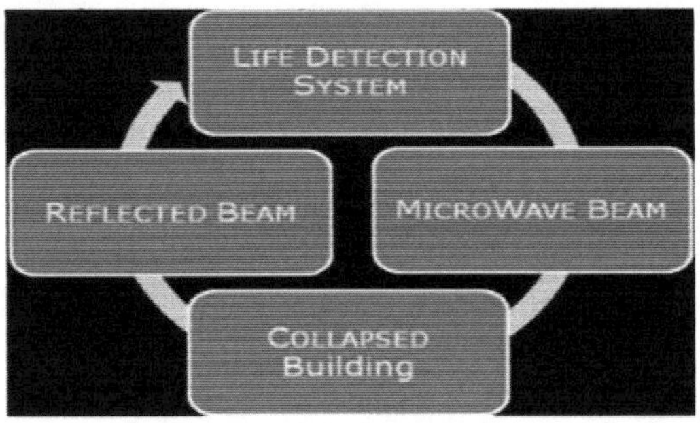

Figure 2. Principle of Operation [3]

3.4.1 Microwave Circuit: This component further consists of phase locked oscillators, directional couplers and oscillators as shown in Figure 3.A Phase-locked oscillator generates a highly stable electromagnetic wave of 2 GHz Range. Power dividers and directional couplers couple part of the transmission power in a transmission line by a known amount, out through another port, often by using two transmission lines set close enough to each other such that the energy passing through one is coupled to the other. In the Microwave life detection system, 10 dB and 3 dB couplers are used. 10 dB coupler is used to divide the power into 1/10th and 9/10th parts while the 3dB coupler divides the power into two equal parts.

The clutter canceller is the soul of the life detection system. It contains the Programmable Phase Shifters, Programmable Attenuator, a Microprocessor based control unit and a RF amplifier which is shown in Figure 4.

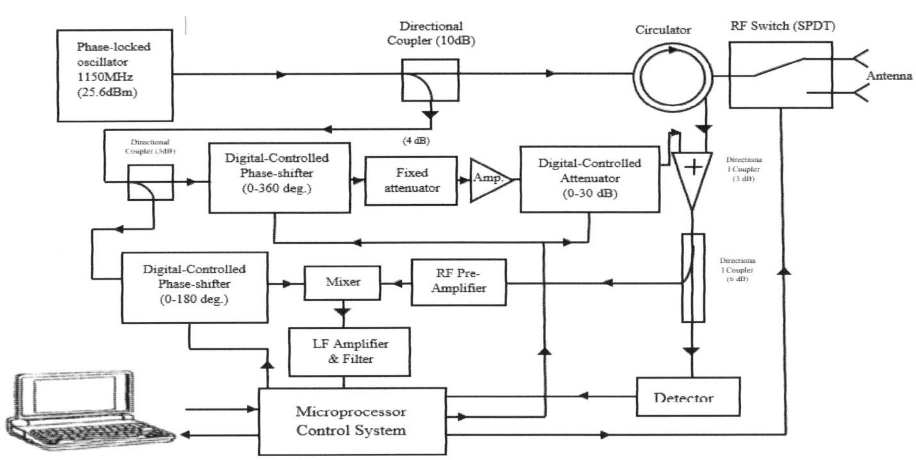

Figure 3. 1150-MHz Microwave Life Detection System. [4]

3.4.2 CLUTTER CANCELLATION CIRCUIT:

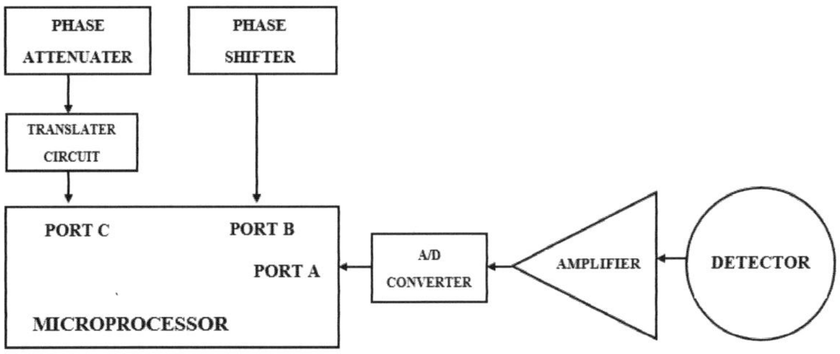

Figure 4. Clutter cancellation Circuit [3]

The clutter canceller is the soul of the life detection system. It contains the Programmable Phase Shifters, Programmable Attenuator, a Microprocessor based control unit and a RF amplifier which is shown in Figure 4.

3.4.3 Working of Life Detection System [5]:The phase-locked oscillator releases an Electromagnetic wave at 1150 MHz and this wave is highly stable. The power output is at 400 mW (25.6 dBm). The EM wave next finds its way through a 10-dB directional coupler and a circulator. The RF switch then energizes the dual antenna system sequentially. The coupler gives off $1/10^{th}$ of the wave amounting to 40 mW and is then divided equally by a 3-dB coupler. One 20 mW part of this output drives the clutter cancellation circuit and the other 20 mW output acts as a local reference signal for the double balanced mixer. The antenna radiates the wave which then penetrates through the rubble for detection. The antenna receives the reflected wave along with the clutter signal from the rubble. The clutter wave is large compared to the reflected wave off the human body. This clutter wave is cancelled by the clutter cancelling signal. The wave emerging of the human body cannot be cancelled by any pure sinusoidal or cancelling signal because of its modulation by the motions on the body. The system consists of two antennas, energized in sequences and it can act as both a transmitter and receiver.

The cancellation circuit contains a phase shifter (0-3600) controlled digitally, an attenuator (4 dB), a RF amplifier (20 dB) and another attenuator (0-30 dB) but controlled digitally. The clutter cancellation circuit gives an output of equal amplitude and opposite phase as that of the clutter.

When this signal is mixed with the signal from the rubble, the clutter signal is cancelled effectively. The output then consists of only the wave reflected off the human body. The output of the 3-dB coupler is made to go through the 6-dB coupler and the signal generated is partially divided. $1/4^{th}$ of the output is amplified by the RF amplifier and is then mixed with a local reference signal. The remaining $3/4^{th}$ of the output is sensed by the microwave detecting system to provide a voltage which ultimately serves as the indicator for the degree or amount of clutter cancellation.

In the double-balanced mixer, the strengthened signal off the human body is mixed with a local reference signal whose phase is controlled by another digitally controlled phase-shifter for better output from the mixer. This output is the motions in the human body like the heartbeat and the breathing, plus some unavoidable noise. This is fed to the low-frequency amplifier (20–40 dB) and a bandpass filter (0.1–4 Hz) and then goes on to be displayed on the monitor. The phase shifter plays the role of controlling the phase of the local reference signal aiming to increase the system sensitivity.

IOP Conf. Series: Materials Science and Engineering **449** (2018) 012005 doi:10.1088/1757-899X/449/1/012005

4.Conclusion

The drone, ANUPADIN 1.0 has found its application in detecting alive human beings buried underneath the earthquake rubble. The entire system is essentially a set of autonomous drones spread in 8 geographical axes which will sense the presence of life trapped, during or after a disaster. The state of the human body is detected through microwaves and that information along with the co-ordinates is passed to the RAF. This technique is compact and as a result, consumes lesser time than manual removal of debris and finding bodies. Adding to that, ANUPADIN 1.0 can reach areas which humans otherwise cannot thereby improving retrieval rates drastically.

Note that this system is automated and does not require any skilled operator. As technology evolves coupled with our research endeavours, ANUPADIN 1.0's capabilities will be enhanced by reducing its size for instance or improving its capability of detection through the debris.

References

[1] Widodo, Iqbal F. Aditama, Khalid Syaifullah, Muthi' a J.Mahya, M.Hidayat, "Detecting Buried Human Bodies using Ground-Penetrating Radar" , Earth Science Research Vol 5,No.2,2016,ISSN: 1927-0542 E-ISSN: 1927-055

[2] Jeong-Soo Lee, Cam Nguyen and Tom Scullion, "Development of a prototype High Frequency Ground –Penetrating Radar system"[2002].

[3] M. D'Urso, "A SIMPLE STRATEGY FOR LIFE SIGNS DETECTION VIA AN XBAND EXPERIMENTAL SET-UP" Centro Ricerche Giugliano, SELEX Sistemi Integrati SpA Via Circumvallazione Esternadi Napoli, zona ASI, Giugliano, I-80014, ItalyProgress In Electromagnetics Research C, Vol. 9, 119{129}, [2012].

[4] Shubhangi C. Deshmukh, Balachandra B.Godbole, Shivaji.M.Sangle, "Detection of Live Human behind the Wall- A Review", IJRAE Vol 1 Issue 11, ISSN:2349-2163,2014

[5] Rajavenkatesan.T, Chinnadhurai. R, Gowthamugan.K.S., Sangeetha.J, Vijayasudha. A, "Detection of Human Beings under building rubble". International Conference on Recent Trends in Engineering Science and Management, ISBN:978-81-931039-2-0,[2015].

Influence of the Earth Rotation on Trajectory of a Returnable Hypersonic Cruise Vehicle

***Kaiqiang Wang, Yanze Hou and Chong Chen**

China Academy of Space Technology, Beijing 100094, China

*wangkaiqiang1988@163.com

Abstract. It is of high difficulty for a hypersonic vehicle to implement a hypersonic cruise flight and return to a scheduled airport without power. Accuracy of mission simulation and analysis is more crucial at this time. Under this circumstance, it is meaningful to discuss whether the earth rotation should be considered in the dynamic model and trajectory design. In this research, the dynamic equations in which the earth rotation is taken into consideration are derived for a hypersonic cruise vehicle. The direct shooting method is utilized for parameterization and discretization of the trajectory, so that the trajectory can be analyzed and simulated through numerical integration. Then, two groups of trajectory simulation results are presented. In each group, the only difference between the two trajectories is that whether the earth rotation is involved, and the two trajectories are compared with each other. For a returnable hypersonic cruise vehicle which returns to the ground without power, whether to involve the earth rotation in the dynamic model is discussed on the basis of the numerical results, and relevant suggestions are given for reference.

1. Introduction

Along with the development of the scramjet, the hypersonic cruise vehicle has been attracting more and more attention. With the use of scramjet, it has better propulsive efficiency and can cruise above the altitude of 20 km with a Mach number larger than 5. After the cruise phase, it can descent without any power until it finally lands at the scheduled airport. Due to its highly fast cruise speed and reusability, the hypersonic cruise vehicle has great potentials in two-stage-to-orbit (TSTO) launch vehicles [1]. Until now, only the US has successfully implemented a few X-43 hypersonic flight tests, in which the hypersonic cruise lasts more than 10s and the Mach number exceeds 7 [2]. However, X-43 was finally controlled to fall into the ocean instead of returning back to the ground.

In the trajectory related researches, some take the earth rotation into consideration, while the others don not. Without considering the earth rotation, Roh and Kim studied on trajectory optimization for a multi-stage launch vehicle [3]; Grant and Braun implemented unpowered trajectory optimization for a hypersonic flight [4]; Manickavasagam et al. optimized the trajectory for long range and air-to-air tactical flight vehicles [5]. As to the reentry trajectory design and optimization, some researches involve the earth rotation into the dynamic model [6-8], while some others don not take it into account [9]. Generally speaking, whether to consider the earth rotation depends on the mission objective, the flight phase considered, total flight range and the landing precision needed in a specific trajectory problem.

However, there are few researches on the similar topic for a returnable hypersonic cruise vehicle. This paper aims to study on the influence of the earth rotation on trajectory and to discuss whether to involve the earth rotation in the dynamic model for the hypersonic cruise vehicle. After the powered climb and cruise phases, it descents and returns to the ground without power. In Section 2, the dynamic equations are derived based on a few basic coordinate systems, and the dynamic model is established in which the earth rotation is taken into account. Section 3 briefly introduces the direct shooting method for parameterization and discretization of the trajectory. In Section 4, two groups of trajectory simulation results are presented, and relevant comparisons and discussions are made with respect to the effects of earth rotation. In Section 5, the research is summarized and suggestions are given.

2. Dynamics modeling

2.1. Coordinates

Five coordinates are used in this paper.

2.1.1. Geocentric coordinate system O_E - $x_E y_E z_E$. The origin O_E is located at the geocenter, axis $O_E x_E$ is in the equator and points to the Greenwich meridian, $O_E z_E$ points to the North Pole, and $O_E y_E$ is determined according to the right-hand rule. The coordinate is fixed with the earth and rotates with it.

2.1.2. Azimuth coordinate system O_1 - $x_A y_A z_A$. The origin O_1 is located at the instantaneous mass center of hypersonic vehicle, axis $O_1 x_A$ is in the local horizontal plane and points to the North Pole, $O_1 y_A$ is perpendicular to the horizontal plane and is straight up, and $O_1 z_A$ is determined according to the right-hand rule which points to the east.

2.1.3. Ground coordinate system O_G - $x_{GL} y_G z_G$. The origin O_G is located at the intersection of the line from the geocenter to O_1 and the ground surface of the earth, axis $O_G x_G$ is in the horizontal plane which contains O_G and points to the direction of the projection of the ground velocity vector $\boldsymbol{v}$ on this horizontal plane, $O_G y_G$ is perpendicular to this horizontal plane and is straight up, and $O_G z_G$ is given according to the right-hand rule.

2.1.4. Body coordinate system O_1 - $x_b y_b z_b$. The origin O_1 is the instantaneous mass center of hypersonic vehicle as explained above, $O_1 x_b$ is along the vertical axis of the hypersonic vehicle and points to its nose, $O_1 y_b$ is in its vertical symmetry plane which is perpendicular to $O_1 x_b$ and sticks straight up, and $O_1 z_b$ is determined according to the right-hand rule. The coordinate is fixed with the hypersonic vehicle.

2.1.5. Trajectory coordinate system O_1 - $x_T y_T z_T$. The origin O_1 is the instantaneous mass center of the vehicle as explained above, $O_1 x_T$ is along the direction of the ground velocity vector $\boldsymbol{v}$, axis $O_1 y_T$, perpendicular to $O_1 x_T$, is located in the vertical plane containing the vector $\boldsymbol{v}$ and sticks straight up, and $O_1 z_T$ is given according to the right-hand rule.

2.2. Coordinate transformations

In this paper, only motion of the hypersonic vehicle in the vertical plane is considered. Then, the following four coordinate transformations are needed.

2.2.1. From geocentric coordinate system to azimuth coordinate system. The direction cosine matrix C_{AE} is obtained through 3 coordinate transformations:

$$C_{AE} = C_z(-\delta)C_y(-90°)C_z(\lambda - 90°)$$
$$= \begin{bmatrix} \cos(-\delta) & \sin(-\delta) & 0 \\ -\sin(-\delta) & \cos(-\delta) & 0 \\ 0 & 0 & 1 \end{bmatrix} \begin{bmatrix} \cos(-90°) & 0 & -\sin(-90°) \\ 0 & 1 & 0 \\ \sin(-90°) & 0 & \cos(-90°) \end{bmatrix} \begin{bmatrix} \cos(\lambda-90°) & \sin(\lambda-90°) & 0 \\ -\sin(\lambda-90°) & \cos(\lambda-90°) & 0 \\ 0 & 0 & 1 \end{bmatrix}$$
$$= \begin{bmatrix} -\cos\lambda\sin\delta & -\sin\lambda\sin\delta & \cos\delta \\ \cos\lambda\cos\delta & \sin\lambda\cos\delta & \sin\delta \\ -\sin\lambda & \cos\lambda & 0 \end{bmatrix} \tag{1}$$

where δ and λ are latitude and longitude of the hypersonic vehicle, respectively.

2.2.2. From azimuth coordinate system to trajectory coordinate system. The direction cosine matrix C_{TA} is given based on the velocity azimuth A_z as well as the flight-path angle γ:

$$C_{TA} = C_z(\gamma)C_y(-A_z)$$
$$= \begin{bmatrix} \cos\gamma & \sin\gamma & 0 \\ -\sin\gamma & \cos\gamma & 0 \\ 0 & 0 & 1 \end{bmatrix} \begin{bmatrix} \cos A_z & 0 & \sin A_z \\ 0 & 1 & 0 \\ -\sin A_z & 0 & \cos A_z \end{bmatrix}$$
$$= \begin{bmatrix} \cos\gamma\cos A_z & \sin\gamma & \cos\gamma\sin A_z \\ -\sin\gamma\cos A_z & \cos\gamma & -\sin\gamma\sin A_z \\ -\sin A_z & 0 & \cos A_z \end{bmatrix} \tag{2}$$

2.2.3. From ground coordinate system to trajectory coordinate system. The direction cosine matrix C_{TG} is acquired through one coordinate transformation based on the flight-path angle γ:

$$C_{TG} = C_z(\gamma) = \begin{bmatrix} \cos\gamma & \sin\gamma & 0 \\ -\sin\gamma & \cos\gamma & 0 \\ 0 & 0 & 1 \end{bmatrix} \tag{3}$$

2.2.4. From body coordinate system to trajectory coordinate system. The direction cosine matrix C_{TB} is obtained based on the angle of attack α, in which sideslip angle is maintained to zero all the time:

$$C_{TB} = C_z(\text{-}\alpha) = \begin{bmatrix} \cos(-\alpha) & \sin(-\alpha) & 0 \\ -\sin(-\alpha) & \cos(-\alpha) & 0 \\ 0 & 0 & 1 \end{bmatrix} \tag{4}$$

2.3. Dynamic equations

It is worth noting that only motion of the hypersonic vehicle in the vertical plane is considered in this paper. According to Newtonian mechanics, the absolute derivative of the ground velocity vector $\boldsymbol{v}$ with respect to time t is given by the following equation, in which the convected acceleration and Coriolis acceleration are taken into account:

$$\frac{\mathrm{d}\boldsymbol{v}}{\mathrm{d}t} = (\frac{\partial \boldsymbol{v}}{\partial t} + \boldsymbol{\omega}_{TG} \times \boldsymbol{v}) + \boldsymbol{\omega}_e \times (\boldsymbol{\omega}_e \times \boldsymbol{r}) + 2\boldsymbol{\omega}_e \times \boldsymbol{v} = \frac{\boldsymbol{F}}{m} + \boldsymbol{g} \tag{5}$$

where m is the mass of the vehicle, $\partial\boldsymbol{v}/\partial t$ is the relative derivative of the vector $\boldsymbol{v}$ in the trajectory coordinate system, $\boldsymbol{\omega}_{TG}$ is the rotational angular velocity of the trajectory coordinate compared to the geocentric coordinate system, $\boldsymbol{\omega}_e$ is the angular velocity vector of the earth rotation, $\boldsymbol{r}$ is the distance vector from the geocenter to the mass center of the vehicle, $\boldsymbol{F}$ represents the external force vector acting on the hypersonic vehicle, and $\boldsymbol{g}$ is the gravitational acceleration vector.

2.3.1. Differential of the ground velocity vector $\boldsymbol{v}$. In the trajectory coordinate system, $\boldsymbol{\omega}_{TG}$ is described as

$$\omega_{TG}^{T}=C_{TA}C_{z}(-\delta)C_{y}(-90°)\begin{bmatrix}0\\0\\\dot{\lambda}\end{bmatrix}+C_{TA}\begin{bmatrix}0\\0\\\dot{\delta}\end{bmatrix}+C_{TG}\begin{bmatrix}0\\-\dot{A}_{z}\\0\end{bmatrix}+\begin{bmatrix}0\\0\\\dot{\gamma}\end{bmatrix} \tag{6}$$

The derivatives of latitude δ and longitude λ with respect to time t are as follows:

$$\frac{\mathrm{d}\lambda}{\mathrm{d}t}=\frac{v\cos\gamma\sin A_{z}}{(r_{e}+h)\cos\delta} \tag{7}$$

$$\frac{\mathrm{d}\delta}{\mathrm{d}t}=\frac{v\cos\gamma\cos A_{z}}{(r_{e}+h)} \tag{8}$$

where v is the flight speed, r_e is the radius of the earth, and h is the altitude of the vehicle. Then, in the trajectory coordinate system:

$$\frac{\partial\boldsymbol{v}}{\partial t}+\omega_{TG}\times\boldsymbol{v}=\begin{bmatrix}\dot{v}\\0\\0\end{bmatrix}+\omega_{TG}^{T\times}\begin{bmatrix}v\\0\\0\end{bmatrix}=\begin{bmatrix}\dfrac{\mathrm{d}v}{\mathrm{d}t}\\ v\dfrac{\mathrm{d}\gamma}{\mathrm{d}t}-\dfrac{v^{2}\cos\gamma}{r_{e}+h}\\ -v\cos\gamma\dfrac{\mathrm{d}A_{z}}{\mathrm{d}t}+\dfrac{v^{2}\cos^{2}\gamma\tan\delta\sin A_{z}}{r_{e}+h}\end{bmatrix} \tag{9}$$

2.3.2. The convected acceleration. It is first expressed in the azimuth coordinate system, and then is transformed in the trajectory coordinate:

$$\begin{aligned}\omega_{e}\times(\omega_{e}\times\boldsymbol{r})&=C_{TA}\begin{bmatrix}\sin\delta\\-\cos\delta\\0\end{bmatrix}\omega_{e}^{2}(r_{e}+h)\cos\delta\\&=\begin{bmatrix}\sin\delta\cos\gamma\cos A_{z}-\cos\delta\sin\gamma\\-\sin\delta\sin\gamma\cos A_{z}-\cos\delta\cos\gamma\\-\sin\delta\sin A_{z}\end{bmatrix}\omega_{e}^{2}(r_{e}+h)\cos\delta\end{aligned} \tag{10}$$

2.3.3. Coriolis acceleration. The angular velocity vector of the earth rotation $\boldsymbol{\omega}_e$ is firstly transformed from the geocentric coordinate to the trajectory coordinate as

$$\omega_{e}^{T}=C_{TA}C_{AE}\begin{bmatrix}0\\0\\\omega_{e}\end{bmatrix}=\begin{bmatrix}\cos\delta\cos\gamma\cos A_{z}+\sin\delta\sin\gamma\\-\cos\delta\sin\gamma\cos A_{z}+\sin\delta\cos\gamma\\-\cos\delta\sin A_{z}\end{bmatrix}\omega_{e} \tag{11}$$

Then, in the trajectory coordinate system, the Coriolis acceleration is formulated as

$$2\boldsymbol{\omega}_e \times \boldsymbol{v} = 2\omega_e^{T\times}\begin{bmatrix} v \\ 0 \\ 0 \end{bmatrix} = 2\omega_e v \begin{bmatrix} 0 \\ -\cos\delta\sin A_z \\ \cos\delta\sin\gamma\cos A_z - \sin\delta\cos\gamma \end{bmatrix} \tag{12}$$

2.3.4. The external force vector $\boldsymbol{F}$. As thrust vector control is excluded in this paper, the thrust is always along the direction of axis O_1x_b.Thus, in the trajectory coordinate system:

$$F^T = C_{TB}\begin{bmatrix} F_t \\ 0 \\ 0 \end{bmatrix} + \begin{bmatrix} -D \\ L \\ 0 \end{bmatrix} = \begin{bmatrix} F_t\cos\alpha - D \\ F_t\sin\alpha + L \\ 0 \end{bmatrix} \tag{13}$$

where F_t is the thrust, D and L are aerodynamic drag and lift, respectively, calculated as follows:

$$\begin{cases} D = q_d S_r C_D \\ L = q_d S_r C_L \end{cases} \tag{14}$$

In the above equation, q_d is the dynamic pressure, S_r is the aerodynamic reference area, C_D and C_L are drag and lift coefficients, respectively. The dynamic pressure is given by

$$q_d = \frac{1}{2}\rho v^2 \tag{15}$$

where ρ is the atmospheric density.

2.3.5. The gravitational acceleration vector $\boldsymbol{g}$. In the trajectory coordinate system, the vector is described as

$$g^T = C_{TG}\begin{bmatrix} 0 \\ -g \\ 0 \end{bmatrix} = \begin{bmatrix} \cos\gamma & \sin\gamma & 0 \\ -\sin\gamma & \cos\gamma & 0 \\ 0 & 0 & 1 \end{bmatrix}\begin{bmatrix} 0 \\ -g \\ 0 \end{bmatrix} = \begin{bmatrix} -g\sin\gamma \\ -g\cos\gamma \\ 0 \end{bmatrix} \tag{16}$$

where g is the gravitational acceleration.

2.3.6. Additional equations. The following additional equations are needed in the meantime:

$$\frac{\mathrm{d}m}{\mathrm{d}t} = -m_c \tag{17}$$

$$\frac{\mathrm{d}h}{\mathrm{d}t} = v\sin\gamma \tag{18}$$

In equation (17), m_c is the mass flow of engine.

2.3.7. The summarized dynamic equations. The dynamic equations are finally shown as follows. The angle of attack α is the only control variable, whilst v, γ, A_z, m, h, λ and δ constitute the state variables.

$$\begin{cases}\dfrac{\mathrm{d}v}{\mathrm{d}t}=\dfrac{(F_\mathrm{t}\cos\alpha-D)}{m}-g\sin\gamma+\omega_\mathrm{e}^2(r_\mathrm{e}+h)\cos\delta\times(\cos\delta\sin\gamma-\sin\delta\cos\gamma\cos A_z)\\ \dfrac{\mathrm{d}\gamma}{\mathrm{d}t}=\dfrac{(F_\mathrm{t}\sin\alpha+L)}{mv}+(\dfrac{v}{r_\mathrm{e}+h}-\dfrac{g}{v})\cos\gamma+\dfrac{\omega_\mathrm{e}^2(r_\mathrm{e}+h)\cos\delta}{v}(\cos\delta\cos\gamma+\sin\delta\sin\gamma\cos A_z)\\ \quad+2\omega_\mathrm{e}\cos\delta\sin A_z\\ \dfrac{\mathrm{d}A_z}{\mathrm{d}t}=\dfrac{v\cos\gamma}{r_\mathrm{e}+h}\tan\delta\sin A_z+\dfrac{\omega_\mathrm{e}^2(r_\mathrm{e}+h)\cos\delta\sin\delta\sin A_z}{v\cos\gamma}+2\omega_\mathrm{e}(\sin\delta-\cos\delta\tan\gamma\cos A_z)\\ \dfrac{\mathrm{d}m}{\mathrm{d}t}=-m_\mathrm{c}\\ \dfrac{\mathrm{d}h}{\mathrm{d}t}=v\sin\gamma\\ \dfrac{\mathrm{d}\lambda}{\mathrm{d}t}=\dfrac{v\cos\gamma\sin A_z}{(r_\mathrm{e}+h)\cos\delta}\\ \dfrac{\mathrm{d}\delta}{\mathrm{d}t}=\dfrac{v\cos\gamma\cos A_z}{(r_\mathrm{e}+h)}\end{cases} \tag{19}$$

2.4. *Geophysical Model of the Earth*

2.4.1. Gravity model. To describe the gravitational acceleration and radius of the earth, the J2 gravity model is used here:

$$g=\frac{\mu_\mathrm{e}}{(r_\mathrm{e}+h)^2}+\frac{3J_2r_a^2\mu_\mathrm{e}(1-3\sin^2\delta)}{2(r_\mathrm{e}+h)^4} \tag{20}$$

$$r_\mathrm{e}=\sqrt{(r_a\cos\delta)^2+(r_b\sin\delta)^2} \tag{21}$$

where μ_e=398600 km^3/s^2 is the gravitational constant of the earth, J_2=1.08264×10^{-3} is the J_2 gravitational constant, r_a=6378.135 km and r_b=6356.912 km are equatorial and polar radiuses of the earth, respectively.

2.4.2. Atmosphere model. The US SA76 model is utilized as the atmosphere model. The atmospheric density and temperature are calculated according to the altitude:

$$(\rho,T)=f_{atm}(h) \tag{22}$$

where f_{atm} is the function to analyze the atmospheric parameters on the basis of the SA76 model.

Then, the sound velocity is given as

$$v_\mathrm{s}=20.0468\sqrt{T} \tag{23}$$

Furthermore, the Mach number is calculated as:

$$Ma=\frac{v}{v_\mathrm{s}} \tag{24}$$

3. Parameterization and discretization of trajectory

The direct shooting method is utilized for parameterization and discretization of the trajectory. First, the continuous time is discretized into a group of time points as follows:

$$t_0 = t_1 < t_2 < \cdots < t_N = t_f \tag{25}$$

where t_0 is the initial flight time, t_f is the final flight time, and N is the number of the discretized time points. These time points can be evenly or non-evenly distributed in the time interval $[t_0, t_f]$.

Then, the corresponding discrete control variables are defined as

$$\boldsymbol{u} = (\boldsymbol{u}_1, \boldsymbol{u}_2, \cdots, \boldsymbol{u}_N) \tag{26}$$

In each time interval $[t_i, t_{i+1}]$, the approximate control variables are obtained through linear interpolation:

$$\boldsymbol{u}(t) = \boldsymbol{u}_i + \frac{t - t_i}{t_{i+1} - t_i}(\boldsymbol{u}_{i+1} - \boldsymbol{u}_i), t_i \le t \le t_{i+1}, i = 1, \cdots, N-1 \tag{27}$$

Then, the trajectory has a limited set of state and control variables, which can be calculated and simulated through numerical integration. In addition, as the number of the discrete time points increases, the precision of the numerical calculation rises.

4. Simulation results and comparison

Table 1. Initial values of the state variables

State variable	Initial Value
Altitude /km	12
Ground velocity /(m/s)	236
Flight-path angle /°	5
Azimuth /°	90
Longitude /°	86
Latitude /°	42

Table 2. Comparison on two trajectory parameters

Group	Earth rotation considered ?	Total range /km	Final altitude /km
1	Yes	1069	4
	No	1055	2.9
2	Yes	985	3
	No	967	2.5

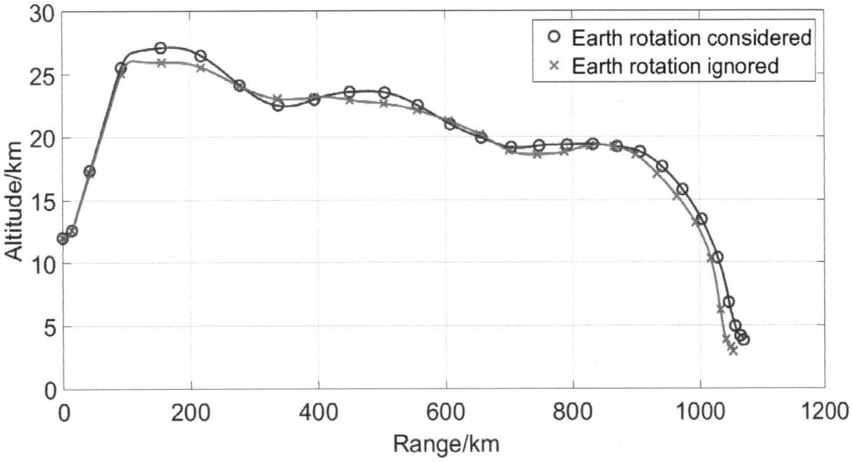

Figure 1. Comparison on trajectories in 1st group.

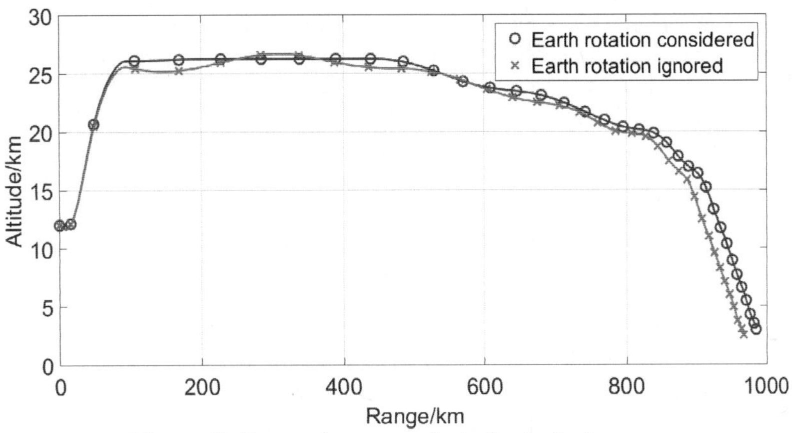

Figure 2. Comparison on trajectories in 2nd group.

In this paper, two groups of trajectory simulation are implemented. It is worth noting that the trajectory is only simulated till the start of approach and landing phase.

In each group, the initial state, the control variable as well as input parameters which includes engine, mass and aerodynamic parameters are all the same. The only difference is whether the earth rotation is considered. In addition, in each simulation, initial values of the state variables are set in table 1. Then, the two groups of results are depicted in figure 1 and figure 2, respectively. The comparison on the total range as well as the final altitude is made in table 2.

It is inferred from the results that the two trajectories in each group are different with each other. When the earth rotation is ignored, the total range is shortened by about 15 ~ 20 km with a percentage of 2%. The distance reduction approximately equals the whole range of approach and landing phase. Meanwhile, the final altitude decreases about 0.5 ~ 1 km with a relatively large percentage of 15% ~25%. Both of these two points may deteriorate the control precision of the unpowered approach and landing flight and may make the final landing fail. As high accuracy is needed for a returnable hypersonic cruise vehicle during the unpowered return flight, it is suggested to take the earth rotation into consideration in order to improve the accuracy of the trajectory simulation and mission analysis.

5. Conclusion

In this paper, the dynamic equations for a hypersonic cruise vehicle are derived, in which the earth rotation is taken into account. The direct shooting method is used for parameterization and discretization of the trajectory. Afterward, based on two groups of trajectories obtained through numerical simulation, the trajectory with the earth rotation included is compared with that without the rotation in each group. It is shown that the earth rotation has effects on the flight range as well as the final altitude. As high accuracy and precision is needed for a returnable hypersonic cruise vehicle, especially in the final unpowered return flight, it is suggested to involve the earth rotation in order to improve the accuracy of the mission analysis and control subsystem design.

Acknowledgments

This work is supported by grants from the Natural Science Foundation of China (NSFC 61403028).

References

[1] Marc A B 2004 *Performance Study of Two-Stage-To-Orbit Reusable Launch Vehicle Propulsion Alternatives* (Ohio: Air Force Institute of Technology Press) p 12–19
[2] McClinton C R, Rausch V L, Nguyen L T and Sitz J R 2005 Preliminary X-43 flight test results *Acta Astronautica.* **57** 266–276
[3] Roh W and Kim Y 2002 Trajectory optimization for a multi-stage launch vehicle using time finite element and direct collocation methods *Engineering Optimization.* **34** 15–32
[4] Grant M J and Braun R D 2015 Rapid indirect trajectory optimization for conceptual design of hypersonic missions *J. Spacecraft and Rockets.* **52** 177–182
[5] Manickavasagam M, Sarkar A K and Vaithiyanathan V 2015 Trajectory optimisation of long range and air-to-air tactical flight vehicles *Defence Science Journal.* **65** 107–118
[6] Zhao J and Zhou R 2013 Reentry trajectory optimization for hypersonic vehicle satisfying complex constraints *Chinese Journal of Aeronautics.* **26** 1544–53
[7] Liu X and Shen Z 2016 Entry trajectory optimization by second-order cone programming *J. Guidance, Control, and Dynamics.* **39** 227–241
[8] Li Z, Hu C, Ding C, Liu G and He B 2018 Stochastic gradient particle swarm optimization based entry trajectory rapid planning for hypersonic glide vehicles *Aerospace Science and Technology.* **76** 176–186
[9] Rahimi A, Kumar K D and Alighanbari H 2013 Particle swarm optimization applied to spacecraft reentry trajectory *J. Guidance, Control, and Dynamics.* **36** 307–310

The Analysis of Compound Control in the Pre-launch Attitude of Internally Carried Air-launched Rocket Based on Simulink

Zhou Huan-Ding*, **Xie Wei-Qi and Cai Yuan-Wen**

Space Engineering University, Beijing101416, China

*651062107@qq.com

Abstract. With the development of modern war and the aggravation of space confrontation, modern battlefield has put forward a strong demand for the hidden, fast and flexible ability to enter space. All countries attach great importance to improving their space emergency launch capability in wartime. Air-launch rocket is based on aircraft, which can be rapidly deployed to any location in the world to achieve full launch. It is very suitable for rapid response to launch performance requirements. Internally Carried air-launched rocket is in the powerless flight state during the period when the main engine is ignited from the carrier, the position, motion and attitude of the carrier rocket are difficult to predict, and the initial position is located. The error of initial condition caused by orientation has a great influence on the precise guidance and navigation of rocket flight process. Prelaunch attitude control not only plays a significant role in the launching process of internally carried air-launched rocket, but also has an important influence on engine ignition and orbit injection accuracy. On the basis of full understanding of the attitude control methods of air launched rockets at home and abroad, also combined with the specific control requirements of pre-launch attitude control, a new control method using grid rudder and RCS and combined of aerodynamic force and direct force is proposed. The simulation of the compound control scheme is carried out by using Simulink. The control characteristics of the scheme are analyzed, and the feasibility of the scheme is verified.

1. Introduction

Internally carried air launched rocket has the characteristics of concealed launch, strong adaptability to payload and low requirements for airframe refitting. Therefore, it is the most potential for military applications in the air emergency launch mode .The study shows that the adjustment of the ignition attitude of the Internally carried air launched rocket directly determines the success or failure of launching the payload and the precision of orbit entry[1]. Therefore, it is necessary to adopt the auxiliary attitude stabilization control method to obtain the appropriate ignition time. After a detailed study of the single control method[2] and the control method of the " stable umbrella +" control method[3] for the forward attitude of the currently internally carried air launched rocket. Although these methods can achieve the purpose of attitude adjustment, there are some problems, such as poor stability, low control efficiency, insufficient anti-jamming ability and so on.

The compound control method of missile (rocket) body attitude has been mature and used in missile and reusable launch vehicle[4][5],but in the field of air launched rocket , there is no research precedent .In this paper, a combined control method of grid rudder and RCS is proposed, which can form a closed loop control by the combined action of continuous aerodynamic force and discrete direct force, thus improving the control efficiency and control precision of the air launched rocket in the air. Improving the stability under the uncertain disturbance in the air.

2. Compound Control Scheme Design

In the design of composite control scheme, the control system is designed into redundant system by referring to the composite control idea of missile and adopting the combination of direct force and aerodynamic force. Through reasonable control allocation, the control efficiency of the control system is improved, and the control ability of the control system under complex external disturbance is improved. The layout design and control ability design of compound control are the key points in the design process.

2.1. Grid rudder design

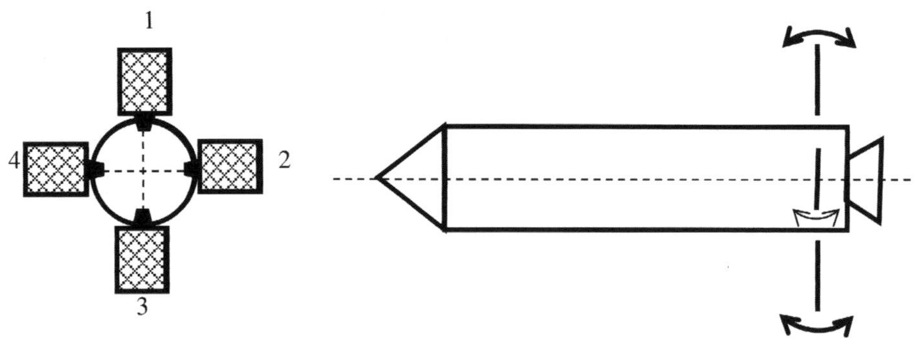

Figure 1. Grid rudder layout and expansion diagram

According to the control characteristics of all kinds of pneumatic rudders and the existing application precedents of pneumatic rudder in carrier rocket, in the design of pneumatic device, a grid rudder of appropriate size is installed at the rear of the rocket to carry out aerodynamic control of the rocket body. In the design, refer to the mature application of the grid rudder on missiles, rockets and spaceships, and select the typical grid shape and layout (Figure.1).However, the effect of grid rudder in this scheme is different from that in other applications, so it is different from other applications in the design of rudder surface rotation. The rudder surface 1/3 is swinging back and forth in the vertical rocket plane of the grid rudder, the rudder surface 2/4 rotates around the symmetrical axis of the rudder plane, and in the process of control and adjustment, the four rudder faces are always in parallel plane.

According to the control characteristics of the grid rudder and the particularity of the attitude control before launching, the grid rudder control device is required to provide the maximum rotation moment for the rocket in the pitch channel, thus helping to realize the attitude adjustment. Therefore, in the selection and use of the grid rudder, it is required to produce as much aerodynamic control torque as possible. At the same time, in order to facilitate the separation of the rocket from the loader, the typical curved grid rudder is used in the design process, and the single rudder surface size is $0.6m \times 0.8m$.

2.2. Side injection device design

On the basis of the research on the RCS control scheme proposed by Northwestern Polytechnic University[6],Combined with the difference of control gain between upwind and leeward surfaces of RCS and the characteristics of pre-launch attitude control of internally carried air-launched rocket[7], the RCS is selected to be arranged on the head of rocket body.According to the literature[8],to reduce rocket costs without affecting rocket payloads, direct force control is provided by satellite attitude control engines.The layout scheme (Figure.2) uses a simple cross distribution,1/3 side jet device to control the pitch channel and the 2/4 side jet device to control the yaw passage.

At the same time, according to the control ability of attitude control engine mentioned in the literature[9], the engine can produce a maximum control force of $3000N$, and at this time it can produce a control torque of about $50000N \cdot m$ to the rocket body in the head.According to the moment of inertia of the rocket, under the action of the torque, the attitude control engine can produce enough control effect to the rocket body, so the control force of the side injection device can be set to $3000N$.

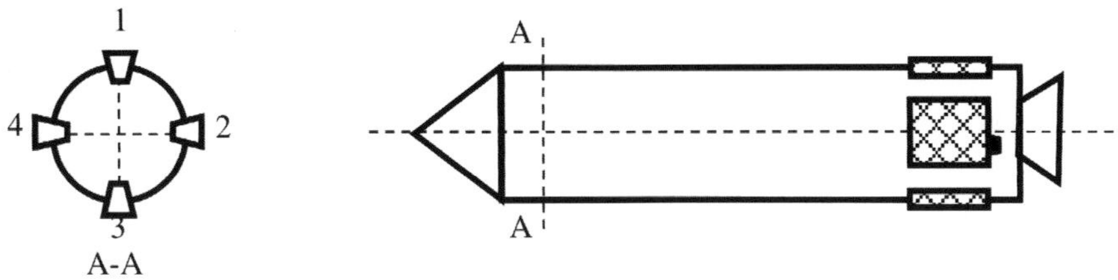

Figure 2. RCS layout and distribution

3. Establishment of Motion Model for Attitude Control

Pre-launch attitude control is a control process to control the internal air-launched rocket in its proper posture,when leaving from the aircraft, so as to help the ignition of the main engine.It is a high angle of attack and low mach control in the atmosphere, which needs to consider various factors such as the influence of gravity comprehensively,loss of altitude and velocity,gusty winds and crosswind effects in atmospheric environment[10].

The aim of this paper is to design a new compound control scheme to improve the rapidity and stability of attitude control before launching.At the same time, the control method is simulated and analyzed.According to the us launch experiment of its internally carried air-launched rocket quick reach[11], the key point for the attitude control of the internally carried air-launched rocket before launch is the change control of the pitch channel at high angle of attack at low mach number.In the control time of 3 to 4 seconds, the change of yaw and roll passage is not obvious. At the same time, adding the spoiler to the QuickReach later basically eliminated the adverse effects caused by asymmetric vortices[12]. Therefore, in the process of establishing the model, only the pitching channel is discussed. In the process of modeling, in order to facilitate simulation analysis, the following reasonable assumptions are put forward.

(1)the change of the height of the control process is less than 100, and the change of atmospheric density is ignored;

(2)the mass of the rocket is constant, and the position of the center of mass is unchanged;

(3)the rocket has axisymmetric aerodynamic shape and stable yawing and rolling channel.

(4)free falling body motion with initial velocity in vertical direction;

(5)the rcs control force is a switch and acts vertically on the longitudinal axis of the rocket.

In the attitude control process, only the change of pitch channel is considered, so the modeling process only needs to consider the dynamic equations of rotation around the center of mass and the kinematics equation of rotation around the center of mass in the rocket coordinate system.

The kinetic equation of rotating around the center of mass:

$$J \bullet \frac{d\omega}{dt} = M \tag{1}$$

To rotate the kinematic equation about the center of mass :

$$\frac{d\varphi}{dt} = \omega \tag{2}$$

J : moment of inertia;

ω : pitch velocity;
M : total moment;
φ : pitch angle.

At the same time, the process of establishing the attitude control model of compound control is as follows:

The combined external force of the rocket consists of four forces:

$$F = Z_a + T + G + \Delta Z_a \tag{3}$$

F : combined external force;
Z_a: aerodynamic force;
T : direct force;
G : gravity;
ΔZ_a :interference caused by tail flow and gusts

Because gravity acts directly on the center of mass, it has no effect on the pitching channel. It only affects the angle of attack of the rocket by influencing the angle of the velocity of the rocket, thus exerting an effect on the variation of the angle of attack of the rocket.

The angle of attack α and the total moment M can be expressed by analyzing the characteristics of attitude control before launching:

$$\alpha = \varphi + \arctan\left(\frac{v_g}{v_z}\right) \tag{4}$$

$$M = M_a + M_t + \Delta M \tag{5}$$

v_g : vertical velocity;
v_z : horizontal velocity;
M_a: aerodynamic moment;
M_t : direct moment;
ΔM : disturbance torque(mainly caused by wake and gusts) .

The expressions of M_a and M_t are as follows:

$$\begin{cases} Ma = q(S_1 L_1 m_\alpha \alpha + S_2 L_2 m_{gr} \delta_{gr}) \\ Mt = TL_3 \end{cases} \tag{6}$$

q, S, L :dynamic pressure, characteristic area, characteristic length;
m_α, m_{gr}: aerodynamic moment coefficient;
δ_{gr} : rudder angle.

The value of disturbance torque ΔM is about 10% of the total torque.

4. Simulink simulation

According to the established system model, the control process is simulated by Simulink (Figure.3) , and the control characteristics and feasibility of the compound control method are analyzed.

The control model module is written according to the formulas (1) and (2), the angle calculation module is written according to the formula (4), and the velocity variation module is based on the aerodynamic characteristics of the initial velocity of the rocket and the different angles of attack. Combined with the effect of external forces to calculate the results. At the same time, the angle of

IOP Conf. Series: Materials Science and Engineering **449** (2018) 012007 doi:10.1088/1757-899X/449/1/012007

velocity inclination is combined. The core module is the interactive realization of two control torque. The following main pneumatic torque and direct force torque control module to explain.

Because the main task of the attitude control engine is to adjust the attitude and orbit of the satellite, its fuel stock is an important factor affecting the satellite's life, so in the process of attitude adjustment before launching, it should be controlled mainly by aerodynamic force. Reduce fuel use of side injection device as far as possible and use direct force as auxiliary control force. In this paper, a control instruction allocation algorithm based on grid rudder saturation is proposed. In the process of attitude adjustment, when the control ability of the rudder is limited, the side spray device will intervene when the control torque can not be provided. Provide direct force to ensure the initiative and rapidity of pre-launch attitude control.

When using the grid rudder to control the rudder, according to the feedback of the rocket angle of attack, the rudder deflection angle is adjusted in real time, so that the grid rudder can be kept at the angle of attack which can produce the maximum pitching moment on the rocket body. The variation of the moment produced by the grid rudder to the rocket body is calculated.

In the use of side spray device for auxiliary control, according to the target state (expected pitch angle 80 ±4 °, expected angular velocity 0 ±4 °/ s) and the control ability of the grid rudder, the side jet device can be used to determine whether the side jet device is involved in the control and produces direct force torque. There are two specific cases:

$$\begin{cases} (1)\ \text{when} \left(\dfrac{2\varphi}{\omega}\right)^{\frac{1}{2}} < \dfrac{J|\omega_e - \omega|}{M}, \quad \text{open side spray device}. \\ \\ (2)\ \text{when} \left(\dfrac{2\varphi}{\omega}\right)^{\frac{1}{2}} \geq \dfrac{J|\omega_e - \omega|}{M}, \quad \text{close side spray device}. \end{cases}$$

In the case of the current aerodynamic torque, the time required to reach the desired angular velocity is longer than the desired angle at the current angular velocity, then the side spray device is opened. On the other hand, the side spray device does not work.

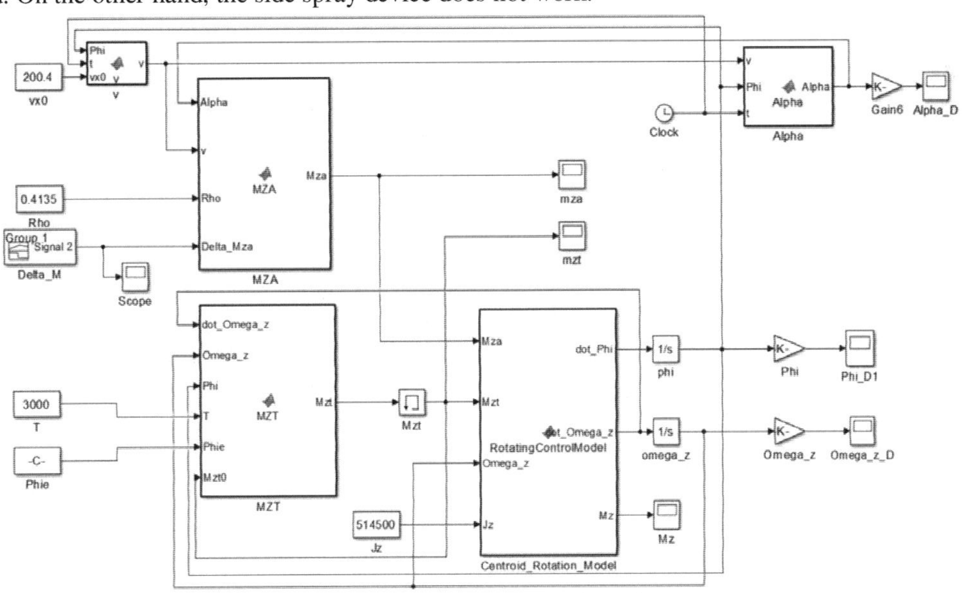

Figure 3. Simulink model simulation diagram

IOP Conf. Series: Materials Science and Engineering **449** (2018) 012007 doi:10.1088/1757-899X/449/1/012007

5. Numerical Example and Analysis

Because of the particularity of attitude control before launching, the control process should reduce the loss of altitude and velocity and at the same time take into account the safety distance of the carrier. Therefore, according to the existing experiments, the control time is required, that is, the desired attitude is achieved at 3.5 seconds. At the same time, the loss of altitude and velocity of the rocket can be minimized under the premise of ensuring the safety of the carrier. In addition, at an altitude of 9000 meters, there are aerodynamic disturbances such as tail current and gusts, so the following disturbance torque is introduced(Figure.4) , because the tail current of the body after it is out of the cabin accelerates the rotation of the rocket's pitching channel, contrary to the control direction. At the same time, in order to verify whether the control ability of the scheme can meet the control requirements, taking the gust disturbance torque is also opposite to the control direction, that is, negative values.

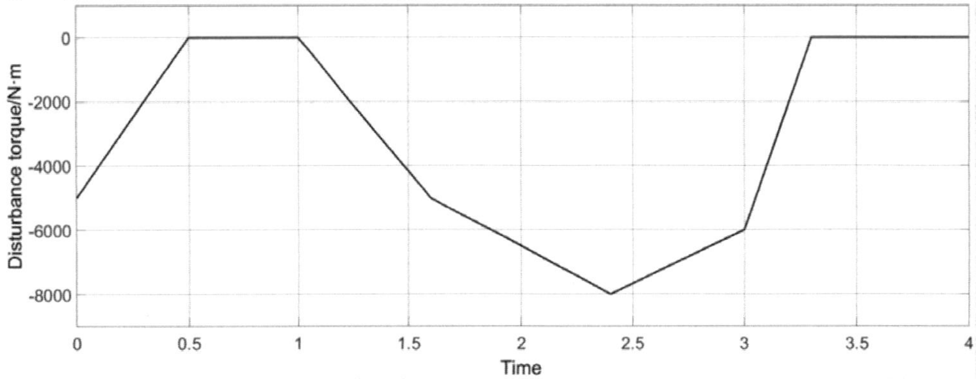

Figure 4. Wake and gust interference

The feasibility of the proposed control scheme is analyzed by using Simulink. According to the experimental data, the initial state and expected state of the rocket are set as follows (Table 1):

Table 1. Initial and expected states of rocket

Parameter	Initial state 1	Initial state 2	Initial state 3	Initial state 4	Expectation
Angle of pitch (°)	20	25	20	25	80±4
Rate of pitch (°/s)	30	30	35	35	0±4

Control of aerodynamic moment and direct force moment of rocket under wake and gust disturbance, and variation of pitch angle and pitch angle velocity in different initial states(Figure.5-Figure.8).

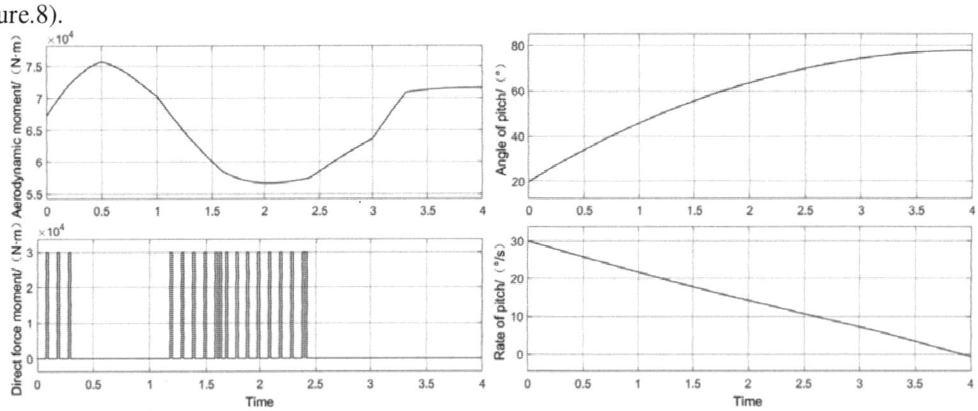

Figure 5. Initial state 1 simulation results

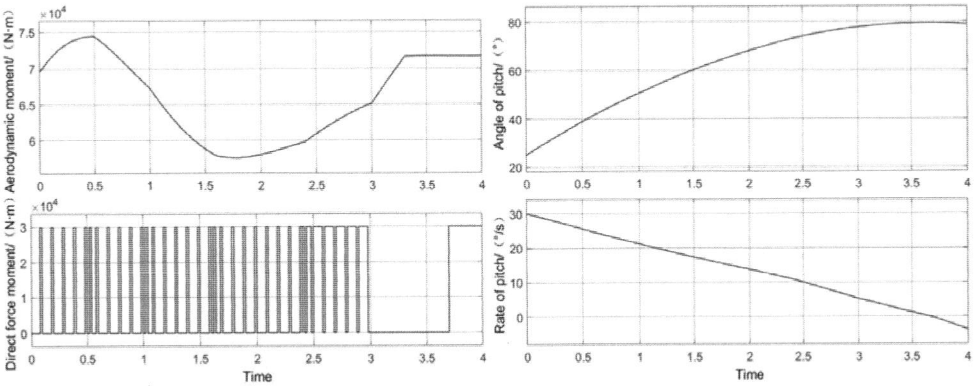

Figure 6. Initial state 2 simulation results

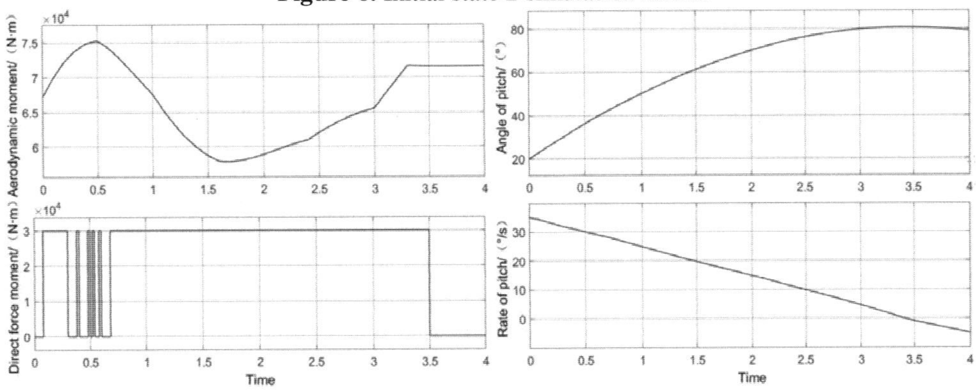

Figure 7. Initial state 3 simulation results

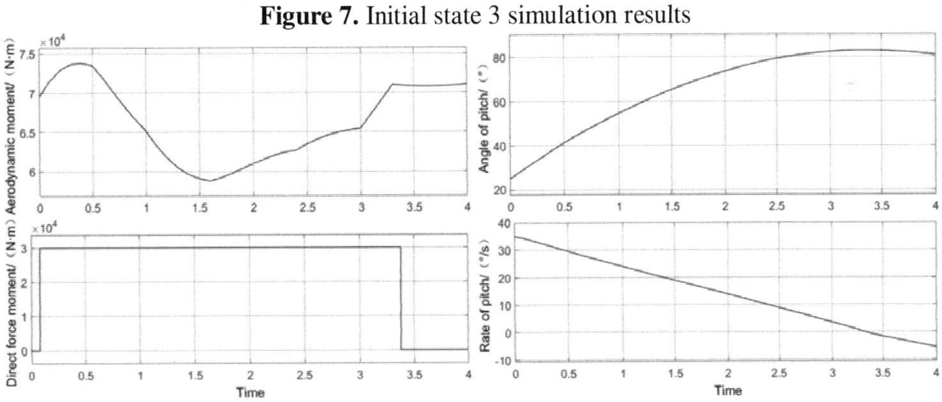

Figure 8. Initial state 4 simulation results

It can be seen from the figure5 to figure 8 that the pitch angle and the velocity of the pitch angle are shown in Table 2 at 3.5 s.

Table 2. Pitch angle and pitch velocity at different initial states at 3.5s

Parameter	Initial state 1	Initial state 2	Initial state 3	Initial state 4
Angle of pitch (°)	72.85	76.08	83.88	89.58
Rate of pitch (°/s)	3.44	2.21	2.01	2.58
Requirements	×	√	√	×

It can be seen from the above simulation results that the control accuracy of the composite control scheme can not reach the expected requirement of the forward attitude of the rocket body in different initial states . The initial state 3 and 4, the control scheme has been saturated control , and the side spray device has always maintained the open state from the beginning to the 3.5 s . Therefore , the simulation study of the next step is carried out , that is , when the initial state is in the range of the control scheme , the control method can complete the forward attitude control task . The simulation results are shown in Table 3.

Table 3. Implementation of control corresponding to different initial states under this control scheme

Initial state parameter		Rate of pitch (°/s)					
		30	31	32	33	34	35
Angle of pitch (°)	20	×	×	×	√	☆	√
	21	×	×	√	√	☆	×
	22	×	×	√	√	√	×
	23	×	√	√	☆	√	×
	24	×	√	☆	☆	×	×
	25	√	√	☆	√	×	×

×:Unable to meet the requirements
√:To meet the requirements normally;
☆:Meet the requirements, and the accuracy is (80±2) ° and (0±2) °/s

The simulation results show that the designed composite control scheme can not meet the expected requirements of the attitude of the rocket body in different initial states. In the initial state 3/4, the control scheme has been saturated and the side injection device has been involved to 3.5s from the beginning. Table 3 shows that the control scheme presents different control effects and control precision in the process of initial state change. In order to ensure better control effect, the pitch angle should be controlled at 23°-25°and the pitch angle velocity should be controlled at 31°/s-33°/s at the time of complete separation of the rocket. In this region, the average control precision of the rocket is the highest.

The simulation results also show that when there is redundancy between direct force and aerodynamic force, this method can achieve the reasonable distribution of control force requirements of two control mechanisms without too complicated control allocation strategy, but there is still room for improvement of control accuracy. In addition, in the subsequent optimization design, the control ability of each control mechanism should be increased appropriately within a reasonable range, so as to ensure that there are redundant control quantities for different initial states, which provides the possibility for increasing the control range of the scheme. At the same time, the adaptive gain value can be set according to different initial states, so that the control range of the system can be improved and the fault-tolerant rate of the system to the initial state can be improved.

6. Tag

In this paper, a method of combining aerodynamic force and direct force is proposed for the pre-launch attitude control of an internally carried air-launched rocket. By means of the grid rudder designed on the rocket body and the satellite attitude control engine, the attitude of the rocket body is fully adjusted to meet the ignition condition in the uncontrolled section between the rocket body and the carrier before ignition. Compared with the stable umbrella and the "stable umbrella" control method carried out in foreign countries, the compound control scheme has obvious advantages: ①. closed loop active control, strong anti-jamming ability. ②. no umbrella surface, so the horizontal velocity loss is small. ③. at the same time of controlling the pitch angle.④. there is no complex rope connection to control the pitch angle velocity. The model is relatively simple and easy to establish in the simulation research, and it is also helpful to improve the actual control efficiency. The simulation results show that the designed scheme fully meets the control requirements and has a strong feasibility in theory and has a certain reference value for the design and research of the air-launched rocket.

Reference

[1] Xu Zhi , Tang Shuo . Parachute Downward Motion Analysis for Air Launch Rocket[J]. (Journal of Projectiles, Rockets, Missiles and Guidance Vol.25 No. 2 Aug 2005)pp:561-563
[2] Sarigul-KlijnM,Sarigul-KlijnN.Trade Studies for Air Launching a Small Launch Vehicle from a Cargo Aircraft[R].AIAA 2005-0621
[3] OKAMURAK, NOGUCHIY, MATSUDA S.ALSET-Japanese Air Launch System Ground Tests and Applications. [C]. SSC.2015-P-20
[4] Zhen Hua-Ping. Study on Composite Aerodynamic Control of Aerodynamic Rudder/Reaction Control System or Hypersonic Vehicle [D]. (BeiJing: Beijing University of Aeronautics and Astronautics, 2013)
[5] Deng Yi-Gao ,Tian Jun-ting,Wang Ya-Feng .Attitude Control Methods of Aircraft Are Reviewed[J].(Tactical missile control technology,2006,2)pp:7-13
[6] Xu Zhi, He Min, Tang Shuo. Research on the Attitude Control Scheme of RCS Separated by Internal Rocket Launcher [J].(Flight dynamics,2011,2) pp:70-73
[7] Li Jian,Li Bin,Liu Xian-Min. The Effect of Jet Interactionon Aerodynamic Characteristicat High Angle of Attack for Missile[J].(Journal of Projectiles, Rockets, Missiles and Guidance Vol. 33 No. 4 Aug 2013)pp:153-156
[8] He Min. Study on Initial Trajectory Stability Method of Interior-mounted Space Launch Vehicle[D].(Xi'An:Northwestern Polytechnical University,2010)pp:37-48
[9] Chang Li-Ping, Tang Shuo, Xu Zhi. Study on Attitude Control Method Before Launch of interior Space Rocket[J].(Science Technology and Engineering,2012,9)pp:2100-2104
[10] Yang Zhou,Chen Jian-Ping, Zhang Hong-Yin. Analysis on the Safety of an Internal Gravity Rocket Moving out of the Capsule[J].(Aeronautical Computing Technique, 2016,3) pp:50-53
[11] FacktorLeporeD, PadavanoJ. Process Toward First Flight of the Quich Reach TM Small Launch Vehicle [C]. SSC,2007-III-3
[12] Sarigul-KlijnM, Sarigul-KlijnN. Flight Testing of a Gravity Air Launch Method to Enable Responsive Space Access [R].AIAA 2007-6146

IOP Conf. Series: Materials Science and Engineering **449** (2018) 012008 doi:10.1088/1757-899X/449/1/012008

Research Progress of Key Technologies for Typical Reusable Launcher Vehicles

Zhong Ya[1*], Liu Danghui and Wang Chen

[1]Department of Aerospace Science and Technology, Space Engineering University, Beijing, China, 101416

[*]zy_930703@yahoo.com

Abstract. In recent years, driven by space X and other commercial companies, the requirement of low cost and fast response to transport, reusable launch vehicle technology is attracting the world's attention. Aerospace agencies in various countries have stepped up research and made some progress in the related technologies of reusable launch vehicles. The development of reusable launch vehicle has become an important direction of space development in the future. First, the development status of reusable vehicles, the Space Shuttle Mode, the Falcon-9 Mode, the Launch Mode on board, and the Airborne Aircraft Mode, are introduced. Second, the structure design, the propulsion system, the recycling scheme, the thermal protection system and the launch costs of each mode is analyzed. Third, some related development suggestions are put forward in terms of the carrier configuration, the design of recycling scheme, the choose of propulsion system and the thermal protection system for reusable launch vehicle designer.

1. Introduction

With the development of Space Vehicles, people's activities extend to space. At present, multi-stage disposable launch vehicle is still the main mode of delivery. Although multi-stage launch vehicle is a reliable way, the high launch cost greatly restricts the development of civil, commercial and military activities in space. Since the 1960s, space agencies have tried to design a fully reusable vehicle to reduce launch costs and increase launch frequencies. Reusable launch vehicle is an effective way to reduce the cost of space transportation system. from the space shuttle to x series testing machines to the falcon - 9 launch vehicle, which has been developing rapidly in recent years, reusable launch vehicle is an important step to capture the commanding heights of space in the future.

In recent years, with the increasing importance of space and the rise of commercial spaceflight, there has been more interest in finding new ways to recycle vehicles to reduce costs. Table 1 shows the current research progress of typical reusable launcher vehicles.

Therefore, by analyzing the current development of various reusable launch vehicles, we can provide references for the development of reusable launch vehicles in our country. According to the different reusable methods, the development of reusable launch vehicle is divided into the space shuttle mode, the falcon-9 mode, the launch mode and the space plane mode. Four modes are further analyzed from aspects of the structural design, the propulsion system, the launch return design, the thermal protection system and the launch and development cost.

Table 1. Research Progress of Reusable Launcher Vehicles

Type	Model	Research Progress
Parachute Recovering	Saturn I	Complete the subscale test
	SRB	Engineering application
	Energy boosters	Propose a concept
	K-1 Rocket	Complete the imitative start
Vertical return	Sub-Mars	Complete the flight test
	DC-X	Complete the low altitude flight test
	Charon	Complete the low altitude flight test
	New Shepard	Complete once reusable flight test
	MOD	Complete the vertical take-off and landing in different locations
	Falcon-9	Complete the recycle multiple times, breakthrough the corresponding technologies
Winged return	Baikal's booster	Propose a concept
	Adeline System	In the research
	XS-1	Phase II flight test
	Leopard Cat III	Propose a concept
	RLV-TD	Complete the first-time flight test
	X-37B	Engineering application, multiple reused
	X-38	Complete the imitative start
	RLV-T1	Complete the multiple hovering flight

2. Space Shuttle Mode

2.1 Structure Design

The space shuttle consists of three main components: an orbiter, an outer fuel tank, and two solid rocket boosters (SRB).The liquid propellant is installed in an independent external fuel storage tank, which does not occupy the space of the orbiter and effectively reduces the size and weight of the orbiter. The outer fuel storage tank burned down in the atmosphere after being separated, and the orbiter and two solid rocket boosters were recycled. The orbiter is divided into front, middle and rear sections. The middle section is the payload compartment. although it can carry up to 29 t of payload at a time, due to the wing, the payload ratio is relatively low. The rear section is equipped with three main engines, two orbital motor engines and a reaction control system. while meeting the power requirements, the rear section is also equipped with elevon, flap, vertical tail, rudder and speed brake. in the process of blindly pursuing the advanced nature of the system and the ability to adapt to various tasks, the space shuttle system has very complicated structure and low reliability. Many new structures are proposed in the space shuttle, such as the design of the tail wing, which provides a good reference for the next space shuttle.

2.2 Propulsion System

The main engine of the space shuttle is SSME (also called RS-25 in the company), as shown in figure 1, which is the largest phase combustion liquid hydrogen / liquid oxygen engine in the world at present. Each main engine provides about 1.8MN [1] thrust during take-off. The main engine and propellant parameters are shown in table 2.

During flight, SSME thrust can be adjusted from 67% to 109%, and the launch usually uses 104.5 % thrust. Although adjustable thrust is achieved, some studies show that when engine thrust exceeds 104.5%, it has obvious influence on engine reliability [2]. Based on SSME, a new type of three-mode heat pipe jet engine (PRRDJE) is proposed, which makes full use of oxygen and hydrogen fuel in the atmosphere and has a fast combustion speed and can meet the thrust requirements of the whole space flight [3]. Its simple structure and light weight are beneficial to the optimal pneumatic layout and good operation economy, and it can be an ideal engine for single-stage on-orbit and horizontal take-off and landing carriers.

Figure 1. SSME engine

Table 2. SSME engine and propellant parameters

Length	4.3m
Diameter	2.4m
Thrust	2090kN/1670kN
Specific Impulse	452s/366s
Thrust-weight Ratio	73.1:1
Cell Pressure	2747 psi
Exit Pressure	1.049 psi
Burning Time	520s
Propellant	Liquid Oxygen/ Liquid Hydrogen
Propellant Mixture Ratio	6.0

2.3 Recycling Scheme

The space shuttle recovery by taking off vertically and landing horizontally, SRB recovery adopts traditional parachute recovery; The orbiter of the space shuttle is landing horizontally and the external fuel tank is not recycled. The SRB's recovered parachutes and payloads are the largest. Each booster USES three main umbrellas to slow down and land at sea. Orbiter adopts the recovery method of horizontal landing, which is intended to be close to aircraft mode in operation, so as to shorten the time between two launches, and enable the recovery and reuse of the spacecraft's expensive equipment, so as to achieve the goal of fast, reliable and cheap. But because of the space shuttle's design of return blindly pursue advanced and the adaptability of each task, its structure is too complex, maintenance costs rose sharply, it takes 14,000 standard hours to replace the surface heat shield alone, the ground operation costs accounted for 45% of the cost [4]. So, compared with the benefits of recycling, the benefits are not worth the loss. In addition, the complex structure leads to the reliability of the shuttle need to be improved.

2.4 Thermal Protection System

In high temperature (1260-1500 °C), mainly is the nosecone, the lower part of the fuselage and the leading edge of a wing which have Carbon/carbon composites and metal materials with good reusability. Carbon/carbon composites have excellent properties of high strength, high modulus and low density. However, Carbon/carbon composite materials do not have excellent thermal insulation performance, so it is necessary to fill the insulation between the outer shell and the body structure of the composite materials. The main metal material is refractory metal, which has the disadvantage of high density, such as Cobalt - and nickel-base high temperature alloys, tantalum, niobium and so on.

High temperature ceramic heat insulation tiles are generally used as heat protection materials on the nose and the lower surface of the fuselage, while low temperature ceramic heat insulation tiles are used on the side and vertical tail of the fuselage. The two kinds of heat insulation tile materials are the same, only there are differences in density, thickness and coating. The advantages of these two kinds of insulation tiles are small thermal expansion coefficient, low thermal conductivity, light weight and good thermal stability. However, due to the brittle ceramic materials and long maintenance period and high cost, they are easy to fall off and have the disadvantage of low reusability. Flexible heat insulation material is used on the upper surface of the fuselage with the temperature lower than 400 °C.

2.5 Development and launch costs

The United States has developed and launched five space shuttles, each of which has been spent $2 billion and has been launched over a hundred times, it costs $500 million every time, and each time it has to go back with a lot of manpower and resources to maintain and repair. In 1985-1988, the shuttle launch costs increased by 85%, to $90 million.

3 "Falcon-9" Mode

In the eighties of last century, McDonnell Douglas's dc-x test machine first proposed a vertical take-off and landing approach in order to develop a reusable vehicle that would be more reliable than the space shuttle and provide lower launch costs and faster responsiveness. The Charon and New Shepard of Blue Origin, the Mod of Armadillo Aerospace, the Xombie of Masten and the RLV-T3 Rocket of Link space in China all have tested the vertical take-off and landing scheme successively. In December 2015, SpaceX launched its Falcon-9 rocket has realized the launch and recovery of the inland landing field for the first time, reclaiming the carrier rocket's sub-stage on the premise of completing a satellite launch. Vertical take-off and landing aircraft developed by global commercial space corporation is shown in figure 2.

3.1 Structure Design

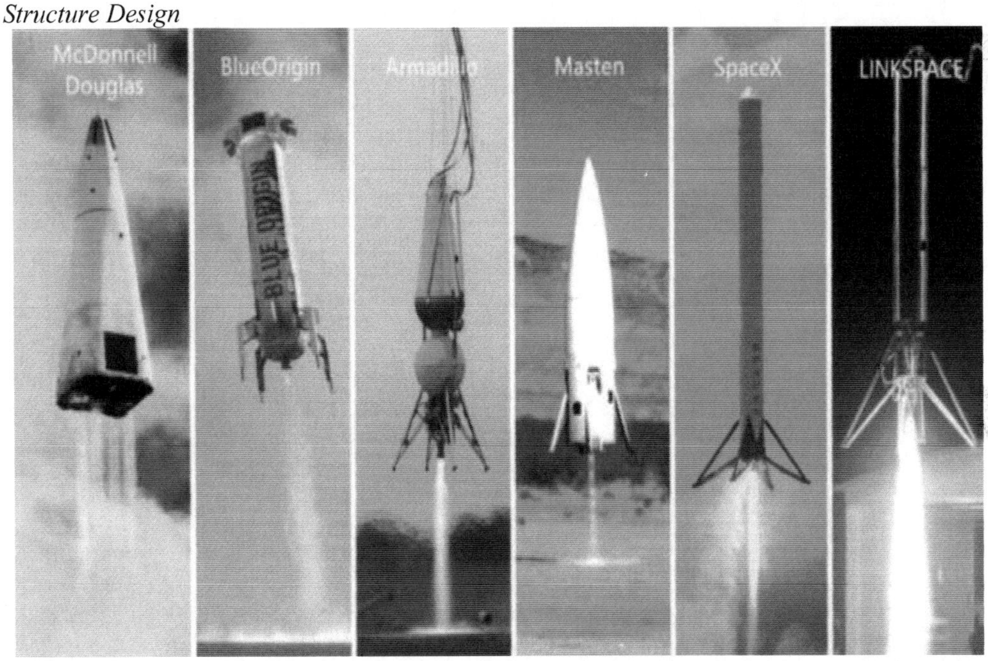

Figure 2. Vertical Take-off and Landing Aircraft Developed by Global Commercial Space Corporation

The reusable launch vehicle using reverse vertical take-off and landing basically adopts the traditional rocket configuration, which is simpler and easier to control than complex structures such as winged and lifting body. "Falcon-9" added four folded x-wing grills to the arrow body at one sub-stage for recovery steering and four landing buffers. The first stage of New Shepard is loaded with six footpads, and any failure of one of them can still ensure a successful landing. It uses a pneumatic rudder to slow down and has a small capacity to recover damage at sea [5].

3.2 Propulsion System

The first stage of "Falcon-9" selects 9 Merlin1D engines, including 1 central engine, and the remaining 8 engines are in a ring around the central engine, as shown in the figure 3 [6]. This structure simplifies the design and assembly of engine parts and simplifies the manufacturing process. The

Merlin 1D engine uses liquid oxygen/kerosene propellant. The main parameters are shown in the table 3. Compared with previous versions of Merlin series engines, the Merlin 1D can achieve throttling control from 100% to 70%. Later, the engine has been improved to work at a minimum of 40% full thrust. In addition, the basic mixing ratio of Merlin engine is controlled by the size of the supply pipeline, and a small amount of fuel flow is regulated by a throttle valve controlled by a brake, which can provide precise mixing ratio control, so as to realize the reverse thrust control of variable thrust vector in the process of return, with good throttling and stopping performance. In order to adapt to the full thrust range of Merlin 1D, the dense treatment of propellant is adopted to increase the mass of propellant.

Figure 3. Falcon-9 Octaweb structure

Table 3. Merlin1D engine parameters [7]

Length	4.3m
Diameter	2.4m
Thrust	2090kN/1670kN
Specific Impulse	452s/366s
Thrust-weight Ratio	73.1:1
Cell Pressure	2747 psi
Exit Pressure	1.049 psi
Burning Time	520s
Propellant	LOX/RP-1
Propellant Mixture Ratio	6.0

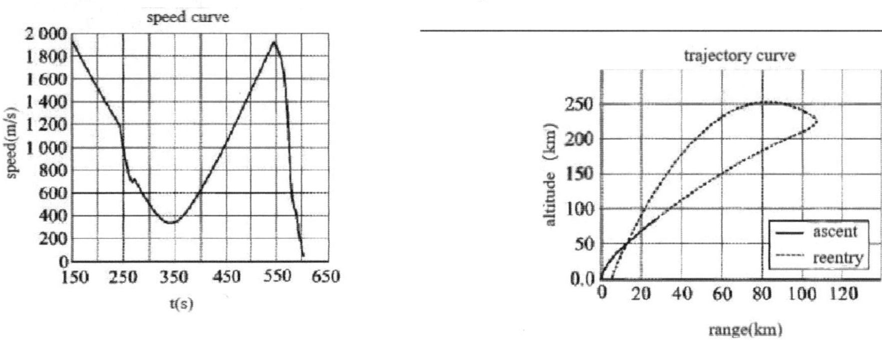

a) First stage recovery trajectory curve **b)** First stage speed curve

Figure 4. Faclon-9 first stage recovery flight path

3.3 Recycling Scheme

During the flight of "Falcon-9", after the first and second stage separation of the rocket at an altitude of about 80km and the inertial impulse reaches the highest point, the flight attitude control is carried out by using the RCS control system, grid wings and so on, and finally the recovery of the land launch site is realized. Figure 4 shows the trajectory of the "Falcon-9" sub-class recovery flight [8].

When landing at the landing site, the primary center engine starts and decelerates to a landing speed of 2m per second. After 10 seconds of landing spray, the four landing supports were opened to complete the landing.

3.4 Thermal Protection System

"Falcon-9" shell and dome are made of aluminum and lithium alloy, with reliable welding technology to ensure the strength of 5 structures. The inter-stage is made of carbon fiber aluminum core composite structure, and the second and the first stages are made of high strength aluminum lithium alloy. In addition, the chief technology officer of Space X announced BLOCK 5 "Falcon-9" will be launched at the end of 2017, hich uses a more heat-resistant titanium grille and a new class of thermal protection coatings, and a reusable thermal protection material for rocket engines. A thermal protection layer is added at the end of the first stage to reduce the ablation of the shell and engine by backward thrust.

3.5 Launch Costs
The cost of the "Falcon-9" launch was $61.2 million. With only one sub-level recovery, the launch price can be reduced to $45.36 million for each re-use, down 25.9%. Two times can be reduced by 36.3%. If the two-stage is also reusable, the cost of a rocket can be reduced to 10 % of its initial cost when it is launched 10 times and 1 % when it is launched 100 times [9].

4. Launch Mode on board

The carrying and launching mode here is not traditional carrying, but uses reusable carriers to carry the orbiter to a certain height and release the ignition and launching. Like small launch vehicle "Bloostar" of Spanish Zero2infinity, which uses warm balloons as its base stage to launch it into the air, then fires to put payloads into orbit. Similarly, "Launcher-One" and " Pegasus " small launch vehicles of virgin galactic use the aircraft as a base class, and after launch, the plane's landing is like its base class recovery.

4.1 Structure Design

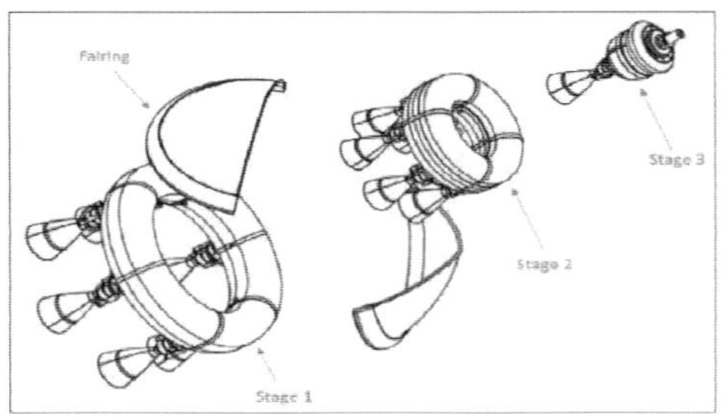

Figure 5. the ring structure of "Bloostar"

"Bloostar" has chosen a unique three-stage parallel configuration. the first and second sub-stages adopt annular tanks. the first sub-stage surrounds the second sub-stage and the third sub-stage is in the center position. as shown in figure 5. This circular configuration makes the rocket compact and easy to assemble. The use of a balloon to rise to the edge of space and reignite the launch is almost impervious to air resistance. Therefore, the design of shape is not restricted by aerodynamics. To avoid the influence of atmospheric environment, the fairing is designed in a collapsible form, with a carbon fiber framework and a multilayer composite surface, the inner layer of which is a gas barrier film.

4.2 Propulsion System
"Bloostar" has six liquid-oxygen/methane engines at one sub-stage, each providing 15kN thrust, and six liquid-oxygen/methane engines at 2kN thrust at the second stage. The engines of the second stage and the third stage are the same, but only one. The first and second stage tanks are made of carbon fiber composite materials wrapped together. The interior is equipped with anti-sloshing plates to prevent the low-frequency shaking of the propellant. The surface tanks is also coated with multi-layer insulation materials to control the evaporation loss of the propellant. In addition, all tanks are pressurized with helium to maintain the overall structural strength of the rocket, as shown in table 4 for the parameters of the methane engine used in the "Bloostar" first stage. The use of an aircraft to carry an orbiter is more restrictive than "Bloostar" uses balloons to bring the orbiter to its intended altitude.

The engine is used in high altitude, low pressure and vacuum environment, so the engine nozzle can be optimized in a single working environment to achieve optimal performance. The engine size is small, and 3D printing technology can be used. In terms of the development of liquid oxygen methane engine, Shanghai aerospace academy plans to develop a new reusable vehicle using liquid oxygen methane engine after 2020, realizing more than 20 times of reuse, and the launch cost is expected to be reduced by about 30%.

Table 4. Liquid Oxygen/Methane Engine Parameters [10]

Vacuum Thrust	15kN
Cell Pressure	10 psi
Vacuum Specific Impulse	345
Total Mass Flow Rate	4.4kg/s
Propellant	Liquid Nitrogen/Methane

4.3 Launch Scheme

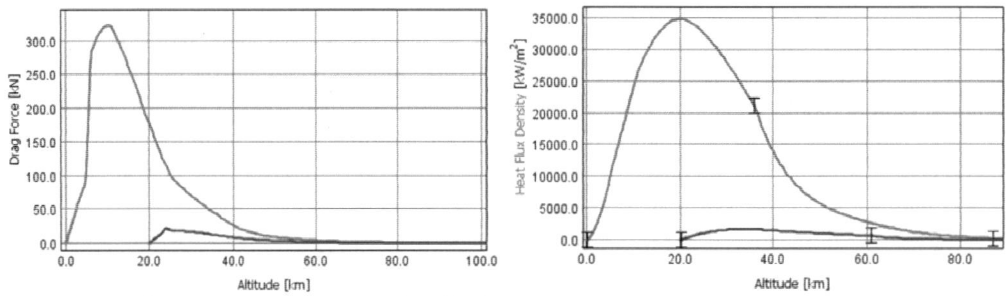

Figure 6. The left picture shows the relationship between ground launch (red) and 20 km launch (blue) drag and altitude. The right one shows the relationship between heat flux and altitude

"Bloostar" uses balloons to reach altitudes of 20-25km, followed by three sub-stage engines firing at the same time, providing 104kN of thrus [11]. Figure 6 shows the trajectory and heat flux heat curve of "Bloostar". Compared with a standard ground launch, the use of a vehicle to transport the orbiter to the stratosphere height can reduce the speed increment needed to enter the orbit. Zero2infinity estimates that, using the ASTOS tools, the balloon is launched at higher altitudes compared to the ground, the speed delta required by the balloon is reduced by 8%, the emission resistance is reduced by about 20 times, and the heat flux is reduced by a factor of 10. In addition, after separation, alloon and other vehicles can be used as high altitude repeater, which saves costs. Using balloon to launch has the advantage of aerial launch, but it has the disadvantage of too long uplift time. Besides, during the flight test, the balloon gas overflow caused by unpredictable gust of wind also occurred, resulting in the balloon damage phenomenon.

4.4 Launch Costs

Zero2infinity plans to launch it in the second half of 2018, a 600-kilometer SSO orbit with a capacity of 75kg, which costs $4.5 million, about $60,000 per kilogram.

5. Airborne Aircraft Mode

The RLV of the space plane model adopts a horizontal landing mode, uses atmospheric resistance to slow down before landing, has excellent rapid response capability and hypersonic flight capability, can develop into a military hypersonic vehicle with strategic deterrence, adapts to the requirements put forward by the new military space application requirements on the space delivery system, and various space powers and mainstream organizations have begun to study their related technologies one after another. DARPA and the Boeing company cooperation project, NASA's XS-1 X-37 b and India RLV-TD are reusable launch vehicles of Airborne Aircraft Mode .

5.1 Structure Design

The recovery part adopts horizontal landing method, but there are great differences in configuration and launching method. The XS-1 is a vertical launch and a two-stage parallel configuration, the first is a skyplane configuration, and the first is hunched over the top; The x-37b and RLV-TD are launched vertically, but they need to be put into orbit by an atlas rocket. As shown in figures 7, XS-1 is partially reusable patterns. In the field of pneumatic layout, the three are similar, it's not a purely lift design, it's designed to be called a hybrid body, a blunt head cone shaped body and a slender wing, and a pair of pointy delta wings. In order to install the upper stage on the back of the fuselage, the XS-1 eliminated the double vertical tail, increased the aspect ratio of the wing, and added the trailing edge with the small wingtip of the control rudder face. The belly of the fuselage is designed to be similar to the shape of a flat plate, forming a semi-lift structure similar to the wing, which increases the lift force in the re-entry stage while reducing the wing area, and enhances the maneuverability and endurance performance.

5.2 Propulsion System

The XS-1 is fitted with two SpaceX Merlin1D engines, which were introduced in the introduction of the falcon 9. The AJ26 (NK33) engine is not reusable at the upper stage. The X-37B is equipped with an AR2-3 rocket engine and uses methylhydrazine and N2O4 as propellants. Although methylhydrazine is highly toxic, the technology is relatively mature.

5.3 Thermal Protection System

Both the XS-1 and the X-37B are made of a composite fuselage, using a TUFROC at the leading edge of the hottest wing and the nose wing area. TUFROC is a new type of thermal protection materials, it has been applied to the high temperature area of aircraft, can withstand the high temperature of 1697 °C, and under the condition of high temperature to maintain strong aerodynamic shape. TUFI/AETB was used in the body's windward and other parts of the body in the sub-high temperature area. In low temperature area leeward side body etc., choose the reusable of CRI Blankets, which can guarantee the durability of the aircraft's aerodynamic shape and thermal protection due to its structural properties [12]. However, there are also disadvantages to heat shield. First, all the equipment is arranged on the surface of the body. Second, it increases the body weight and reduces the payload mass. In addition to the heat shield, the XS-1 has added an insulating layetabler on the outside of the fuselage, using a hard, smooth coating on the outside.

Figure 7. conceptual XS-

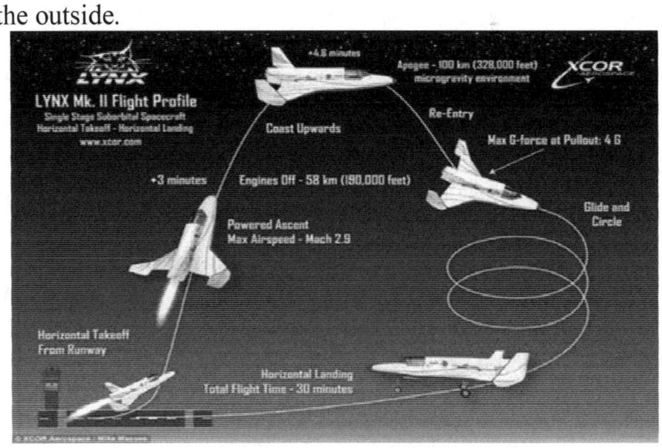

Figure 8. “Lynx III” running track

5.4 Recycling Scheme

The return phase of XS-1 is the same as the “bobcat III” horizontal landing mode proposed by XCOR, as shown in figure 8. It uses a horizontal landing approach originating from the space shuttle. Although the horizontal landing recovery method can recycle the expensive space equipment, it still faces the test of autonomous landing technology, re-entry atmospheric return technology and thermal protection system. And it has to lower the cost of recovery maintenance, or it is going to go the way of the shuttle. Just like the X-38 recovery is also horizontal landing, but it has no power to taxi in the re-entry process, and then uses the wing umbrella to slow down the landing. Compared with the horizontal landing of the wing, the use of the wing umbrella can reduce the consumption of propellant, but also lead to the reduction of the payload carrying capacity.

5.5 Launch Costs

In 2014, DARPA applied for $10 million in research grants for the first phase of the XS-1 project, with a $140 million budget for the second phase. The XS-1 is designed to be a reusable military space jet, with a payload of 1360kg, 10 consecutive satellite launch missions in 10 days, and each launch costs $5 million [13].

6. Summary

Based on the above analysis and taking reliability and low cost into consideration, the following Suggestions are put forward for the research of reusable launch vehicle:

1) Throughout the development history of RLV, the mainstream technological approach for development is to develop new technologies and construct new configurations and design a vehicle that is very different from the traditional one-time carrier rocket. The pursuit of advanced technology will also bring great risks. Like America's space shuttle, excessive pursuit of technology advancement, finally do more harm than good. The focus of development should be shifted from the pursuit of advanced technology to the goal of high reliability and low cost.

2) At the present stage, RLV should take rocket power as the main body and inhale combined power as the future development direction. In the 1990s, after the failure of many countries' aspirated single-stage in-orbit aircraft program due to the great technical difficulties, they all chose the gradual development model successively. So at this stage, we need to develop a more mature rocket power, and on the basis of the original launch of the rocket-propelled rocket, the development of a propulsion system that is more suitable for RLV.

3) Thermal protection system has always been a technical difficulty that restricts the development of RLV and is also the core part of RLV. In order to solve the heat matching problem in high temperature area, X-37B put forward the idea of thermal insulation integrated design. In this regard, the core is to improve the practicability and economy of reusable thermal protection materials for carriers, upgrade the performance of thermal protection structures and materials in different temperature zones, and enhance the reusability and structural lightness of the thermal protection system.

Reference

[1] Thompson J R. Space shuttle main engine[J]. Test Plan Technology Test Bed Program Rockwell International Corp September, 1980, 70(Suppl 1):S75.

[2] Report of the SSME assessment team[J]. Nasa Sti/recon Technical Report N, 1993, 93.

[3] Zhang guozheng. Thermodynamic Duct Jet Engine of Three Model—New Type of Space Plane Main Engine[J]. Journal of Propulsion Technology, 1988(1):69-73+97.

[4] Wang Zhenguo, Luo Shibing, Wu Jianjun. Research progress on reusable launch vehicle [M]. National defense university of science and technology press, 2004.

[5] Yang Kai, Qu Jing, Cai Manrui. Analyse of World Launch Vehicle Development in 2017[J]. Missiles and Space Vehicle, 2018(1).

[6] SPACEX. OCTAWEB [EB/OL]. SPACEX NEWS http://www.spacex.com/news/2006/01/01/octaweb (2013-07-29) [2018-03-16].

[7] Lindsey, Clarks S. SpaceX Merlin 1D [EB/OL]. Webcite. http://www.webcitation.org/6BKKlzLoQ (2011-06-08) [2018-3-18].

[8] Mou Yu, Chen Yu, Yu Guangxue. Enlightenment of innovative technology of American SpaceX falcon rocket [J]. Winged Missiles Journal, 2016(06):3-8+35.

[9] Liu Bo, Shen Lin. Success and cost analysis of " falcon" 9 rocket's first stage marine recovery test [J]. Aerospace China, 2016(5):22-25.

[10] Yang Kai, Xue Jingning. Preliminary Analysis of Spanish Bloostar Rocket Launched by Balloon[J]. Space International, 2017(6):38-42.
[11] Herbert J. Kramer. Zero 2 Infinity: Introduction of a game-changing launch technology for small satellites [EB/OL].EoPortal Directory.https://directory.eoportal.org//eoportal/satellite–missio ns/v-w-x-y-z/zero-2-infinity [2018-03-18].
[12] Zhou Zhiyong, Ma Bin, Zhang Cui. Reusable Thermal Protection System for Orbital Test Vehicle X-37b[J]. Spacecraft Engineering, 2016, 25(4):95-101.
[13] Chen Youlong, U.S. space weapon XS - 1 space plane [J]. Tank & Armoured Vehicle, 2016(4).

IOP Conf. Series: Materials Science and Engineering **449** (2018) 012009 doi:10.1088/1757-899X/449/1/012009

Simultaneous Transmission and Reception Transponder for Nano-satellite

Yuan Fu

College of Aerospace Science and Engineering, National University of Defense Technology, Changsha, Hunan, P. R. China

yuanfu@nudt.edu.cn

Abstract. The feasibility of simultaneous transmission and reception (STAR) in the same frequency for space transponder in nano-satellite is discussed. The demand of the self-interference cancellation for the transponder is analysed by space-ground link budget, and a STAR superheterodyne architecturebased transponder is also presented. Based on the conventional space TT&C frequency scheme, a feasible frequency scheme and digital signal processing process for the STAR transponder are expatiated. A FFT based carrier acquisition method is modified to adapt to such STAR transponder.The implementation of the STAR transponder is simulated. The simulation results show that with properly deployingthe passive and active self-interference cancellation modules, the self-interference can be cancelled to meet with the demand of demodulation which also implies that the STAR transponder presented in this paper is feasible.The simulation results also show that the radio frequency self-interference cancellation(RFSIC) does matter in such STAR transponder by realizing 29dB self-interference cancellation, and a total self-interference cancellation of about 68dB can be realized by the combination of passive suppression and RFSIC. The numerical results show that the digital self-interference cancellation is not necessary for the transponder when the transmission power of ground-satellite uplink is high enough, and the phase noise power is increased by the digital self-interference module.

1. Introduction

The simultaneous transmission and reception (STAR) technology is based on the transceiver working at the same carrier frequency in the same time to achieve two-way data transmission. Compared with the traditional FDD, the communication capacity of the system can be doubled in theory, and the spectral resources can be relieved [1, 2]. As an alternative to 5G mobile communication technology, STAR has attracted wide attention in the field of civil and commercial applications.

The current research on STAR is mainly aimed at the wide-bandwidth self interference cancellation and related performance based on OFDM modulation using the direct conversion transceiver architecture. The application scenes are mostly focused on indoor short distance wireless transmission. The research on STAR in satellite communication scene is less investigated. Ref [3] explored the military applications of STAR, and gives different modes of operation for listening and jamming. Ref [4]discussed the feasibility of STAR for the low orbit satellite communication system, but did not consider the space TT&C communication system. Ref [5] designed a test platform for narrow band STAR and validated its performance. Ref [6] discussed the application of STAR in personal communication in the satellite communication environment, but there was no specific discussion for

IOP Conf. Series: Materials Science and Engineering **449** (2018) 012009 doi:10.1088/1757-899X/449/1/012009

TT&C system either. By far, there were no research results on STAR technology in the field of TT&C system. In this paper, based on the standard unified carrier TT&C system and the widely used transceiver architecture, a preliminary implementation scheme of the STAR TT&C transponder is given and simulated.

2. The Design of STAR Transponder

2.1. Feasibility Analysis

Whether the STAR is feasible for TT&C communication system is mainly determined by the high performance self interference algorithms which can cancel the self interference signal and its distortion under the receiver noise floor so as to ensure the demodulation performance of the receiver.The noise floor of the receiver can be expressed as:

$$F_t(\text{dBm}) = -174 + NF + 10\log B, \tag{1}$$

where NF is the noise figure, B is the signal bandwidth.

Assuming that the bandwidth of a receiver is 2MHz and the noise figure is 3dB, the noise floor of the receiver can be calculated to be -108dBm. Assuming the transmit power is 27dBm, and then the self interference cancellation capability should be at least 135dB, which will satisfy the request of transponder demodulation. According to the research results of the self interference cancellation for 20MHz OFDM signal in [7], the RF self interference cancellation of 45dB and the digital self interference cancellation of 55dB were achieved, thus extra 35dB interference cancellation is required. If a three port network (circulator / duplex) with 20dB isolation is used to transmit and receive signal with single antenna in the RF front end of the transceiver, then the total interference cancellation can reach 120dB which still not meet the self interference suppression demand of 135dB. In addition, according to [5], the interference suppression ability and the signal bandwidth are closely related, i.e.

$$SI_{RF,\max} = 10\log_{10}\frac{\Delta\tau}{T_s} = 10\log_{10} B\Delta\tau, \tag{2}$$

where B is the symbol rate, $\Delta\tau$ is the time delay between two taps in the RFSIC. Therefore, using the analog domain interference cancellation method and the implementation structure in [7], when the signal bandwidth is reduced from 20MHz to 2MHz, an extra 10dB self interference cancellation can be obtained. Then the total interference suppression ability should be 20+45+10+55=130dB, and the self interference signal can theoretically cancelled to the noise floor of the receiver. The demodulation performance of the receiver is basically guaranteed.

A low orbit satellite with a height of 600km is taken as an example. Assuming that the spaceflight transponder for the satellite works in the S band, then the farthest transmission distance (the elevation angle of the ground antenna is 5 degree) is about 2330km, and the corresponding maximum path loss is 168dB. Suppose the ground station uses a parabolic antenna with 10m caliber to transmit signals, and the transmit power is 100W. According to the link budget in Table 1, the minimum receiving power of the satellite receiver is -79dBm.

Table 1. Link budget for a nano-satellite

Frequency (MHz)	2400
Transmit power (dBm)	50.0
Ground Station Gain（dB）	45
Ground Cable Loss（dB）	3
EIRP (dBm)	92
Ground Station Point Loss（dB）	0.5
Ground Station Polarization Loss（dB）	0.5
Ground Station Path Loss（dB）	168
Satellite Point Loss（dB）	0.5
Satellite Polarization Loss (dB)	0.5
Satellite Cable Loss dB)	1
Satellite Gain（dB）	0
Minimum Received Power(dBm)	-79

According to this minimum received power, if the receiver bandwidth is 2MHz, the noise figure is 5dB, then according to

$$P_{r\min} = -174 + NF + 10\log B + SNR_{\min} \tag{3}$$

the minimum signal to noise ratio of the receiver is obtained of 27dB. Based on the BPSK modulation, the demodulation signal-to-noise ratio of 10.5dB can achieve a bit error ratio of 10^{-6} theoretically.

Considering the demodulation loss, the signal to noise ratio for demodulating the uplink signal can set by 13dB, then there are still about 14dB allowance for degradation the self interference cancellation ability. Thus the self interference and noise power just needs to less than -108+14 = -94dBm to guarantee the demodulator work properly. Then, the demand of total interference suppression capability is 114dB, which reduces the requirement of 135dB self interference suppression capability of the system. To sum up, by designing appropriate suppression strategy and interference suppression algorithms with the actual application background, the STAR exploiting for space transponder is feasible.

2.2. The Implementation Scheme of STAR Transponder

Superheterodyne architecture is the most popular architecture used in the FDD transponder. In a superheterodyne transceiver, impacts of the leaked or reflected transmitted signal to the local oscillator are weakened by adopting twice frequency translation processes. However, additional filters are needed to filter out the unwanted sideband signals in the RF channel. The receiver can also reduce the requirement of high Q value filter in frequency selection, and assign the total gain of the entire receiving channel to radio frequency, intermediate frequency and baseband, and also increase the stability of high gain amplifier [8]. The disadvantage of superheterodyne architecture is that such transceiver can induce interference such as inter-module interference (parasitic interference), especially image interference. But this can be done by designing a high Q bandpass filter at the front end of the mixer to filter out the image frequency signal. Compared with the superheterodyne architecture, the direct conversion architecture does not have mirror frequency interference, and the receiver only contains high frequency LNA and mixer, which decrease the device quantity of the transceiver. However, there are some problems such as local oscillator leakage, DC bias and I/Q mismatch existing in the direct conversion architecture. Most of the literatures on STAR are focused on the performance and interference suppression ability with direct transformation architecture, such as the interference suppression ability of I/Q imbalance [9] or the leakage of local oscillator and so on.

Although the superheterodyne architecture is complex, there are no problems such as local oscillator leakage, DC bias and so on. In addition, combining the characteristics of the unified carrier TT&C, the carrier acquisition needs the upper and lower band information of the signal, and the transceiver assigns the gain to the baseband, the intermediate frequency and the radio frequency blocks, which can greatly guarantee the linear characteristic of the nonlinear components (power amplifier, mixer and so on). A superheterodyne architecture for STAR transponder is shown in Figure 1.

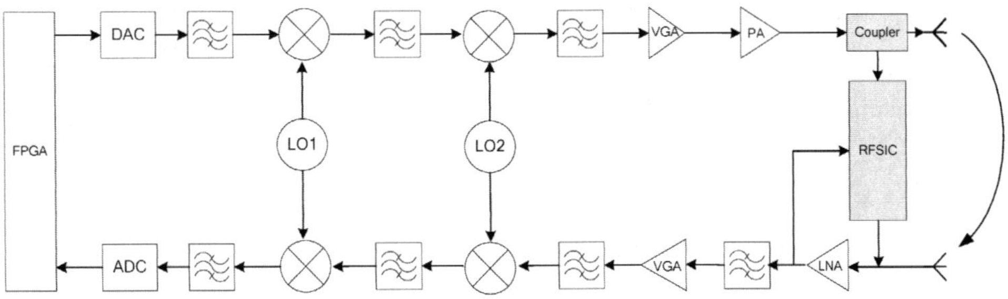

Figure 1. A superheterodyne architecture for STAR transponder implementation.

In Figure 1, the modulation signal firstly pass through DA converter, low pass filters, band-pass filters and mixers, VGA, driver amplifier and power amplifier, then the RF signal are coupled to two separate RF signals. One RF signal is sent to the RFSIC module, and the other is transmitted by the antenna. The received signal passed through the low noise, band-pass filter and VGA, down frequency conversions, and then accesses into the ADC for sampling. In Figure 1, the RFSIC is implemented as the scheme depicted in figure 2, in which the RFSIC contains N taps, and each tap includes a power splitter, a I/Q mixer, a vector modulator and ADC circuits[10]. The samples of the ADCs are used to calculate the control parameters of the vector modulator with LMS algorithms in MUC/FPGA. Assuming a sample of the PA output is $d(k)=d_i(k)+id_q(k)$, N taps outputs are $\boldsymbol{x}(k)=[x_1,\cdots,x_n,\cdots x_N]$ and the control parameters are $\mathbf{w}(k)=[w_1,\cdots w_n\cdots,w_N]$. Then based on the complex LMS algorithms in [11], the calculation procedure of $\mathbf{w}(k)$ can be expressed as follows:

STEP1: Initialized $\mathbf{x}(0)=\mathbf{w}(0)=[0\ 0,\cdots,0]^T$;

STEP2: If $k\geq 0$, calculate $e(k)=d(k)-\mathbf{w}^H(k)\mathbf{x}(k);\mathbf{w}(k+1)=\mathbf{w}(k)+\mu_c e^*(k)\mathbf{x}(k)$.

The real part and imaginary part of $\mathbf{w}(k)$ calculated by the LMS algorithm are used as the I/Q control input of the vector modulator after the DAC circuit, and then adjusting the amplitude and phase of the self interference signal in RFSIC[12, 13]. The specific adjustment process can refer to the data sheet of the selected vector modulator chip.

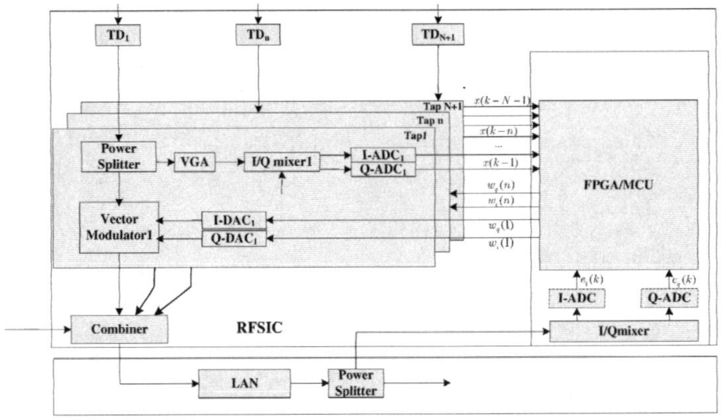

Figure 2. Implementation scheme of a RFISC.

In Figure 1, FPGA implements modulation and demodulation and digital domain interference cancellation algorithm. For low earth orbit nano-satellite systems, the Doppler frequency shift of S band transponder is generally less than 100kHz. In order to overcome the influence of telemetry tone to the demodulation of telecommand tone in the STAR transponder, the frequency of telemetry tone in the STAR transponder is selected as 250kHz with data rate 4096bps, the frequency of telecommand tone is 8kHz with data rate 2000bps. According to the frequency scheme, even considering the 100kHz Doppler shift, the telecommand tone and the telemetry tone still have a frequency interval of 150kHz, which provides convenience for design the demodulation filter in the digital signal processing procedure. In addition, the maximum frequency of the range tone is 100kHz, which still have 50kHz frequency interval to the telemetry tone even with 100kHz Doppler frequency shift. It is convenient for designing the digital FIR filters in demodulation of the range tone.

The digital processing flow within FPGA is shown in Figure 3. In Figure 3, the DSIC block is followed by a 256 order FIR bandpass filter to filter the telemetry tone and its harmonic components. The passband of the FIR filter is 100kHz. The implementation of the FFT capture block needs to eliminate the influence of the DC offset, which can be realized by setting the value of the DC offset to zero after the transformation. The order of FFT is 4096. Following the DDC1 block is a FIR filter with passband of the telecommand symbol rate, and the filter order is 512. Then the real part or imaginary part of the FIR's output are selected to mixing with local telecommand subcarrier in DDC2. The output of the DDC2 block is then filtered by a 2048 order FIR with passband of symbol rate.

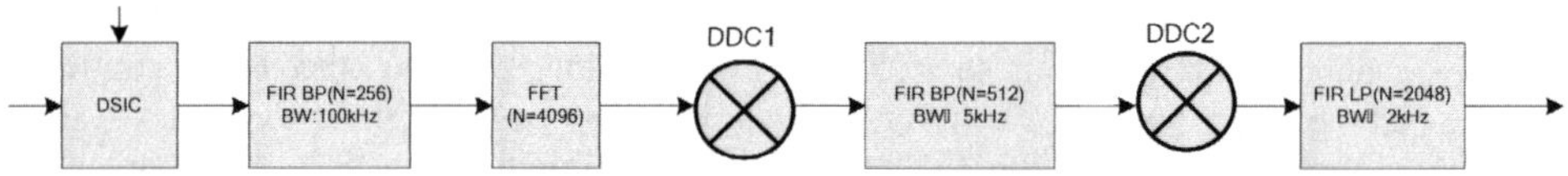

Figure 3. The DSP process in FPGA

2.3. Signal Model

According to the STAR transponder architecture in figure 1, the signal model of the whole system can be expressed as follows if the VGAs and the filters are working in the linear zones. PCM-BPSK-PM modulation is exploiting in the transponder, the complex baseband signal can be expressed as:

$$\theta_s(t)=\sum_{k=0}^{\infty} m_x d(t-kT)\cos(w_0 t), \tag{4}$$

where m_x is the modulation index, $d(t-kT)$ is the NRZ PCM symbol, w_0 is the angle frequency of the telemetry tone or telecommand tone. With two converters, the signal can be up-converted to the radio frequency, expressed as

$$x_{m2}(t)=\exp\{j((w_1+w_2)t+\theta_s(t)+\phi_1(t)+\phi_2(t))\} \tag{5}$$

where the $\phi_1(t)$ and $\phi_2(t)$ are the phase noises, w_1 and w_2 are the angle frequency of the local oscillators. Then the signal is amplified by the power amplifier, which can be expressed as [14]

$$x_{PA}(t)=a_1 x_{m2}(t)+\sqrt{2}a_2|x_{m2}(t)|^2-3/2\,a_3 x_{m2}(t)|x_{m2}(t)|^2 \tag{6}$$

where a_2, a_3 are the gain coefficients of the second and third order non-linear components. By using the path loss between the transceiver antennas as passive suppression, the self-interference signal enters into the receiver antenna is

$$x_{SI}(t)=h_L(t)*x_{PA}(t-\tau) \tag{7}$$

where $h_L(t)$ is the path loss. Then the received signal is

$$y(t)=y_{SOI}(t)+x_{SI}(t)+n_{th}(t), \tag{8}$$

where $y_{SOI}(t)$ is the signal of interested from the far end, $n_{th}(t)$ is the thermal noise. Assuming the coupled signal of the PA output is $\tilde{x}_{PA}(t)$, then the signal at the input port of the LNA is

$$\begin{aligned} y_{RF}(t) &= y(t) - \sum_{i=1}^{N} \omega_i * \tilde{x}_{PA}(t - i\tau) \\ &= y_{SOI}(t) + \left[x_{SI}(t) - \sum_{i=1}^{N} \omega_i * \tilde{x}_{PA}(t - i\tau) \right] + n_{th}(t) \\ &= y_{SOI}(t) + x_{SI}^{\text{Res}}(t) + n_{th}(t) \end{aligned} \tag{9}$$

where ω_i is the control parameter, and $x_{SI}^{\text{Res}}(t)$ is the residual self interference signal. Considering the LNA is a nonlinear device, the output of the LNA can be expressed as

$$y_{LNA}(t) = \left(k_1 y_{RF}(t) + \sqrt{2} k_2 \left| y_{RF}(t) \right|^2 - 3/2\, k_3 y_{RF}(t) \left| y_{RF}(t) \right|^2 \right) + n_{LNA}(t) \tag{10}$$

where k_2, k_3 are the gain coefficients of the second and third order non-linear components. Then with two converters, the signal frequency is down-converted to the baseband expressed as

$$y_{m2}(t) = y_{LNA}(t) \exp\left\{ -j\left(w_2 t + w_1 t + \phi_1(t) + \phi_2(t) \right) \right\} \tag{11}$$

where w_1 and w_2 are the angle frequency of the local oscillator. Then the ADC converts the amplified baseband signals into digital signals for further processing in the digital base-band.

3. Simulation Results

Based on the above implementation scheme, the STAR transponder is simulated. The simulation parameters are set as follows: the local oscillator phase noise is modelled as a Gauss random variable with a mean of 0.45 degree, and the power amplifier is modelled based on the SSPA Sale model. For a complex baseband signal $x(t) = a(t)\exp\{j\phi(t)\}$, the AM-PM effect can be expressed as [14].

$$\text{AM}\left(a(t)\right) = \frac{\alpha_G \left| x(t) \right|}{1 + \beta_G \left| x(t) \right|^2},\ \text{PM}\left(a(t)\right) = \frac{a_\phi \left| x(t) \right|}{1 + \beta_\phi \left| x(t) \right|^2}, \tag{12}$$

where $\alpha_G = 1, \beta_G = 0.25, \alpha_\phi = 0.26, \beta_\phi = 0.25$. In the simulation, the power amplifier chip selects HMC454 of AD Inc., its gain is 12.5dB and the 1dB compression point is 27.5dBm. The gain coefficient of the second and the third nonlinear components can be obtained by [8]

$$\text{OIP}_m = \text{G}_1 + \text{IIP}_m, \text{OIP}_m = \text{G}_m + m\text{IIP}_m, \tag{13}$$

where G_m is the mth gain coefficient. The low noise amplifier selects HMC669 of AD Inc., its gain is 17dB with the noise coefficient of 1.4dB, and the 1dB compression point is 12dBm. According to (13), the gain coefficients of the second and third non-linear components of the low noise amplifier can be obtained.

For RF interference cancellation module, vector modulator can choose MAX2045. It is assumed that the error in the phase and amplitude adjustment of the self interference signal can be ignored in the vector modulator, DAC and ADC circuits. The number of taps of the RFSIC is 32, and the baseband sampling frequency is 16.3842MHz after frequency down conversion, then the sample time interval is 0.061us which means that each tap experienced a total time delay of 0.061us. The adaptive filter is set to 1024 iterations.

In the simulation, the satellite orbit height is 500km, the transmission power of the ground station is 17dBW, the antenna gain of the ground station is 45dB, and the path loss is 165dB when the antenna elevation is 5 degrees. The transmission power of the transponder is 500mW, the distance of the transceiver antennas of the transponder is 1 meters, the Doppler frequency shift is 40kHz, and the transmitter noise figure is assumed to be 10dB.

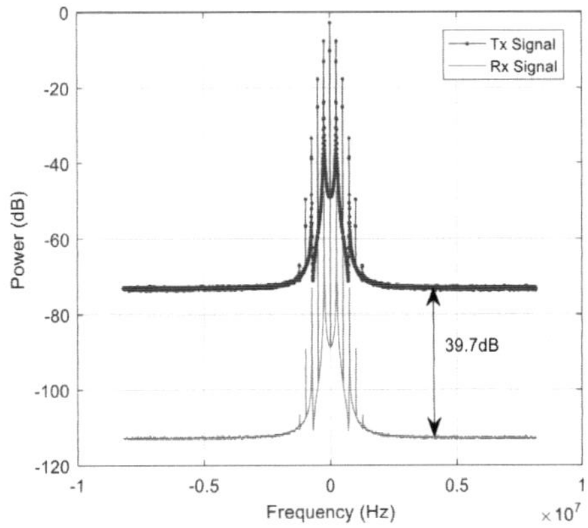

Figure 4. Powers comparison of transmitted and received signal.

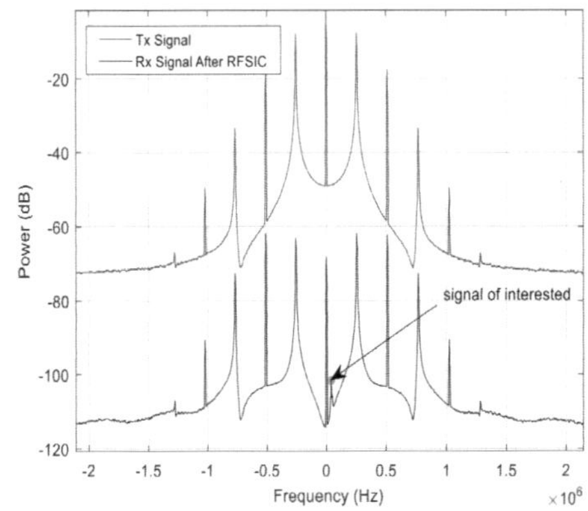

Figure 5. Powers comparison of transmitted signal and received signal after RFSIC.

Figure 4 shows the powers of the reception and transmission signals of the transponder. It is seen that the signal attenuates 39.7dB by the space isolation of the transceiver antennas. The telemeter tone and its harmonics are respectively at 250kHz and 500kHz etc. when the PCM-PM modulation is exploited. It can be seen that the received signal of interested (telecommand tone) is submerged in the received noise. Figure 5 shows the power comparison of the transmitted signal with received signal after RFSIC. It can be seen that after RFSIC, the power of the transmitted carrier signal is attenuated to -68.2dB, and RFSIC achieves the attenuation of about 29dB, while the harmonics and noises with in bandwidth $\pm$500kHz are also attenuated. It can be seen that the signal of interested is appeared at the 40kHz offset frequency of transmitted carrier with power of about -102dBW(-72dBm).

Figure 6 shows the signal power after the DSIC, in which the order of the adaptive filter is set to 128 with 1024 iterations. From the figure, the harmonics of the telemetry tone is further reduced about 10dB after the DSIC, but the power of the transmitted signal within bandwidth of carrier frequency±250kHz is raised about 8dB. This is because, after the RFSIC, the harmonics of the telemetry tone has larger power than the carrier, and the DFSIC mainly cancels the harmonics. It also depicts that the power of received signal are all amplified about 17dB which is accordant with the gain of LAN. Due to the raised power of in-band noise, the performance of demodulation is weakened.

Thus, we considered to remove the DSIC and filter and demodulate the received signal directly. Figure 7 gives the demodulation waveform of the telecommand symbols. In the simulation, the EbN0 is set to 15dB and the symbol rate is 2000bps. It can be seen from the graph that the telecommand PCM symbols are demodulated correctly.

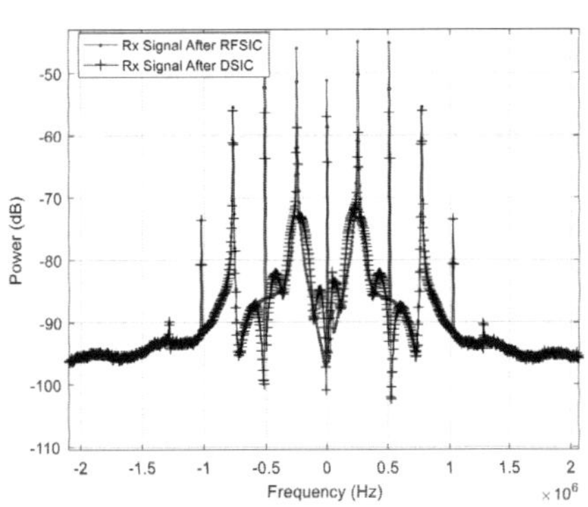

Figure 6. Power comparison of the residual interference signal after RFSIC and DSIC.

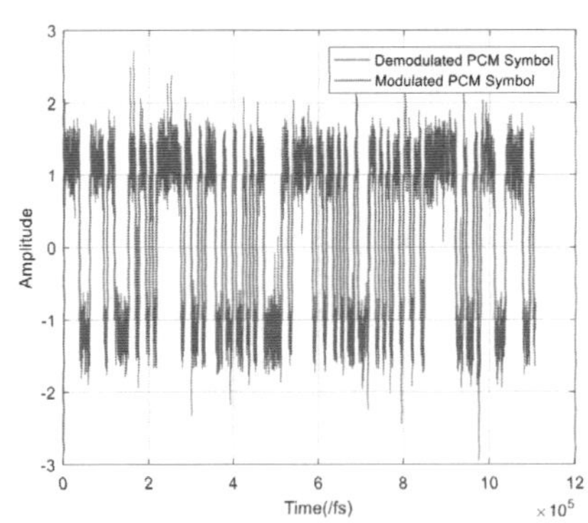

Figure 7. Comparisons of demodulated symbols with modulated symbols.

4. Conclusion

In this paper, the feasibility of the STAR TT&C transponder is analyzed, and the transceiver architecture is presented. The scheme of the RFSIC and the process of digital signal processing are described. The key chips are selected and the system simulation is carried out. The simulation results show that the self interference cancellation of about 29dB can be realized by the RFSIC. The total self interference cancellation of about 68dB can be realized by the combination of the passive suppression and RFSIC. The simulation results also showed that the STAR implementation scheme is feasible and the demodulation of telecommand PCM symbols can be realized. It is necessary to point out that the taps of the RFSIC scheme is 32, which leads to extra power consumption of the RF circuit compared with the traditional transponder, and the circuit is more complex, which needs to be solved in the future.

Reference

[1] Jain, M., Choi, J. I., Kim, T. M., Bharadia, D., Seth, S., Srinivasan, K., Levis, P., Katti, S., & Sinha, P. 2011, Practical, Real-time, Full Duplex Wireless, Las Vegas, Nevada, USA.),

[2] Bharadia, D., McMilin, E., & Katti, S. 2013, Full Duplex Radios, Hong Kong, China.),

[3] Riihonen, T., Korpi, D., Rantula, O., & Valkama, M. 2017, On the prospects of full-duplex military radios, pp.1-6

[4] Grayver, E., Keating, R., & Parower, A. 2015, Feasibility of Full Duplex Communications for LEO Satellite, pp.1-7

[5] Grayver, E., Parower, A., Khadge, G., Logue, K., & Tarasov, K. 2017, Experimental Results for a Narrowband Full Duplex Communications System, pp.1-10

[6] Collins, G. D., & Treichler, J. 2015, Practical Insights on Full-Duplex Personal Wireless Communications Gained From Operational Experience In The Satellite Environment.

[7] Choi, Y., & Shirani-Mehr, H. 2013, IEEE Transactions on wireless communications , 12, pp. 5992-6010

[8] Gu, Q. 2005, *RF System Design of Transceivers for Wireless Communications.* (Springer)

[9] Askar, R., Zarifeh, N., Schubertt, B., Keusgent, W., & Kaiser, T. 2014, I/Q Imbalance Calibration for Higher Self-Interference Cancellation Levels in Full-Duplex Wireless Transceivers,

[10] Huusari, T., Choi, Y., Liikkanen, P., Korpi, D., Talwar, S., & Valkama, M. 2015, Wideband Self-Adaptive RF Cancellation Circuit for Full-Duplex Radio: Operating Principle and Measurements, pp.1-7

[11] Diniz, P. S. R. 2013, *Adaptive Filtering Algorithms and Practical Implementation (*Fourth Edition). (Springer)

[12] Tamminen, J. 2016, Digital Control of RF self-interference canceller in Full-duplex Radio, Tampere University of Technology,

[13] Huusari, T. 2014, Analog RF Cancellation of Self Interference in Full-duplex Transceivers, Tampere University of Technology,

[14] Smaini, L. 2012, *RF Analog Impairments Modeling for Communication Systems Simulation: Application to OFDM-Based Transceivers. (*Wiley)

Efficient Evaluation of Mars Entry Terminal State Based on Gaussian Process Regression

A Gao[1], G Y Wang[2], S S Wu[3] and T Song[4]

[1] Beijing Institute of Technology, 5 South Zhongguancun Street, Beijing, China, 100081;
[2] Beijing Institute of Technology, 5 South Zhongguancun Street, Beijing, China, 100081;
[3] XD Group (XD) and General Electric (GE) Automation Co., Ltd. Xi'an, Shanxi province, china, 710075;
[4] Shanghai Aerospace Control Technology Institute, Shanghai, China, 201109;

wanggaoyue@bit.edu.cn

Abstract. Trajectory optimization technology used for Mars entry is one of the key technologies for planetary exploration. Evaluation of the performance of the entry trajectory under conditions of complex atmospheric dynamics, various vehicular design parameters, and multiple constraints in the process of entry, are important issues pertaining to the design of trajectories. In this study, an efficient evaluation approach of the terminal state for Mars entry is proposed based on Gaussian process regression to evaluate the maximum terminal altitude for different entry velocities, terminal Mach numbers, and vehicular parameters. Additionally, the influences of entry flight-path angle, lift-drag ratio, and ballistic coefficient, on the maximum terminal altitude are analyzed. A genetic algorithm is used in the optimization solver to avoid local minima and to guarantee the data quality of the training samples used for Gaussian process regression. The mean function, kernel function, and hyperparameters are selected as the optimization parameters for Gaussian process regression to describe the correlation between samples, and the maximum terminal altitude prediction model is then established. Numerical simulations demonstrate that the proposed method can achieve the evaluation of more than 3000 group scenarios within tens of seconds with a mean relative error that is less than 4%.

1. Introduction

Planetary exploration is one of the main fields of deep space exploration. Mars, the closest earth-like planet in the solar system to earth is usually the preferred target for human planetary exploration. However, unlike the earth's environment, Mars entry encounters many challenges given the planet's physical environment, with its thin atmosphere and increased gravity, thereby making it difficult to decelerate the vehicle [1,2]. When the vehicle is decelerated to meet the deployment constraints imposed for parachutes or other deceleration devices, the terminal altitude can affect the maneuver time of the landing process and the range of the target landing site that directly affects the success of the entire mission. Therefore, the approach used to evaluate the terminal altitude is an important issue in the design of Mars exploration missions.

IOP Conf. Series: Materials Science and Engineering **449** (2018) 012010 doi:10.1088/1757-899X/449/1/012010

However, when designing the entry trajectory the process involves large-scale work of trajectory design because different entry modes and control mechanisms lead to a variety of flight plans. Considerable amount of research work has been conducted on the entry trajectory design. Carman, Ives, and Geller et al. introduced the guidance algorithm of constant-bank angle profile that was adopted by the Apollo mission for the study of Mars entry vehicles [3,4]. Mendeck and Carman et al. emphasized the importance of the variable-bank profile for the optimization of Mars-entry trajectories [5]. Grant and Shuang et al. considered the terminal altitude as the performance index for the assessment of the maximum terminal altitude that could be reached by the variable-bank profile design in specific application scenarios [6,7]. Lafleur and Jacob et al. comprehensively studied the maximum terminal altitude that a Mars vehicle could reach with combinations of different vehicular parameters and entry process constraints [8,9]. However, all the existing research studies have focused on the complex modeling of the entry process with different vehicular parameters, depending on a large number of numerical calculations, or comprehensively studied the possible entry scenarios of different vehicular types, entry velocities, and deceleration devices that required considerable computation time in an inefficient manner, or simplified the constructed model to improve the solution's efficiency. It is difficult to design an approach that is comprehensive, optimal, and efficient.

Gaussian process regression can use part of the observed values to establish the mapping relationship of the input-output model, so that when the new input values are given, the corresponding output can be estimated. Since the prediction method based on GPR does not need a large number of complex modeling functions, its computational cost is considerably reduced. The idea of GPR was first proposed by Krige for the estimation of the distribution of gold in mines [10]. Thus far, the GPR method has been applied to the field of aerospace research, and includes prior work on the aircraft's airfoil design [11], aircraft's aerodynamic coefficient assessment [12], and for the assessment of the main belt that asteroids can reach [13]. Inspired by this thought, this study presents an efficient evaluation method based on Gaussian process regression to evaluate the terminal state of Mars entry.

This study is organized as follows. Section 2 introduces the Mars trajectory optimization algorithm proposed herein, uses it to generate a large number of simulation results, and provides sample data for the prediction model based on Gaussian process regression. In Section 3, a maximum terminal altitude prediction model of Mars entry is established based on Gaussian process regression, using the existing sample data. Section 4 discusses the effectiveness of the maximum terminal altitude prediction model of Mars entry, and uses the prediction model to analyze the characteristics of the trajectory, and provide the design references under a given mission scenarios.

2. The optimal trajectory for Mars entry

In order to provide sample data for the GPR-based prediction model, and to ensure that the sample data has sufficient empirical information, this section uses the terminal altitude as the performance index, and explores the maximum terminal altitude that the optimal trajectory of the Mars entry can reach under different combinations of entry velocities, terminal Mach numbers, lift-drag ratios, ballistic coefficients, and process constraints, in a comprehensive manner.

2.1. Dynamics model for Mars entry

Considering the rotation of Mars, the following dynamics model is applied.

$$\frac{dr}{dt} = V\sin\gamma \tag{1}$$

$$\frac{d\theta}{dt} = \frac{V\cos\gamma\sin\psi}{r\cos\varphi} \tag{2}$$

$$\frac{d\psi}{dt} = \frac{1}{V}\left[\begin{array}{l}\dfrac{L\sin\sigma}{\cos\gamma} + \dfrac{V^2}{r}\cos\gamma\sin\psi\tan\varphi + \dfrac{\Omega^2 r}{\cos\gamma}\sin\varphi\cos\varphi\sin\psi \\ -2\Omega V(\tan\gamma\cos\psi\cos\varphi - \sin\varphi)\end{array}\right] \tag{3}$$

$$\frac{d\gamma}{dt}=\frac{1}{V}\left[\begin{array}{l}L\cos\sigma+\cos\gamma(\frac{V^2}{r}-g)+2\Omega V\cos\varphi\sin\psi\\ +\Omega^2 r\cos\varphi(\cos\gamma\cos\varphi+\sin\gamma\cos\psi\sin\varphi)\end{array}\right] \tag{4}$$

$$\frac{d\varphi}{dt}=\frac{V\cos\gamma\cos\psi}{r} \tag{5}$$

$$\frac{dV}{dt}=-D-g\sin\gamma+\Omega^2 r\cos\varphi(\sin\gamma\cos\varphi-\cos\gamma\sin\varphi\cos\psi) \tag{6}$$

Herein, r is the distance between the vehicle and the center of Mars, and Ω is the rotation speed of Mars, $g=mu/r^2$ is the local gravitational acceleration, mu is Mars gravitational constant. Equivalently, $\theta,\varphi,\gamma,\psi,\sigma$ are longitude, latitude, flight-path angles, heading angles, and bank angles [14]. The variables L,D are the lift acceleration and drag acceleration. In accordance with the method in [8], 10 bank angles were provided in the velocity domain, and the bank angle profiles are then obtained by interpolation. Assume that the aerodynamic parameters can be fully expressed by the ballistic coefficient and lift-drag ratio, as stated in equations (7) and (8).

$$\beta=\frac{m}{C_D S} \tag{7}$$

$$L=kD \tag{8}$$

Among them, β is the ballistic coefficient, k is the lift-drag ratio, C_D is vehicle drag coefficient, and S is the reference area, and $D=\frac{1}{2\beta}\rho V^2$, ρ is the Mars atmospheric density. Based on the observation data of the Viking, a simplified model of the atmospheric index is adopted herein equation(9), whereby $\rho_0=1.474\times10^{-2}$ kg/m^3, $hs=8.8057\times10^3$ m ,and h is the vehicular altitude from the Mars surface [15].

$$\rho=\rho_0 e^{(-h/hs)} \tag{9}$$

2.2. Vehicular parameters and Mars entry scenarios

Entry vehicles are classified into manned and unmanned, and they have three different shapes, which respectively correspond to the ballistic, ballistic lift, and lift models.

At present, most of the vehicular lift-drag ratio values are less than unity, except for lift-body vehicles, such as the space shuttles. All known Mars missions have used unmanned vehicles with ballistic coefficients of approximately 100 kg/m^2. Considering the need of manned vehicle landing on Mars, and the increase of the required load, future vehicular ballistic coefficients will be correspondingly larger. Based on the use of manned vehicles in other missions, the ballistic coefficient mainly ranges between 200 and 1000 kg/m^2. At present, there are mainly three types of vehicular entry velocities: a) 3.3 km/s for entries from a 500 km circular orbit, b) 4.7 km/s for entries from a 1-sol elliptical orbit, and c) direct entry velocities, which usually vary between 3-7 km/s [1,8].

Therefore, referring to the typical distribution of characteristic vehicular parameters, the ballistic coefficient used in this study was selected to range from 100 kg/m^2 to 1000 kg/m^2 in increments of 100 kg/m^2. The values of the lift-drag ratio were selected to be in the range of 0.1 to 0.9, at 0.1 increments. Moreover, the terminal Mach numbers were equal to 2.0, 3.5, and 5.0, corresponding to the Mach number of deployed parachutes, inflatable decelerators, and propulsive decelerators. The entry velocities were equated to 3.3 km/s, 4.7 km/s, and 5.5 km/s. The maximum acceleration limit of manned vehicles was 4.5 g_0 and the maximum acceleration limit of unmanned vehicles was 30 g_0, where $g_0=9.8066$m/s^2 and represents the Earth surface gravity acceleration. A total of 1620 groups were evaluated with different combinations of vehicular parameters and Mars entry scenarios. The convective heat rate was constrained to 1000 W/cm^2 [8].

2.3. Optimization model

In this study, the entry flight-path angle and the bank angle are considered as the optimization variables. The entry flight-path angle and 10 bank angles evenly spaced within the velocity range are then optimized [9]. The optimal bank angle profile is obtained through interpolation, and the performance index is then calculated. Using a genetic algorithm-based optimization solver, the local minimum is avoided by virtue of its global goodness, so as to ensure the quality of the optimization results.

2.3.1. The objective function. Mars entries are expected to have a higher terminal altitude h_{term}. Correspondingly, this study selected the terminal altitude as the optimization goal and performance index to maximize terminal altitude. As shown in equation (10), r_{term} is the distance from the planet's center to vehicle at the instant when the deceleration device is deployed, and R_m is the radius of Mars.

$$h_{term} = r_{term} - R_m \tag{10}$$

2.3.2. Process constraints. The maximum acceleration limit reflects the deceleration acceleration F that the vehicle or astronaut can withstand. The convective heat rate constraint reflects the heat rate Q that the vehicle can withstand.

$$F = \sqrt{L^2 + D^2} \le F_{\max} \tag{11}$$

$$Q = Q_z \left(\frac{V}{V_e} \right)^{3.08} \sqrt{\rho / \rho_0} \le Q_{\max} \tag{12}$$

The minimum altitude during the process of entering can be classified in two situations. When restrictions are imposed, $h_{\min} \ge 10\text{km}$, and when no restrictions are imposed, $h_{\min} \ge -5\text{km}$ is the simulation condition.

2.3.3. Terminal constraints. Terminal Mach number indicates the velocity at which the deceleration devices can be deployed, and the minimum terminal altitude limits the altitude of the vehicle when the deceleration devices is deployed.

$$M_{term} = M_f \tag{13}$$

$$h_{term} \ge 0\text{km} \tag{14}$$

Therefore, 3240 groups of simulations were conducted with combinations of different Mars entry scenarios, vehicular parameters, and process constraints. A total of more than 1400 groups of optimal solutions were obtained.

3. Design of terminal altitude evaluation method based on Gaussian process regression

In this study, Gaussian process regression is used to design the prediction model of the optimal terminal altitude for Mars entry. Firstly, use the combination of entry velocity, terminal Mach number, ballistic coefficient, lift-drag ratio, acceleration limit, and the minimum altitude as the input x of the prediction model. Use the maximum terminal altitude as the output y of the prediction model.

$$x = \left[V_0 \quad M_{term} \quad \beta \quad \frac{L}{D} \quad F_{\max} \quad h_{\min} \right]^T \tag{15}$$

$$y = h_{term} \tag{16}$$

Secondly, select the appropriate mean function and kernel function to design the framework of the prediction model. In this study, the zero mean function and the Gauss kernel function are selected, the corresponding expressions are in equations (17) and (18). Thus, the prediction data and the training data are subject to a joint normal distribution distribution as in equation (19). Therefore, the mean and variance of the predicted data can be calculated by equations (20) and (21).

$$m(x) = 0 \tag{17}$$

$$K(x, x') = \sigma_f^2 \exp\left(-\frac{1}{2l^2}(x_p - x_q)^2\right) \tag{18}$$

$$\begin{bmatrix} y \\ y_* \end{bmatrix} \sim N(0, \begin{bmatrix} K(x,x) & K(x,x_*) \\ K(x_*,x) & K(x_*,x_*) \end{bmatrix}) \tag{19}$$

$$\overline{y_*} = K(x_*, x)[K(x,x) + \sigma_n^2 I]^{-1} y \tag{20}$$

$$\operatorname{cov}(y_*) = K(x_*, x_*) - K(x_*, x)[K(x,x) + \sigma_n^2 I]^{-1} K(x, x_*) \tag{21}$$

Subsequently, optimize the model hyperparameter ω using the training data. Suppose that the distribution of ω is Gaussian distribution. The Bayesian formula can estimate the likelihood function of the hyperparameter $p(\omega \mid y, x)$,as in equation (23). The joint probability distribution in accordance to the dataset is shown in equation (24) and the marginal probability is calculated in equation (25).

$$\omega = \lfloor l \quad \sigma_f \quad \sigma_n \rfloor \tag{22}$$

$$p(\omega \mid y, x) = \frac{p(y \mid x, \omega) p(\omega)}{p(y \mid x)} \tag{23}$$

$$p(y \mid x, \omega) = \prod_{i=1}^{n} p(y_i \mid x_i, \omega) = \prod_{i=1}^{n} \frac{1}{(2\pi\sigma_n^2)^{\frac{n}{2}}} \exp(-\frac{(y_i - y_*)^2}{2\sigma_n^2}) \tag{24}$$

$$p(y \mid x) = \int p(y \mid x, \omega) p(\omega) d\omega \tag{25}$$

The optimal hyperparameters can be obtained by maximizing the logarithm of the edge likelihood of the training data, as shown in equation (26). By optimizing the hyperparameters, the probability of the model training output is maximized, and the prediction error is controlled within a reasonable range.

$$\log p(y \mid x, \omega) = -\frac{1}{2} y^T (K + \sigma_n^2 I)^{-1} - \frac{1}{2} \log\left|K + \sigma_n^2 I\right| - \frac{\mathrm{n}}{2} \log 2\pi \tag{26}$$

Finally, using the hyperparameters solved by optimization, the prediction model of the maximum terminal altitude is determined. Thus, the use of specific parameters of the exploration mission allows the prediction model to be used to quickly and accurately evaluate the maximum terminal altitude that the vehicle can reach.

4. Numerical results and discussion

The simulations were conducted in MATLAB 2015b using an I5-2520 processor at 2.50 GHz, and a memory of 8 GB.

Based on the Gaussian process regression, the prediction model established in Section 3 was used to predict the maximum terminal altitude of Mars entry. Part of the sample data was selected as training data and the remaining data were used as test data to evaluate the maximum terminal altitude and validate the prediction model. In order to ensure that proper and adequate information can be extracted under different entry velocities, terminal Mach numbers, vehicle ballistic coefficients, and lift-drag ratios and process constraints, the training samples need to have data under different input parameters, and be distributed as evenly as possible. Therefore, in this study, the training samples were randomly and uniformly extracted under different input parameters to ensure that each acquired information type can be learned.

4.1. Performance verification of prediction model for Mars maximum terminal altitude

To validate the prediction model proposed herein, it was trained to evaluate the maximum terminal altitude and analyze the prediction errors of the test data. A total of 138 sets of data were extracted as test data, and 213, 276, 345, 460, and 690 sets of data were evenly extracted from the rest of the data as training data to predict the terminal altitude and analyze the prediction errors. The error performance of the prediction results is shown in Table 1, and the relationship between the prediction error and the size of the training data is shown in Figure 1.

Figure 2 and Table 1 indicate that the use of the GPR-based prediction model to evaluate the terminal altitude yields a mean relative error of 3.7% for the predicted results, a relative error within 7% for nearly 95% of the test data, and a maximum relative error that does not exceed 10%.

Table 1. Performance evaluation of prediction model

number of test data	number of training data	Mean absolute error/*km*	Mean relative error	Maximum relative error	Training,test time-consuming/*s*
138	213	3.50	22.4%	36.8%	0.81
138	276	2.25	11.9%	28.4%	1.50
138	345	1.72	7.8%	19.6%	3.79
138	460	0.74	4.2%	13.5%	18.4
138	690	0.63	3.7%	9.4%	25.7

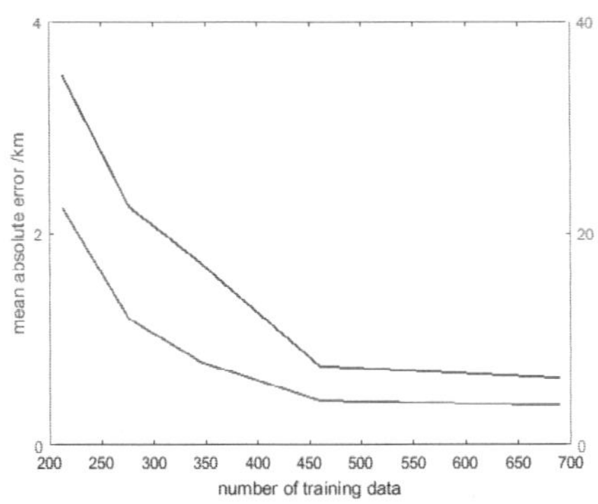

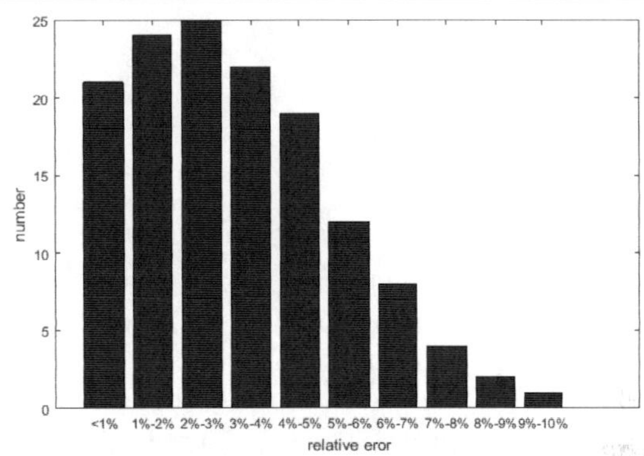

Figure 1. Prediction error and the training data size. **Figure 2.** Relative error distribution of test data.

As shown in Figure 1, when the training data becomes more than 500 groups, the performance of the prediction model exhibits a minor increase, and the mean relative error of the prediction results stabilizes near 4%. Figure 2 shows the relative error distribution with 138 groups test data and 690 groups training data.

4.2. Application to an elliptical orbit entry with a parachute decelerator

The prediction model based on the Gaussian process regression can evaluate the terminal altitude with different Mars entry scenarios, and analyze the characteristics of the entry trajectory with the entry flight-path angle obtained by the optimization algorithm. Due to the length of the space, this section takes the elliptical orbit entry with parachute decelerator as an example to show the numerical results.

Manned Mars exploration is the focus of future Mars explorations. Compared with unmanned vehicles, the atmospheric entry process of manned vehicles imposes strict constraints for the maximum acceleration limit, thereby affecting the maximum terminal altitude. Figure 3 show the maximum terminal altitude and optimal entry flight-path angle of unmanned vehicles. Figure 4 show the maximum terminal altitude and optimal entry flight-path angle of manned vehicles.

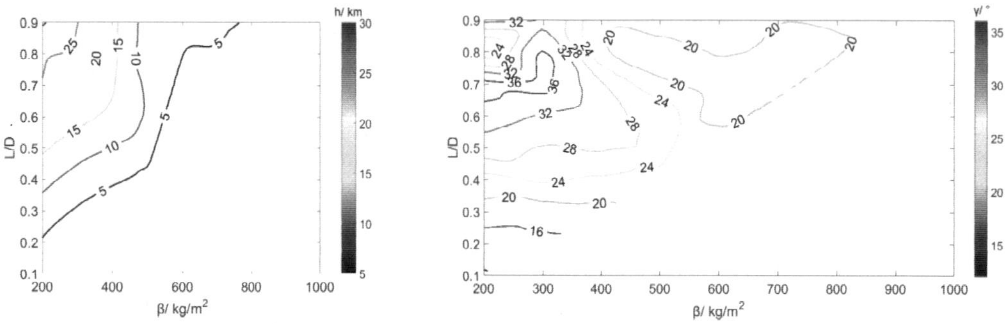

a) b)

Figure 3. Maximum terminal altitude and optimal entry flight-path angle of unmanned vehicle.

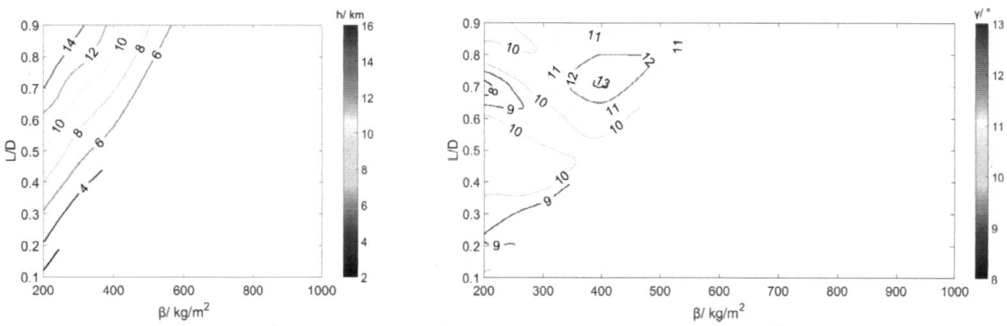

a) b)

Figure 4. Maximum terminal altitude and optimal entry flight-path angle of manned vehicle.

As it can be observed from Figure 3 a) and Figure 4 a), in order to reach a larger terminal altitude, the vehicle is expected to have a higher lift-drag ratio and a lower ballistic coefficient, and the unmanned vehicle can reach a much higher terminal altitude than the manned vehicle. Moreover, the manned vehicle can hardly meet the deployment conditions of the parachute when the ballistic coefficient is more than $600\,\mathrm{kg/m^2}$. Figure 3 b) and Figure 4 b) show that for a manned vehicle with an elliptical orbit entry and uses the parachute as the decelerator, the optimal entry flight-path angle is approximately 10°. Compared to the unmanned vehicle, the optimal entry flight-path angle is much smaller. At the same time, when the vehicle has a lower lift-drag ratio, a relatively small entry flight-path angle should be used to enter the Mars atmosphere. As the lift-drag ratio increases, the optimal entry flight-path angle become larger. However, when the vehicle has a higher lift-drag ratio, the optimal entry flight-path angle fluctuates as the ballistic coefficient and lift-drag ratio change. The reason for these outcomes may be that a larger entry flight-path angle provides the vehicle with more time to decelerate in the dense atmosphere, and the increase of lift-drag ratio can ensure that the vehicle decelerate without impacting the Mars surface. However, as the lift-drag ratio increases, a larger entry flight-path angle will cause an acceleration overload that the vehicle cannot withstand, so when the maximum constraints are reached, the optimal entry flight-path angle decrease instead.

5. Conclusions

In this study, an efficient evaluation approach of the Mars entry terminal state was proposed, based on Gaussian process regression, to evaluate the maximum terminal altitude for unmanned and manned vehicles of different entry scenarios, vehicular parameters, and process constraints. The maximum terminal altitude that the vehicle can reach under different combinations of parameters was optimized using a Mars entry trajectory optimization algorithm that provided the sample data for the GPR-based prediction model. The proposed method was used to evaluate the maximum terminal altitude data from more than 3000 groups. The evaluation process took less than 25.7 s. In contrast to traditional trajectory optimization algorithms, which solve one group of data within tens of minutes to hours, the assessment efficiency was greatly improved. Compared with the optimal solution obtained using the optimization algorithm, the mean relative error of the evaluated results was 3.7%, and 7% for almost 95% of the predicted results, based on Gaussian process regression. Using the GPR-based method proposed herein, the Mars maximum terminal altitude prediction model can provide valuable insights to the design of the Mars exploration mission. Because of its high-efficiency calculation attributes, it is expected to provide support as an auxiliary means for online optimization of the Mars entry trajectories design.

Acknowledgments
This work was supported in part by the following grant: Young Elite Scientist Sponsorship Program by CAST (YESS) No. YESS20160163.

References
[1] Braun, R. D., and Manning, R. M., "Mars Exploration Entry, Descent and Landing Challenges," 2006 IEEE Aerospace Conference. Big Sky, MT, USA, Mar. 2006.
[2] Wells, G., Lafleur, J., Verges, A., Manyapu, K., Christian, J., Lewis, C., and Braun, R., "Entry, Descent, and Landing Challenges of Human Mars Exploration," AAS Guidance and Control Conference, American Astronautical Society Paper 06-072, Feb. 2006.
[3] Moseley, P. E., "The Apollo Entry Guidance: A review of the mathematical development and its operational characteristics," TRW Note No. 69-FMT-791, Dec. 1969.
[4] Carman, G., Ives, D., and Geller, D., "Apollo-Derived Mars Precision Lander Guidance," AIAA 1998-4570.
[5] Mendeck, G. F., and Carman, G. L., "Guidance Design for Mars Smart Landers Using the Entry Terminal Point Controller," AIAA Atmospheric Flight Mechanics Conference and Exhibit, AIAA Paper 2002-4502, Aug. 2002.
[6] Grant, M. J., and Mendeck, G. F., "Mars Science Laboratory Entry Optimization Using Particle Swarm Methodology," AIAA Atmospheric Flight Mechanics Conference and Exhibit, AIAA Paper 2007-6393, Aug. 2007.
[7] Shuang, L., and Yuming, P., "Mars Entry Trajectory Optimization Using DOC and DCNLP," Advances in Space Research, Vol. 47, No. 3, 2011.
[8] Jarret, M. L. and Christopher, J. C., "Mars Entry Bank Profile Design for Terminal State Optimization," Journal of Spacecraft and Rockets, November. 2011.
[9] Geethu, L. J., Geethu, N., and Ramanan, R. V., "Mars Entry Mission Bank Profile Optimization," Journal of Guidance, Control, and Dynamics, Vol. 37, No. 4, Jul-Aug. 2014.
[10] Krige, D. G., "A Statistical Approach to Some Basic Mine Valuation Problems on the Witwatersrand," Journal of the Chemical, Metallurgical and Mining Society of South Africa, Vol. 52, No. 6, 1951, pp. 119-139.
[11] Liu, X., Zhu, Q., and Lu, H., "Modeling Multiresponse Surfaces for Airfoil Design with Multiple-Output-Gaussian-Process Regression," Journal of Aircraft, Vol. 51, No. 3, 2014, pp. 740-747.
[12] Robin, D., Muelenaere, J., and Elham, A., "Trajectory Driven Multidisciplinary Design Optimization of a Sub-Orbital Spaceplane Using Non-Stationary Gaussian Process," Structural and Multidisciplinary Optimization, Vol. 652, No. 4, 2015, pp. 755-771.
[13] Haibin, S., Yuxin, L., "Assessing Accessibility of Main-Belt Asteroids Based on Gaussian Process Regression," Journal of Guidance Control and Dynamics, Vol.40, No.5, May 2017.
[14] Pillai, J. R., John, A., and Ramanan, R. V., "Design and Analysis Tool for Mars Atmospheric Entry Missions," Journal of Aerospace Innovations, Vol. 5, No. 1, Mar. 2013, pp. 29-40.
[15] Benito, J., Mease, K., "Characterizing the controllable entry states and reachable sites for planetary landing," International Planetary Probe Workshop, 2008, Atlanta, Georgia.

IOP Conf. Series: Materials Science and Engineering **449** (2018) 012011 doi:10.1088/1757-899X/449/1/012011

The Effects of Tailplane Ice Accretion on Flight Stability of Commuter Category Aircraft for High Terrain Remote Areas Flight Operation

A B Hakim[1], W Nirbito[1,*] and M Adhitya[1]

[1] Department of Mechanical Engineering, Universitas Indonesia, Depok, West Java 16424, Indonesia

[*] Corresponding author : wahyu.nirbito@ui.ac.id

Abstract. The effect of ice accretion on the surface of the horizontal tail on aerodynamic characteristics and stability of commuter category aircraft is reviewed using predicted ice shapes and polar predictions using CFD software. Ice accretion prediction is carried out in various atmospheric conditions already listed in Appendix C FAR25 within remote high terrain flight operation condition. The predicted results of ice shapes are categorized according to the main shapes which will then be simulated using XFOIL. The resulting polar data then combined with data obtained from wind tunnel test to give lift coefficient predictions and the overall moment of the aircraft at zero flap condition. The most severe impact of ice accretion on the surface of the horizontal tail is found in the decreasing values of stalling angle of attack and maximum lift coefficient due to flow separation, with stalling angle of attack and maximum lift coefficient as low as 9.54 degrees and 0.765. The most visible decrease in static longitudinal stability and the static margin point are 6.88% and 6.34% from the normal condition.

1. Introduction

Commuter category aircraft development is important for short distance flight. While using commuter category aircraft can maintain the operational cost, it is also important to use commuter category aircraft that capable of taking off and landing on the short runway operating in remote and mountainous areas, especially on islands or Papua Region. The problem in high terrain remote areas flight operation is the likelihood of ice accretion because of certain weather [1]. To ensure the safety of an aircraft during flight, certification for flight in icing condition is required. Before undergoing the certification process, it is better to make predictions in advance to save the cost of developing the aircraft, such as the wind tunnel test. Prediction must be in accordance with applicable regulations with results as accurate as possible from the actual conditions. The purpose of this study was to obtain the degradation value of aerodynamic characteristics of horizontal tail commuter category aircraft and the aircraft flight stability parameters that experienced ice accretion, that can be used as a reference before undergoing the certification process.

2. Methodology

To obtain stability and static point margin of an aircraft, certain procedures have to be done starting from data retrieval of an aircraft. The flowchart shown in Figure 1 represents an overview of those procedures starting from data collection to analysis and discussion.

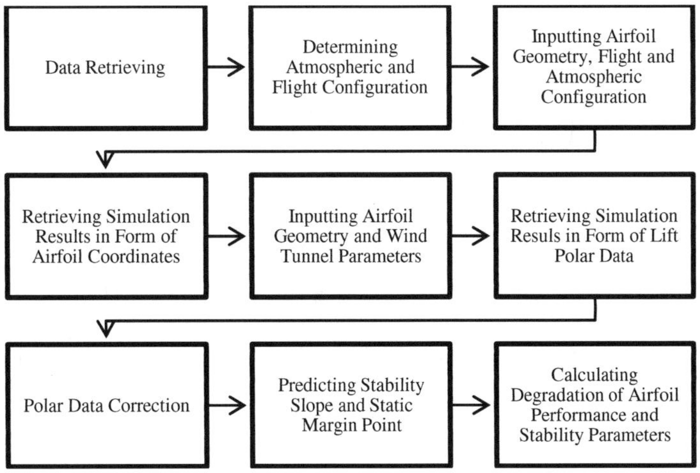

Figure 1. Methodology of Stability Prediction

2.1. Determining Atmospheric and Flight Configuration

Ice accretion simulation using LEWICE requires atmospheric and flight configuration [2]. Atmospheric configuration parameters required are Liquid Water Content (LWC), Median Volume Diameter (MVD) of water droplet, and static ambient temperature. The flight configuration of a commuter category aircraft displayed in the following Table 1.

Table 1. Flight Configuration

Altitude (ft)	**Weight (kg)**	**Airspeed (knot)**	**P (kg/m^3)**	**α (deg.)**
10000	7031	153.04	0.86	3.04
8000	7031	157.09	0.91	2.32

In addition to these geometry variables, the coordinates of the airfoil to be simulated must also be prepared first. In this study, the airfoil that is used is symmetrically modified trailing edge of NASA/LANGLEY MS (1) -0313 airfoil as shown in Figure 2.

Figure 2. NASA / LANGLEY MS (1) -0313 airfoil

After obtaining the input parameters, the simulation of ice accretion is done by using LEWICE code. The result of ice accretion simulation then simulated using XFOIL [3] with the same test conditions with the actual wind tunnel test, with the Reynolds number around 2.9×10^6 and Mach number of 0.2.

2.2. C_L and C_M Prediction

XFOIL simulation result is in the form of infinite wing lift polar, therefore, it is necessary to correct the lift polar into 3-dimensional or finite wing lift polar so that the simulation results of XFOIL can be used to predict the overall lift polar of the aircraft. The first thing to do is to get the lift slope of infinite wing lift polar, then the finite wing lift coefficient correction can be done using Prandtl's Lifting Line theory [4].

Prediction is done by calculating the degradation of lift and moment coefficient on the horizontal stabilizer and substitute the degradation value into the wind tunnel results. To find the degradation of lift coefficient, the following equation is used [5]

$$\Delta C_{L_h} = C_{L_{h_{clean}}} \eta_h \frac{S_h}{S} - C_{L_{h_{ice}}} \eta_h \frac{S_h}{S} \tag{1}$$

After obtaining the degradation value of lift coefficient on the horizontal stabilizer, the next thing to do is predict the lift coefficient of the aircraft by using the following equation

$$C_{L_{ice}} = C_L - \Delta C_{L_h} \tag{2}$$

The moment experienced by the aircraft is influenced by wing moment $C_{M_{AC_w}}$ and the horizontal tail $C_{M_{AC_h}}$ and the moment generated by the lift from the wing and horizontal tail with c.g. as the reference point, however, $C_{M_{AC_h}}$ does not have such a large influence that it can be ignored.

$$\Delta C_{M_h} = -\left(C_{L_{h_{clean}}} \eta_h \frac{S_h}{S} \frac{(x_{AC_h} - x_{cg})}{\bar{c}} - C_{L_{h_{ice}}} \eta_h \frac{S_h}{S} \frac{(x_{AC_h} - x_{cg})}{\bar{c}}\right) \tag{3}$$

After obtaining the coefficient of moment degradation on the horizontal stabilizer, the next step is to predict the overall moment coefficient of the aircraft by using the following equation

$$C_{M_{ice}} = C_M - \Delta C_{M_h} \tag{4}$$

2.3. Stability and Static Margin Calculation

The stability characteristic can be examined taking into consideration the increased pitching moments by changing the angle of attack in trim conditions and calculated using the following formula

$$C_{M_\alpha} = \frac{dC_M}{d\alpha} \tag{5}$$

The slope of the moment coefficient C_{M_α} is used to assess how stable a plane is. The calculation of the static position of the plane margin is done by using equation as follows

$$SM = -\frac{C_{M\alpha}}{C_{L\alpha}} \tag{6}$$

3. Results and Discussion

Accreted ice from simulation results can be categorized into 2 types based on the aerodynamic effect [6], namely double horn and streamwise. At a temperature of about -5 ^{0}C, the ice accreted is double horn while the ice at -10 ^{0}C may change depending on the atmospheric configuration.

Table 2. Simulation results at -5 ^{0}C

Name	T (^{0}C)	MVD (mic.)	LWC (g/m^3)	Ice Shape
10kHW-5111	-5	15	0.7	Double Horn
10kHW-5112	-5	25	0.4	Double Horn
10kHW-5113	-5	40	0.13	Double Horn
8kHW-5111	-5	15	0.7	Double Horn
8kHW-5112	-5	25	0.4	Double Horn
8kHW-5113	-5	40	0.13	Double Horn

At a temperature of -5 ^{0}C, the resulting ices shown in Table 2 are always in the form of a double horn. The temperature conditions allow the droplets to hit the surface of the airfoil to flow aft stagnation point before freezing. Under the -5 ^{0}C temperature conditions, atmospheric configuration variations may affect ice thickness. Atmospheric configuration's effect on ice thickness is due to the amount of air content in the air affecting the mass of water that can be frozen on the surface of the airfoil. The resulting ice shapes at -5 ^{0}C are shown in Figure 3.

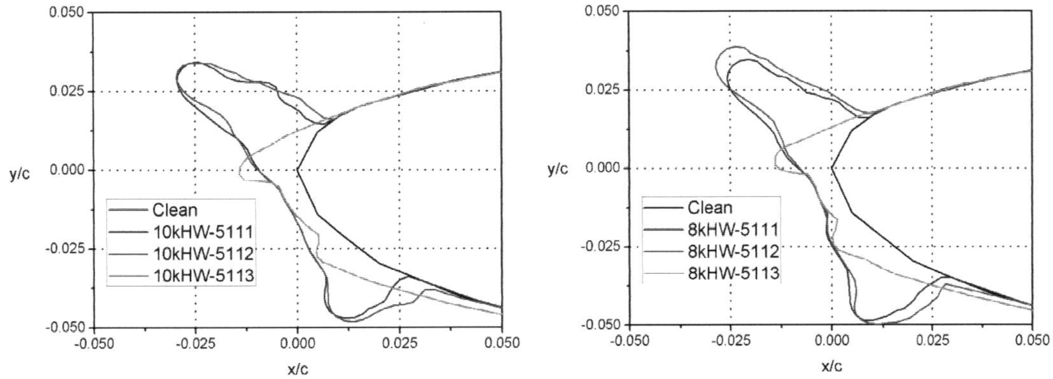

Figure 3. Ice shapes at -5 ^{0}C

At -5 ^{0}C temperature conditions, there is a significant difference. The resulting ice simulation 8kHW-5113 and 10kHW-5113 has a smaller thickness than other simulation results and still follow the leading edge contours, although there is still an ice horn around the stagnation point.

Table 3. Simulation results at -10 ^{0}C

Name	T (^{0}C)	MVD (mic.)	LWC (g/m^{3})	Ice Shape
10kHW-10111	-10	15	0.6	Double Horn
10kHW-10112	-10	25	0.3	Double Horn
10kHW-10113	-10	40	0.1	Streamwise
8kHW-10111	-10	15	0.6	Double Horn
8kHW-10112	-10	25	0.3	Double Horn
8kHW-10113	-10	40	0.1	Streamwise

In the streamwise case at -10 ° C shown in Table 3, ice formation is affected by variations of LWC and MVD. This is evidenced by simulation 10kHW-10113 and 8kHW-5113, the resulting ice is precisely streamwise, unlike the others that produce double horn shapes as shown in Figure 4.

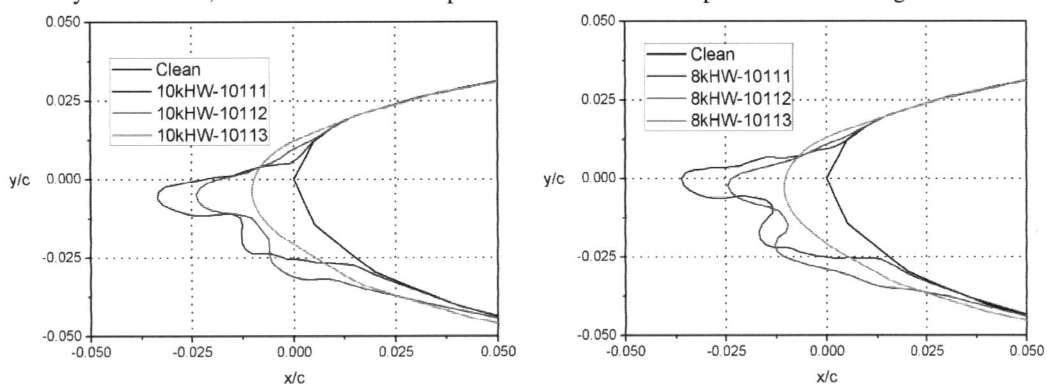

Figure 4. Ice shapes at -10 ^{0}C

Of the 12 ice shapes produced, only 5 ice shapes alone can be simulated using XFOIL software. The method used XFOIL has limitations in doing airfoil simulations are covered with ice, especially ice in the form of double horn. The simulation results are shown in Figure 5.

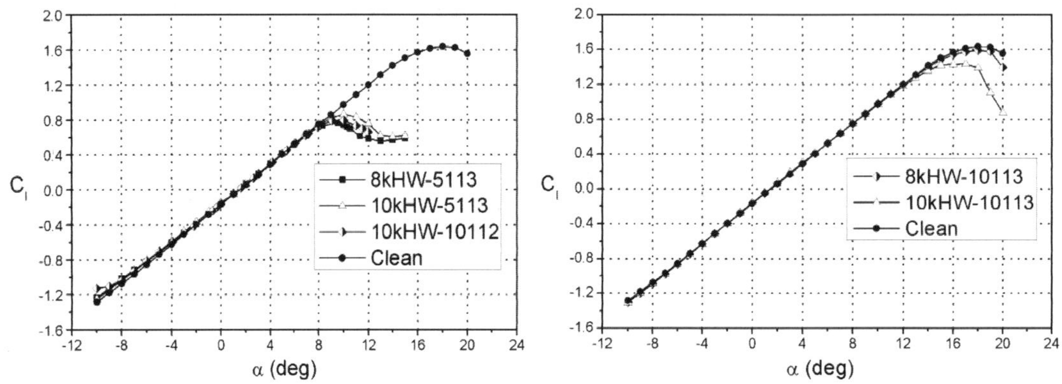

Figure 5. Aerodynamic performance results of various ice shapes

The graph above shows that the ice shape on the leading edge affects α_{stall} and $C_{l_{max}}$ of the airfoil. According to research conducted by Lee and Bragg [7], the decreasing value of α_{stall} and $C_{l_{max}}$ on ice-covered airfoils can be caused by separation bubble aft of the horn, one example is a double horn ice shape. The greatest effect is produced by ice in the form of double horn, however, the formation of ice does not exert too much influence on the slope lift of the airfoil. This means the effect of ice accretion on stability will not be dominant.

Horn ice's main characteristic is the presence of large separation bubbles generated around the tip of the ice. The stagnation point usually lies in the formed ice, and the boundary layer cannot overcome the large pressure gradient loss on the ice tip, thus, the location of the flow separation is usually relatively fixed at the tip of the ice over a large α range. A separated shear layer undergoes a transition to a turbulent stream and then usually returns to the surface of the airfoil. The existence of a separation bubble causes a considerable change in the pressure distribution causing the decrease of the pitching moment and the lift generated.

In this study, predictions were performed using polar data from XFLR5 simulation results and wind tunnel test data. From the prediction calculation of C_L and C_M values, the value of C_L and C_M of an aircraft are presented in Figure 6.

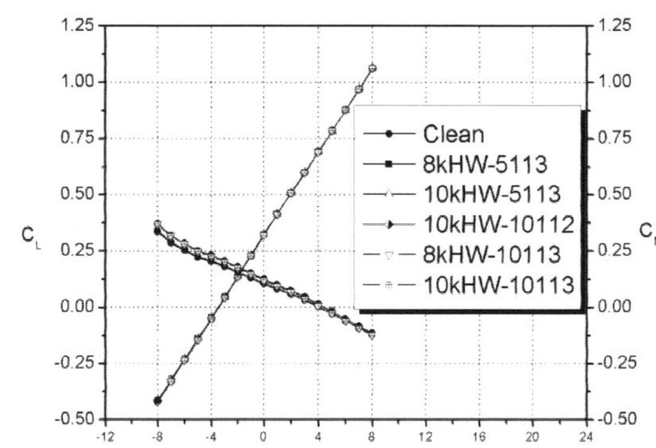

Figure 6. Stability characteristics of various ice shapes

As seen from the graph above, the C_L and C_M values of the aircraft from the wind tunnel test are used as comparators with the predicted C_L and C_M values. The biggest slope lift change occurred on ice simulation result 8kHW-5113. Stability prediction results can be viewed in Table 4.

Table 4. Stability prediction results of various ice shapes

Name	V (Mach)	Re ($\times 10^6$)	C_{L_α}	C_{M_α}	SM	$\%C_{M_\alpha}$	$\%SM$
Clean	0.2	5	0.0927	-0.0286	0.3087	0	0
8kHW-5113	0.2	5	0.0922	-0.0266	0.2891	-6.88	-6.34
10kHW-5113	0.2	5	0.0923	-0.0271	0.2931	-5.47	-5.03
10kHW-10112	0.2	5	0.0923	-0.0270	0.2927	-5.62	-5.17
8kHW-10113	0.2	5	0.0928	-0.0291	0.3130	1.55	1.42
10kHW-10113	0.2	5	0.0928	-0.0291	0.3139	1.86	1.70

In the streamwise case, all the simulation results show an increase in the slope lift. Increased slope lift is caused by the effect of streamwise ice that is spread evenly on the leading edge tends to improve airfoil performance by increasing the length of the cross section of the airfoil to produce lift. Although the streamwise ice shapes in this study increased the slope lift of an airfoil, the increased slope lift produced was insignificant. Since the slope lift is directly proportional to the static longitudinal stability or C_{M_α}, then in the streamwise case, all the simulated ice results in increased stability.

4. Conclusion

The most significant aerodynamic performance degradation shown by various ice cases lies in the decrease of α_{stall} and $C_{l_{max}}$, while the lift slope does not change significantly but its effect on static longitudinal stability and Static Margin is still visible, with the most significant degradation shown by simulation 8kHW-5113 with α_{stall} and $C_{l_{max}}$ values are 9.5 degrees and 0.76 in zero elevator deflection condition. Prediction of flight stability parameters results showed that the most significant decrease in C_{M_α} was caused by simulation 8kHW-5113 around 6.88%, with the reduction of SM equal to 6.34% from normal condition.

5. Acknowledgements

The authors wish to thank Universitas Indonesia for funding this research through PITTA grant program with contract number 2553/UN2.R3.1/HKP.05.00/2018. PT. Dirgantara Indonesia (Indonesian Aerospace) have supported this research providing facilities, guidance, and opportunity to contribute in commuter category aircraft development, thus, deserves our thanks.

References

[1] K. Mortensen, "CFD Simulations of an Airfoil With Leading Edge Ice Accretion," no. August, pp. 1–117, 2008.
[2] W. B. Wright, "User's manual for LEWICE version 3.2," 2008.
[3] M. Drela, "XFOIL: An analysis and design system for low Reynolds number airfoils," in *Low Reynolds number aerodynamics*, Springer, 1989, pp. 1–12.
[4] J. D. Anderson, *Aircraft Performance and Design*. McGraw-Hill, 1999.
[5] T. R. Yechout, "Introduction to Aircraft Flight Mechanics: Performance, Static Stability, Dynamic Stability, and Classical Feedback Control," p. 700, 2003.
[6] M. Bragg, "Aircraft aerodynamic effects due to large droplet ice accretions," in *34th Aerospace Sciences Meeting and Exhibit*, American Institute of Aeronautics and Astronautics, 1996.
[7] S. Lee and M. B. Bragg, "Experimental Investigation of Simulated Large-Droplet Ice Shapes on Airfoil Aerodynamics," *J. Aircr.*, vol. 36, no. 5, pp. 844–850, Sep. 1999.

IOP Conf. Series: Materials Science and Engineering **449** (2018) 012012 doi:10.1088/1757-899X/449/1/012012

Determining the Horizontal Tail Optimum Dimension of Civil Transport Class Aircraft Based on the Previous Model for Upgrading the Passengers Number

N Aditya[1] and W Nirbito[1,2]

[1]Department of Mechanical Engineering, Faculty of Engineering, Universitas Indonesia, Depok, 16424, Indonesia

[2]Corresponding author email : wahyu.nirbito@ui.ac.id

Abstract. As is well known, the horizontal tail of the plane has a function to produce a force to provide stability in the longitudinal spell of the plane. The determination of the contribution of the horizontal tail to longitudinal stability on the plane can be determined by simple methods such as using Digital DATCOM software. The simulations predicted by this method can save both time and cost. Nevertheless, the determination of the horizontal tail volume ratio can still be taken into further calculation so that the horizontal tail of the aircraft to work optimally. Using the analytic calculation of the aircraft equation of motion, the value of the horizontal tail volume ratio on the plane can be determined. The value of this horizontal tail volume ratio can be used as a reference to modify the area of the horizontal tail dimension of the aircraft. The calculation results show that the area of the horizontal tail of the aircraft can be reduced to 13.16 m2 and the dimensions for the new horizontal tail shape can be determined by considering the aspect ratio and taper ratio of the reference design.

1. Introduction

Nowadays, aircraft has become one of the most important transportation to be invented. the number of commercial flight has grown rapidly throughout the year. This also happen in developing country like Indonesia ,However based on research almost 70 % airport in Indonesia can't be landed by aircraft that has more than 50 passenger capacity. Therefore, the Indonesian state-owned enterprise, Indonesian Aerospace Inc is developing their best selling aircraft which was the result of their cooperation with CASA, the CN-235. The develop aircraft will have larger capacity than its predecessor which is 50 passenger, and will have a T tail configuration for its horizontal tail. Now ,this aircraft is in preliminary design phase. The existing configuration of the horizontal tail is an estimation from civil transport aircraft which have the same class category and passenger number as this aircraft based on FAR 25. Therefore in order to design the aircraft properly, the dimension area of the horizontal tail need to be calculated. The purpose of this study is to predict the dimension of the horizontal tail of the aircraft based on the horizontal tail volume ratio that will be earned by calculating the aerodynamic data from the DATCOM simulation.

2. Methodology

2.1. Aircraft Profile

As said before, This aircraft is developed from the CN-235. The input data of the aircraft is calculated using analytical approach based on the equation motion of the aircraft. From the three view drawing of the aircraft , the configuration of the plane can be seen from 3 point of view. This geometry data will be be used as an input into DATCOM software to simulate. This geometry data can also determine the arm length of moments that can be used to review the design results of the aircraft. In addition there are some supporting data for those used for calculation that can be seen in Table 1:

Table 1. The Supporting Data

No.	Spesification	Value
1	Air Density	1,225kg/m^3
2	MTOW.	18700kg
3	Inersia gear	1274203kg.m^2

2.2. Calculation of the Aerodynamics data of the Aircraft

Before calculating the horizontal tail volume ratio , there are geometry data that can be used for data processing contained in three view drawing of the aircraft. From these data we get values such as aircraft fuselage length, airplane wings, wing chord length, aspect ratio, taper ratio and others. The data functions as follows :

- range of center of gravity : The allowable shift of the center of gravity so that the horizontal tail can give stability to the aircraft
- Mean aerodinamycs chord : The average distance of the wing chord that serves as the data input data counting characteristics of aerodynamics
- Equivalent area : Wingspan and tailspan of the aircraft projected on the top view
- Tail volume : Obtained from the configuration data of the previous versions of the aircraft. The function of the previous tail volume to be the reference the determines whether the newtail volume is enlarged or minimized.

The data can then be used as an input for aircraft aerodynamic data calculations. The aircraft aerodynamic data is then processed using Digital DATCOM software. The results of the data processing of the aerodynamic characteristics using DATCOM are dimensionless number of coefficients of lift (C_L), coefficient of drag (C_D), coefficient of pitching moment (C_M) to the value of the angle of attack (α).[1] As for several stages to process aerodynamic data obtained from the configuration data can be seen in the flow chart on figure. 1.

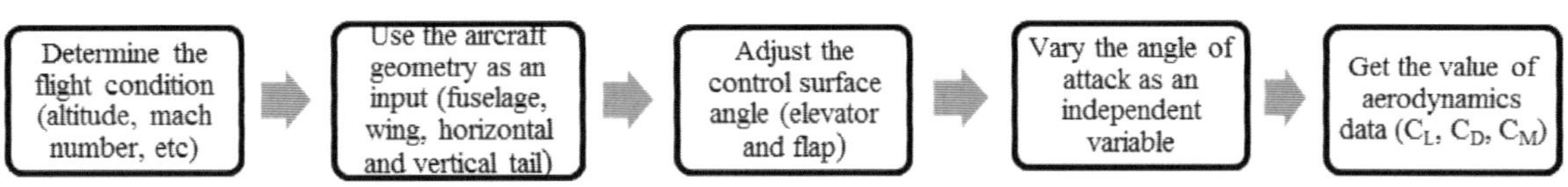

Figure.1. Flowchart of The DATCOM Simulation

3. Discussion

Here are the result of the research obtained from the simulation of Digital DATCOM software and from analytical result of shifting point value of center of gravity to the value of horizontal tail volume ratio.

IOP Conf. Series: Materials Science and Engineering **449** (2018) 012012 doi:10.1088/1757-899X/449/1/012012

3.1 Aerodynamics Characteristic Data

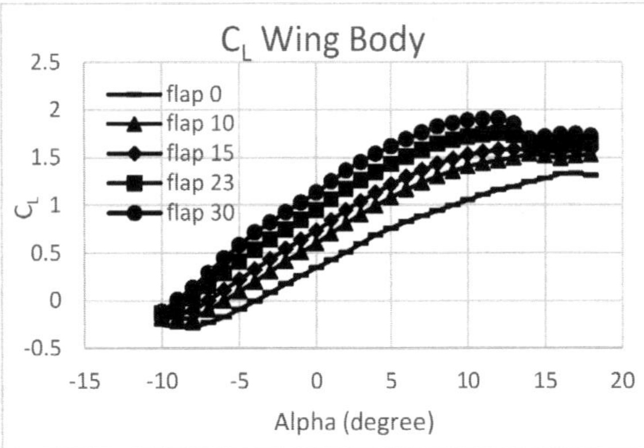

Figure.2. C_L vs α Graph on Wing Body

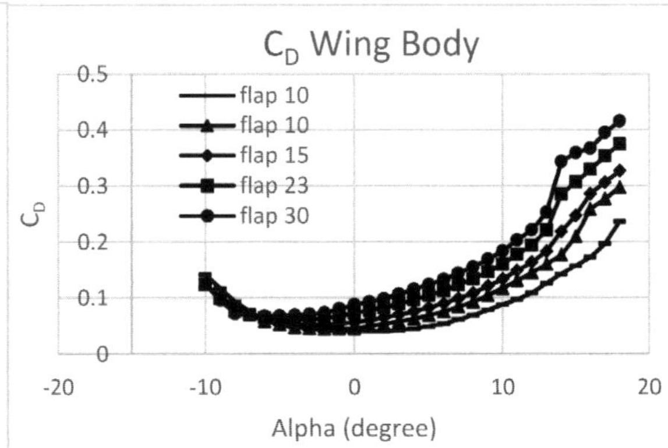

Figure.3. CD vs α Graph on Wing Body

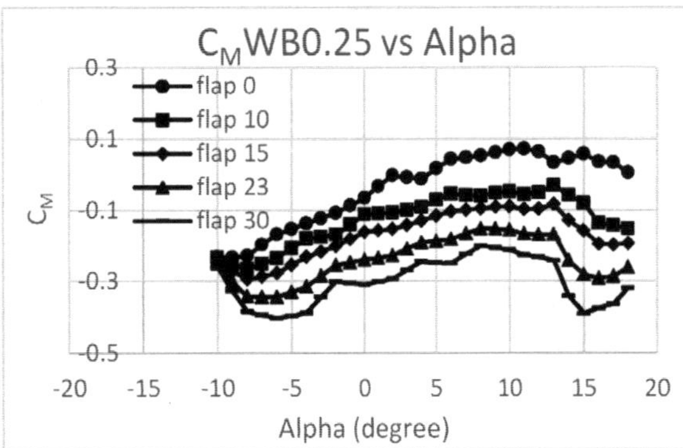

Figure.4.CM vs α Graph on Wing Body

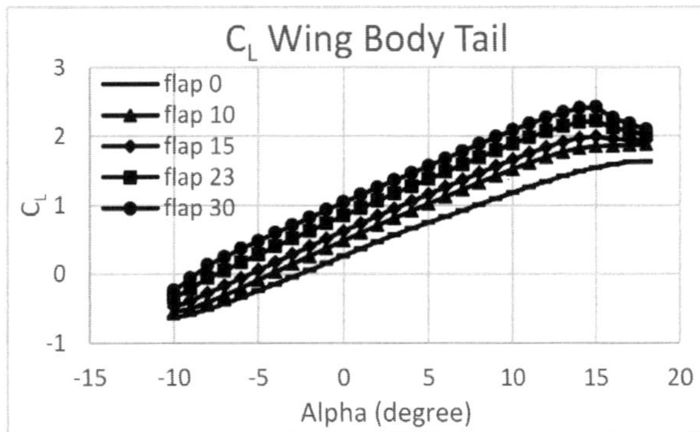

Figure.5. CL vs α Graph on Wing Body Tail

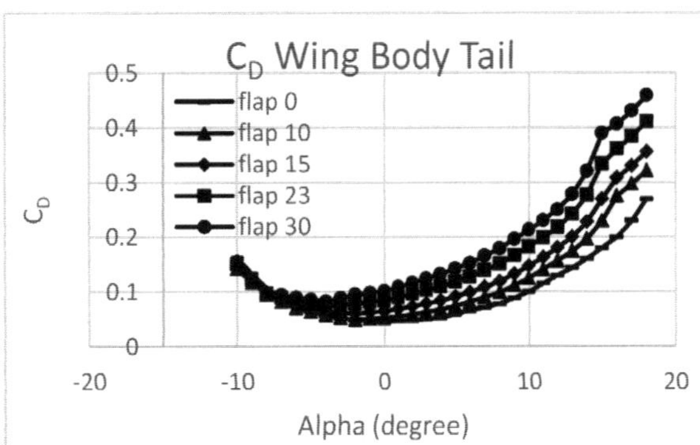

Figure.6. CD vs α Graph on Wing Body Tail

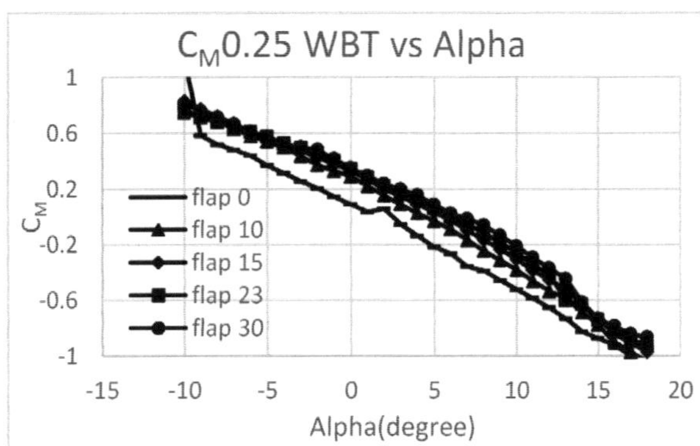

Figure.7. CM vs α Graph on Wing Body Tail

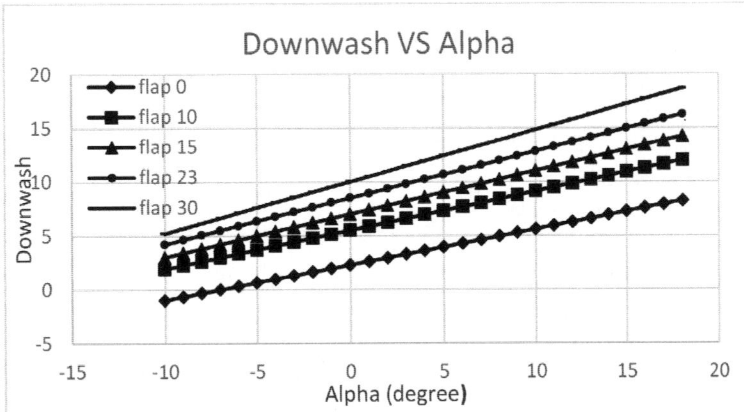

Figure.8. Downwash vs α Graph

3.2 Calculation of Horizontal Tail Sizing

The aerodynamics data from the simulation in figure. 2. until figure. 8. Are used for the input of the equation of motion. To get the the right value of the horizontal tail volume ratio, the flight condition of the aircraft is also become the parameter of the calculation.[2] This has to be done so that the horizontal tail of the aircraft can work optimally due to the various flight conditions that the aircraft is going to perform. The flight condition are like take off, approach, stall, go arroung at 1.3 VS, and go around at stick shacker. [3]

- Calculation on Take Off Condition

The first condition is take off rotation condition. At the time of take off, the aircraft deflect the elevator as a horizontal tail control surface. The equation can be seen in equation 1.

$$XCG = A'-B'-C'+D'-E'+F' \tag{1}$$

Where :

$A' = \{(C_{LH}/C_{Lo})[(X_H - X_G) - \quad Z_{CG}] - (C_{DH}/C_{Lo})(Z_H - Z_{CG})\} \quad (\quad {}_H V_H)/(X_H - X_{AC})$

$B' = (C_{LWB}/C_{Lo})[(X_G - X_{AC}) + Z_{CG}]$

$C' = (C_{DWB}/C_{Lo})(Z_{AC} - Z_{CG})$

$D' = T_{XENG}/W(Z_E - Z_{CG})$

$E' = T_{ZENG}/W[(X_G + X_E) + Z_{CG}]$

$F' = X_G + \mu Z_{X\Gamma} X_{M\Omega B.25}/X_{Ao} + (I_{\Psi\Psi\Gamma EAP}/(\Omega\chi)).\theta$

After obtaining the necessary data, then the value of VH and XCG can be obtained. The values of VH and XCG are obtained in the form of the line equation (y = -1.9189x + 1.4952) and can be seen on graphic.

- Calculation on Approach, Stall, Go Arround at 1.3 VS, and Go Arround at Stick Shacker Condition

In the conditions of approach, stall, and go arround phases, the aircraft will change its altitude by deflecting the elevator so that the angle of attack will have a positive value. When the center of the gravity is at forward, the needs of the elevator to change the altitude of the aircraft is at the highest point ,so it must be designed for those criteria so that aircraft can perform optimally.[4] aerodynamic data in each of these phases is also different. The equation that is used can be seen in equation 2.

IOP Conf. Series: Materials Science and Engineering **449** (2018) 012012 doi:10.1088/1757-899X/449/1/012012

$$\text{XCG} = (A - B)VH + 0.25 - \frac{CMWB0.25}{C'LO} - C - D + E \tag{2}$$

Where :

$A = (C_{LH}/C'_{LO}) [\cos\alpha + \frac{ZHo+\Delta CG}{LHo} \sin\alpha]$ $\quad C'_{LO} = C_{LO} \cos(\gamma + \alpha)$

$B = (C_{DH}/C'_{LO}) [\sin\alpha + \frac{ZHo+\Delta CG}{LHo} \cos\alpha]$ $\quad W' = W \cos(\gamma + \alpha)$

$C = (T_{XEng}/W') (\frac{\Delta ZCG-Heo}{c})$

$D = (T_{ZEng}/W') (\frac{\Delta ZCG-Heo}{c})$

Using the equation above , the values of VH and XCG for certain flight conditions can be obtained. The value is obtained either in the form of equations of lines or in the form of VH vs. XCG graph that can be seen in graph .

- Calculation of Neutral Point & Maneuver Point

Neutral point is the point where the moment of the entire aircraft is equal to zero.[5] The maneuver point is location where the aircraft moment is also equal to zero during maneuvering flight.[5] If the center of gravity is at this point, the entire aircraft will have neutral static stability.[5] The location of the maneuver point usually located aft of the the neutral point, but both the neutral point and maneuver point on the aircraft must be behind the CG. Those point can be determined using the equation 3 and 4.

$$X_{NP} = 0,25 - \frac{\frac{D_{CMWB}}{D_{CLWB}} C_{LaWB} - C_{LaH}(1-\varepsilon\alpha)V_H}{C_{LaW} + \frac{V_H}{LH/c} C_{LaH}(1-\varepsilon)} \tag{3}$$

$$X_{\text{MP}} = X_{\text{NP}} + 0.55\rho g \frac{l_H}{W/S} C_{L_{\alpha H}} \bar{V}_H \tag{4}$$

After use the aerodynamic data, the value of V_H and XCG can be obtained in the form of the equation of the line that is equal to (y = 2.6825x - 0.3472) or in the form of graph VH against XCG. The position of the neutral point usually located on the aft of the center of gravity. But, if the location of neutral point is to far aft, the stick force thet felt by the pilot is to small. Based on the FAR 25, the stick force for the pilot has to be greater than 1 lbs/6 knot .[6] So , when designing an aircraft, usually the position of the neutral point is assumed to be closer to the center of gravity by reducing the distance of the neutral point by 5 %.

The value of VH and XCG can be seen in figure. 9 .[6]

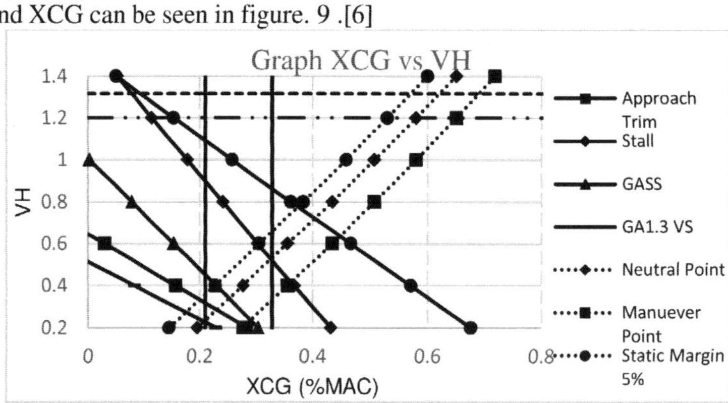

Figure.9. The XCG vs α Graph

3.3 The V_H vs XCG Graph Analysis

VH is the horizontail tail volume ratio that can be obtained from the equation 5 :

$$V_{HT} = \frac{L_h S_h}{cS} \tag{5}$$

From the VH vs XCG graph above the position of the CG forward position is 0.21% of the MAC, and the position of the AFT CG position is 0.33% of the MAC. In order for the aircraft to fly properly , the required VH value must be outside the center of gravity boundary, either forward or in AFT CG. It is necessary that the tail of the aircraft provide a force to stabilize the aircraft even though the center of gravity lies beyond the center of gravity.[7] Based on the graph on figure. 9. ,it can be determined that the minimum value of VH to be used is approximately 1,176. To get a safe value, the number is estimated V_H obtained is smaller than the value of VH obtained previously which is 1.317. After reviewing these aspects it can be said that the shape of the horizontal tail of the calculated result will be the same as the reference horizontal tail but has a smaller scale. [7]

3.4 The Horizontal Tail Configuration and Dimension Analysis

From the results obtained from the VH vs XCG graph, the optimum horizontal tail volume ratio value to be used is 1.2. From that value it can be concluded that the horizontal tail are can be reduced from 14.45 m ^ 2 to 13.16 m ^ 2. The value can be determined from the equation. After obtaining the area from the horizontal tail, the dimensions for the horizon of the aircraft can be determined by considering the following aspects like the geometry of the tail, the aspect ratio, and the taper ratio.[7]

4. Conclusion

Based on various of calculation, the value of horizontal tail volume ratio is 1.2 . This value is smaller than the previous estimation from civil transport aircraft which have the same class category. This means that the horizontal tail area of the aircraft is smaller than the preivous model and the tail configuration can be reduced . By using the equation ,the area of horizontal tail can be reduced from 14.45 m^2 to 13.36 m^2. After determine the area, the dimension of the horizontal tail can also be determined. In order to do this, there are a few aspect that need to be consider such as configuration, aspect ratio, and taper ratio of the horizontal tail. The new dimensions obtained are the root chord length is 2,090 meters, the chord tip length is 1,100 meters, and the tail span length is 8,110 meters. These value is smaller than the preivous model. The new dimension was better than the previous model because, while the tail still produce enough lift to stabilze the aircraft, the structrural and material cost is reduced. This cause the new horizontal tail is lighter than the one before and sufficient for T tail configuration

5. Acknowledment

The authors would like to thank the Indonesian Aerospace Inc that has given the opportunity to study and facilitate this engineering research program and also the PITTA grant programme of University Indonesia for funding this research with grant number 2553/UN2.R3.1/HKP.05.00/2018

References

[1] F. Nicolosi, P. Della Vecchia, and D. Ciliberti, “An investigation on vertical tailplane contribution to aircraft sideforce,” *Aerosp. Sci. Technol.*, vol. 28, no. 1, pp. 401–416, 2013.
[2] Dr. Jan Roskam (Ackers Distinguished P, *Airplane_Aerodynamics_and_Performance_Nav.pdf.* 1981.
[3] R. C. (Departmen. of A. and M. E. Nelson and University of Notre Dame), “Nelson Flight Stability and Automatic Control Second Edition.” WCB McGraw-Hill, p. 452, 1998.
[4] U. S. D. of Transportation and Federal Aviation Administration, *Weight And Balance Handbook.* 2016.
[5] Thomas R. Yechout (American Institute of Aeronautics and Astronautic, “Intoduction of Flight Mechanics,” p. 633, 2003.
[6] J. (Aerospace E. U. of K. Roskam, *Airplane Design.* 1985.
[7] J. D. Anderson, *Aircraft Performance and Design.* McGraw-Hill, 1999.

An Estimation of the Maximum Take-off Weight Commuter Category Aircraft Based on Airstrip Condition on Remote High Terrain Area

L L Sunara[1] and W Nirbito[1,2]

[1]Department of Mechanical Engineering, Faculty of Engineering, Universitas Indonesia, Depok 16424, Indonesia.

[2]Corresponding author's e-mail : wahyu.nirbito@ui.ac.id

Abstract. The conditions of the runway on remote high terrain area with mountain conditions are the problem for the aircraft. The short of runway length and the higher location can affect to the take-off performance of the aircraft. Thus the determination of the maximum take-off weight is critical in order for the aircraft to take-off safely. The purpose of this research is to determine the maximum take-off weight for 19 passengers commuter category and double engine propeller driven aircraft based on runway condition at ISA +20 on remote high terrain area. The intended of runway conditions are the altitude of the runway location, the slope of the runway, the surface condition of the runway and the wind speed around the runway. The take-off distance approach based on aerodynamic characteristics, geometry, and power plant is described. The results obtained that the aircraft can take off at remote high terrain area with an altitude above 5000 feet and runway length of 393 m to 600 m for all engine operative conditions.

1. Introduction

Air transportation is one of the solutions to connect one area with other areas. The remote area with geographical conditions in the mountains is a difficult location to access. The aid of food and health are needed for people in remote areas, therefore the access of the transportation is essential as a bridge to aid delivery. The runway conditions that are located in mountainous areas with an altitude above 5000 feet and there are gaps and valleys such as Papua region in Indonesia is one of the problems. Thus the appropriate calculation is needed for the take-off operations in that area. The maximum take-off weight is important in flight operations because it includes the amount of passengers and fuel that the aircraft can carry. The commuter category with 19 passenger seats and double engine propeller driven aircraft is an aircraft that is still in development stage and aims to operate in remote areas. Thus it is necessary to predict data for the maximum take-off weight on remote areas. At the take-off phase, the aircraft speed is arranged regularly in accordance with the regulation [1]. The regulation uses Civil Aviation Safety Regulation (CASR) part 23 for commuter category aircraft as a reference.

The take-off distance approach is used to determine the maximum take-off weight based on runway conditions. There are several terms of take-off distance including take-off distance available (TODA), take-off run available (TORA) and accelerate-stop distance available (ASDA) [2]. Engine failure may occur during the take-off process. if an engine failure occurs, then the pilot must determine whether the aircraft should continue take-off with one of the engine failures or rejected take-off. The maximum speed at which a pilot decides to rejected take-off when an engine failure occurs is at speed V_1. If the engine fails before V_1, the pilot can rejected take-off (RTO) and brake until the aircraft stops for the rest of the available runway [2]. The engine failure at the time of take-off process is undesirable. Thus it is necessary to estimate the maximum take-off weight when the engine failure occurs to pevent the accident.

The required distance from the aircraft starts moving to a halt in the rejected take-off condition is ASDA. The required take-off distance for all engine operative (AEO) and continue take-off in case of the engine failure occurs until reach the screen height 35 feet is TODA [1]. Each runway is described by the relevant distances available, referred to as TORA, TODA, and ASDA respectively [2]. Runway conditions such as altitude location, wind velocity around the runway, runway surface conditions and runway slope are some conditions that affect the take-off performance of the aircraft. Knowing the maximum take-off weight (MTOW) for each runway, it can be a reference for a pilot or crew to determine the amount of payload in order to take-off safely on the runway.

2. Methods

The aircraft move following the regulated speed during the take-off process for safety purpose [3]. In the regulation has been set about the speed limits on take-off conditions by considering that the speed is the minimum safe speed.

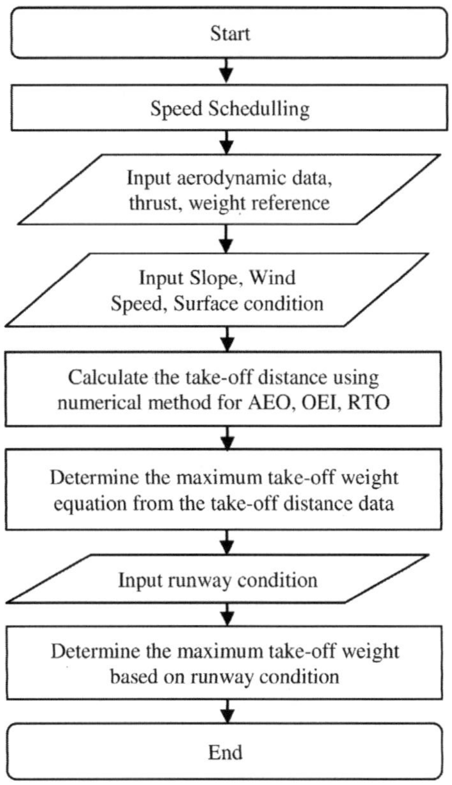

Figure 1. Flowchart of calculating process

Figure 1 shows the flowchart of calculating process. The take-off distance is calculated using numerical method [4]. The length of distance traveled by aircraft depends on the acceleration of the aircraft in conditions with different altitude, the slope of runway, runway surface conditions, weight and wind speed [1]. On the calculation of take-off distance using 3 weight as a reference, including 4310 kg, 6000 kg and 7030 kg. The weight of 4310 kg is the empty weight of the aircraft without payload. The weight of 7030 is the maximum take-off weight based on design structure limit, while the weight of 6000 kg is the middleweight between 4310 kg and 7030 kg. Take-off distance is measured using ground speed, therefore the wind speed will affect ground speed [5], [6]. The calculation of the take-off distance includes the condition of all engine operative (AEO), one engine inoperative (OEI) and rejected take-off (RTO), where for the case of RTO one of the engines does not operate the same as in the case of OEI. In this calculation, the case of engine failure occurs for the OEI dan RTO condition is the same. Thus the pilot decides whether continue take-off or rejected take-off at V1 for each case. Acceleration of the aircraft during on the ground refers to the equation below [5] :

$$a = \frac{g}{W}[T - \mu_{\mathrm{R}} W\cos\theta - (C_D - \mu C_L)qS - W\sin\theta] \tag{1}$$

The take-off distance data is used to determine the maximum take-off weight. The take-off distance result is take-off distance available (TODA). The result from a numerical calculation is the take-off distance chart. Based on the take-off distance chart there are a curves of the relationship between the required take-off distance with correction factors such as wind speed (V_w), runway slope, runway surface conditions (S_c) and altitude (Alt) of the runway location [5]. The multiple regression method with least squares approximation is used to determine the maximum take-off weight equation based on each curve in the take-off distance chart [4], [7]. A suitable equation is obtained based on the best fit of the curve. Thus the maximum take-off weight is $W_{TO} = f(Alt, RL, V_w, Slope, S_c)$. In each case, there are 4 equations with each factor affecting the required take-off distance in determining the maximum take-off weight.

$$S_1 = f(RL, V_w) \tag{2}$$

$$S_2 = f(S_1, Slope) \tag{3}$$

$$S_3 = f(S_2, Sc) \tag{4}$$

$$W_{\mathrm{TO}} = f(S_3, Alt) \tag{5}$$

The distance of S_1 is the effect of wind speed on the available runway length (RL). S_2 is the distance at S_1 which is affected by the slope of the runway while S_3 is the distance of S_2 which is affected by the runway surface condition. W_{TO} is the maximum take-off weight with the distance S_3 which is affected by the altitude of the runway. The results of the equations comes from multiple regression methods. Thus the equations for all engine operative conditions are as follows :

$$S_{1,\mathrm{AEO}} = 0.201 + 1.004(RL) + 0.891(V_{\mathrm{w}}) + 0.018(RL)(V_{\mathrm{w}}) \tag{6}$$

$$S_{2,\mathrm{AEO}} = 0.1197 + 1.003(S_1) + 14.35(Slope) - 0.073(S_1)(Slope) \tag{7}$$

$$S_{3,\mathrm{AEO}} = -31.84 + 1.145(S_2) + 31.84(Sc) - 0.144(S_2)(Sc) \tag{8}$$

For the engine failure conditions, there are two cases that is take-off with one engine or rejected take-off. The equations for one engine inoperative (OEI) conditions are as follows :

$$S_{1,\mathrm{OEI}} = 0.167 + 1.002(RL) + 2.608(V_{\mathrm{w}}) + 0.013(RL)(V_{\mathrm{w}}) \tag{9}$$

$$S_{2,\mathrm{OEI}} = 21.842 + 0.969(S_1) + 30.384(Slope) - 0.116(S_1)(Slope) \tag{10}$$

$$S_{3,\mathrm{OEI}} = -46.454 + 1.167(S_2) + 46.454(Sc) - 0.167(S_2)(Sc) \tag{11}$$

The equations for rejected take-off (RTO) conditions are as follows :

$$S_{1,\mathrm{RTO}} = 1.309 + 1.003(RL) + 4.983(V_{\mathrm{w}}) + 0.015(RL)(V_{\mathrm{w}}) \tag{12}$$

$$S_{2,\mathrm{RTO}} = 2.028 + 0.99(S_1) + 24.225(Slope) - 0.031(S_1)(Slope) \tag{13}$$

$$S_{3,\mathrm{RTO}} = -70.291 + 1.142(S_2) + 70.291(Sc) - 0.1428(S_2)(Sc) \tag{14}$$

IOP Conf. Series: Materials Science and Engineering **449** (2018) 012013 doi:10.1088/1757-899X/449/1/012013

S_3 is the required take-off distance for a given runway length that has been corrected by wind speed, runway slope and runway surface conditions. In determining the maximum take-off weight using the distance of S_3 which is corrected by the altitude of the runway location. Here are the maximum take-off weight (MTOW) equations for AEO, OEI and RTO conditions in kilogram :

$$W_{\text{TO,AEO}} = 9774.4 - 9.38\,x\,10^6(S_3{}^{-1.32}) - 0.111(Alt^{1.04}) - 20.861(S_3{}^{-1.32})(Alt^{1.04}) \quad (15)$$

$$W_{\text{TO,OEI}} = 9282.16 - 380816(S_3{}^{-0.76}) - 0.166(Alt^{0.99}) + 1.729(S_3{}^{-0.76})((Alt^{0.99}) \quad (16)$$

$$W_{\text{TO,RTO}} = -68038 + 50339(S_3{}^{-0.06}) + 2.026(Alt^{0.97}) - 1.514(S_3{}^{-0.06})(Alt^{0.97}) \quad (17)$$

3. Discussion

The airstrip data in the remote area are presented in table 1. The runway conditions include the runway length in meters, altitude of the runway in feet, runway slope in percent and wind speed around the runway in knots. Figure 2 and 3 shows the maximum payload can carry by aircraft for each runway.

Table 1. The maximum take-off weight data for each airstrip on remote area

Airstrip	RL	ISA	Alt	Slope	Surface	Vw	MTOW		
							AEO	OEI	RTO
A	600	20	7975	-4	Asphalt	1	7120.54	5912.34	4702.80
						-1	7036.48	5841.90	4478.25
B	590	20	6950	0	Asphalt	1	6862.5	5567.81	4923.94
						-1	6759.80	5492.61	4725.41
C	500	20	6227	-2	Asphalt	1	6776.32	5554.16	4199.98
						-1	6653.58	5462.39	3960.44
D	415	20	6100	-2	Grass	1	5894.97	4905.29	3452.52
						-1	5721.57	4791.84	3224.46
E	444	20	5150	-7	Grass	1	6867.13	5683.08	3375.00
						-1	6729.78	5572.73	3081.25
F	393	20	5200	-1	Grass	1	5669.49	4764.64	3436.00
						-1	5478.27	4647.35	3210.14
G	500	20	5900	-6	Grass	1	7006.42	5805.05	3978.83
						-1	6891.67	5713.38	3726.10

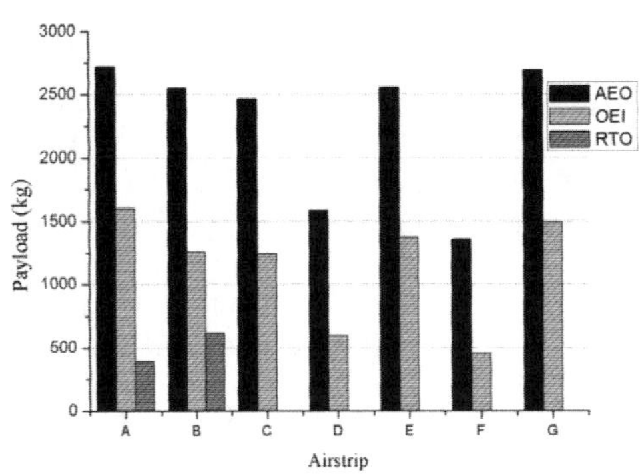

Figure 2. Maximum payload for various airstrip with headwind condition.

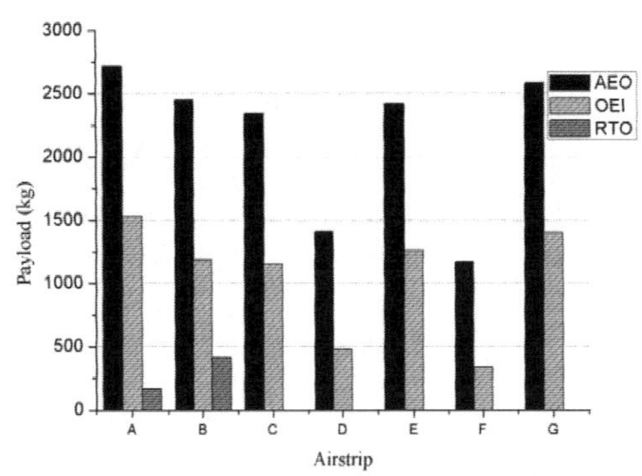

Figure 3. Maximum payload for various airstrip with tailwind condition.

The variation of the runway conditions for each runway will affect the take-off distance. Due to the runway length on the remote area is short. Thus if the required take-off distance exceeds the runway length, it will impact the reduction of the weight of the aircraft in order to take-off with the available runway length. The aircraft can operate with the maximum take-off weight over 7030 kg in A airstrip. Because the weight of 7030 kg is the maximum structural limit weight for the aircraft. Thus the aircraft must operate with 7030 kg even though the aircraft can operate over that weight.

The maximum take-off weight on AEO case is greater than OEI and RTO case. This is because when the engine failure occurs, the available thrust only comes from one engine and hence, the take-off distance will be farther for the same weight. The maximum payload for OEI is greater than RTO, therefore determining the maximum take-off weight on OEI condition is preferred for operational purpose. Thus if the engine failure occurs, the aircraft can still take-off safely with one engine fails. Based on the equation for each condition, the results obtained between the effect of the runway condition to the maximum take-off weight are shown in Figures 4 to 7. Figure 4 is the effect of runway altitude to the maximum take-off weight. Increasing altitude of the runway will decrease the maximum take-off weight. This is because in the higher altitude the air density more lowest and thus the required take-off distance will be farther.

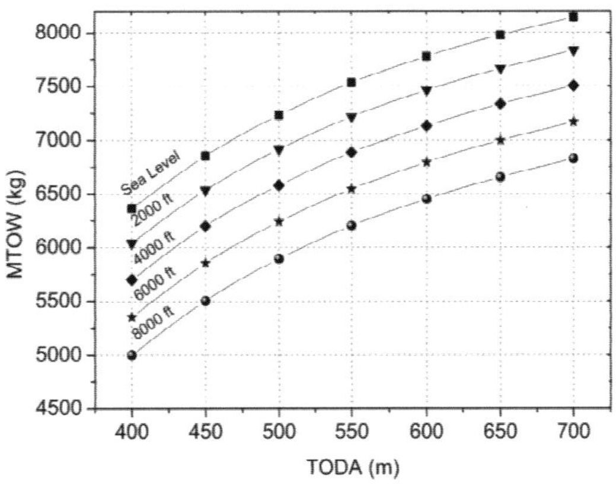

Figure 4. MTOW vs TODA for variation altitude with AEO condition.

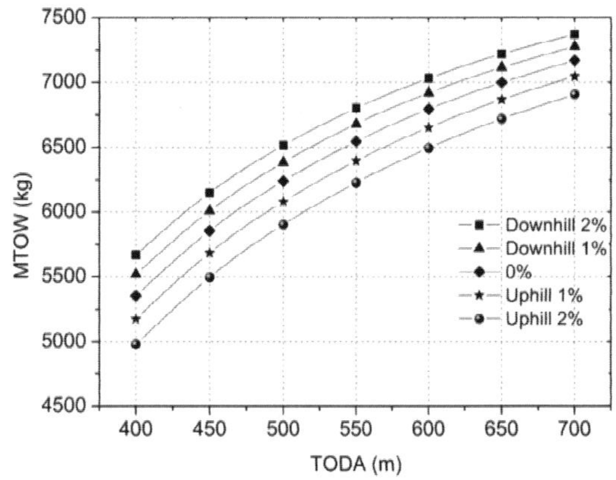

Figure 5. MTOW vs TODA for variation slope with AEO condition at 6000 feet.

The effect of the runway slope to the maximum take-off weight is presented in figure 5. In the runway with a downhill condition, the maximum take-off weight will be greater rather than an uphill condition. This is because the down slope will enhance the acceleration of the aircraft during the ground roll phase, and thus the required distance becomes shorter. The effect of wind speed to the maximum take-off weight is presented in figure 6. The direction of the headwind is opposite to the direction of the aircraft motion and it will enhance the relative velocity between the aircraft and the air. The greater of relative velocity between the aircraft and the air will enhance the lift force of the aircraft and the take-off distance becomes shorter. Thus the rest of the available runway length can be used to enhance the weight of the aircraft. Figure. 7 represents the effect of the runway surface condition on the maximum take-off weight. The coefficient of rolling between asphalt and grass is different. Asphalt has the rolling coefficient of 0.025 while grass is 0.05. The greater of the rolling coefficient value will decrease the acceleration of the aircraft. Thus the required take-off distance will be farther and impact to reduce of the aircraft weight.

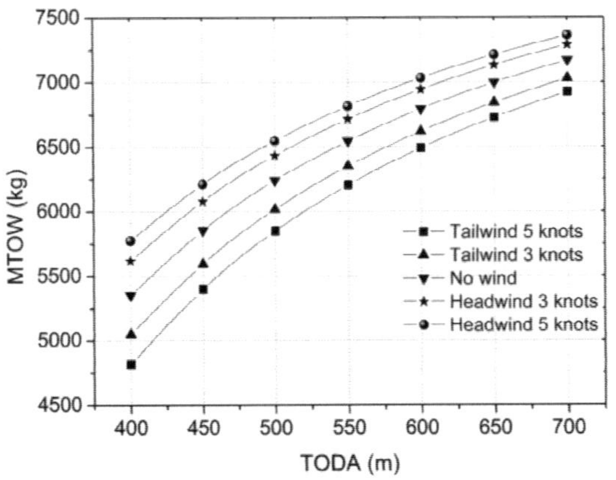

Figure 6. MTOW vs TODA for variation wind velocity with AEO condition at 6000 feet.

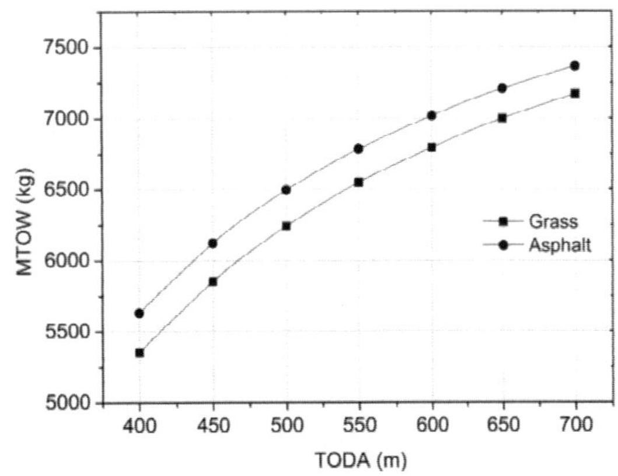

Figure 7. MTOW vs TODA for variation surface condition with AEO condition at 6000 feet.

4. Conclusion

The maximum payload with RTO case is lower than OEI case in engine failure condition. There are 5 airstrips that are not possible to rejected take-off (RTO) such as C, D, E, F and G airstrip. The maximum take-off weight on C, D, E, F and G airstrip is below the empty weight of the aircraft for RTO case in engine failure condition. Thus the required take-off distance will exceed the runway length even if the aircraft does not carry any payload. The most critical airstrip is an F airstrip with a maximum take-off weight only 5669.49 kg under headwind conditions and 5478.27 kg under tailwind conditions for AEO conditions. The maximum payload on AEO condition above the empty weight of the aircraft. Thus for normal operation without engine failure occurs, this aircraft can still operate safely at all of the runways.

5. Acknowledgements

The authors would like to thank PITTA grant programme of Universitas Indonesia for funding this research with grant number 2553/UN2.R3.1/HKP.05.00/2018. The authors also grateful to Indonesian Aerospace company for the facilities, guidance, and opportunities in the development of commuter category aircraft in this research.

Reference

[1] Y. Zhu, J. Wang, Y. Chen, and Y. Wu, "Calculation of Takeoff and Landing Performance Under Different Environments," *Int. J. Mod. Phys. Conf. Ser.*, vol. **42**, p. 1660174, 2016.

[2] D. Zammit-Mangion and M. Eshelby, "Simplified Algorithm to Model Aircraft Acceleration During Takeoff," *J. Aircr.*, vol. **45**, no. 4, pp. 1090–1097, 2008.

[3] C. T. E. Roskam, Jan and Lan, *Airplane Aerodynamics and Performance*. Kansas: DARcorporation, 1997.

[4] M. Arjomandi and N. K. Liseytsev, "Simplified method for estimating the take-off weight for short-haul transports," *Aircr. Des.*, 2000.

[5] Boeing Commercial Airplanes, "Jet Transport and Performance Methods," *Flight Oper. Airplanes, Boeing Commer.*, no. March, p. 742, 2009.

[6] I. Vasile and A. Nicolae, "Take-off and landing of aicraft," *Incas Bull.*, vol. **4**, no. 2, pp. 53–60, 2012.

[7] M. K. Greenway, S. J. Koob, and A. Force, "Relating Take-Off-Gross-Weight and Mission Aircraft AIAA 16TH AEROSPACE," pp. 78–98, 1978.

IOP Conf. Series: Materials Science and Engineering **449** (2018) 012014 doi:10.1088/1757-899X/449/1/012014

Design, Modeling and Performance Optimization for Propulsion of a Small Launch Vehicle

Layachi Hemza[1, a], Boudjemai Abdelmajid[2] and Houhou Hatem[3]

[1,3] Algerian Space Agency, Algeria
[2] Centre de développement des satellites, ORAN, ALGERIA

[a] layachi-hamza@hotmail.com

Abstract. In this study, we will establish a platform for design modeling and performance analysis for a propulsion module of a small launch vehicle. After giving it the suitable mathematical reformulation, we will use MATLAB-SIMULINK as a numerical tool for calculating the specification for the needed factors. The validation of each parameter has to be sequential to reach the optimized model, our simulation basics are established on tracking and a regulation methodology, after defining the design variables, giving the mathematical formulation, the simulation procedure set up on the interaction of the designated parameters to reach the suitable scenario. The integration of a regulation tool was in aim to set the output of each sub-system and avoid possible dysfunction in the whole system.

1. Introduction

Space launch systems are composed of a great number of components assembled in a hierarchy of subsystems. The performance of the vehicle depends on the specific performance of each subsystem [1, 2], which dependent on the design sets and parameters [3, 4]. Adjustments in design parameters are propagated in the hierarchy of subsystems and components, the flight path and the payload capacity [3, 5, 6]. The design of launch vehicles is fundamentally a task of great complexity. Thus, to manage such complexity, its design is generally divided into several disciplines, such as trajectory, propulsion, mass and aerodynamics [7, 8]. Since the minor inaccuracy can lead to disastrous economic, material and human consequences, we also consider the cost and reliability of the disciplines just as important [2, 9]. Conventionally, design tools and computer codes have been developed to treat these disciplines separately. Unfortunately, the disciplines are not isolated from each other, but are all interconnected in a system of units.by this project, the primitive goal is to classify the disciplines with a direct impact to the overall performance in a first order then, analyze the behavior of such vehicle in critical load conditions by multiple modeling and simulation techniques to reach the optimized model [9].

The design of a space launcher is a growing field of research; It is used for transporting satellites or payloads generally [10, 11]. As launching, is a complex operation, requiring highly qualified personnel and huge infrastructure [12, 13], in a difficult security context. The development of such a launcher must meet very high specifications in terms of reliability and performance to ensure the delivery of the payload in a precise position [13].

Small launch vehicle propulsion systems are interesting topic of research, in order to ensure the deliverance into a specified orbit successfully, a couple of design variable must be analyzed by respect of mathematical respective formulation. In this article, we will introduce a mathematical model to analyze the mission of insertion of payloads to specific orbit and respective matlab-simulink platform for visualization and analysis.

2. Space launch vehicle propulsion specifications

2.1. Trajectory and rockets mission

For this rocket, a payload of 100KG will be carried along the mission and its final destination will be the low earth orbit 200 KM altitude. The rocket is composed of two stages as it is shown in figure 1, so the mission is composed of two phases. As a first phase, the whole rocket will be launched to a given distance to earth at 100KM, then the first stage is separated the calculation of propellants tanks volume is based on:

$$\Delta V = C \ln \{(m_p + m_l)/ m_p\} \quad (1)$$

$$m_p = m_l (exp(\Delta V / C) - 1) \quad (2)$$

After the separation of the stage 1, the rocket will continue its mission to deliver the payload to the low earth orbit altitude as specified before. For this paper, the mission is to deliver the payload into the specific altitude with a precise velocity.

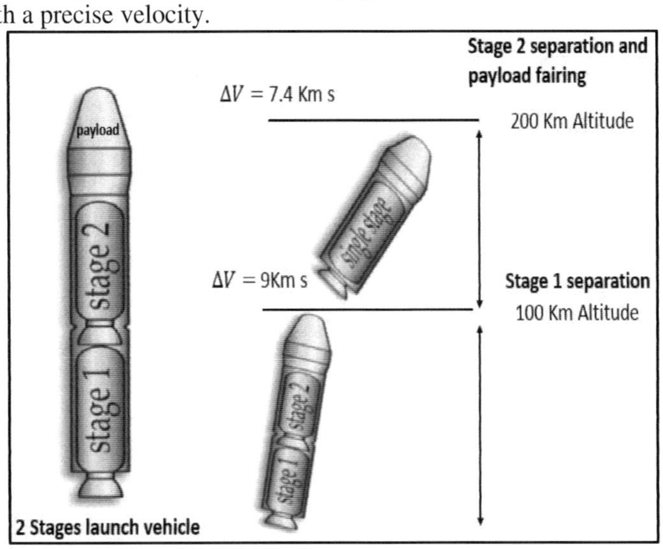

Figure 1. Rocket mission and trajectory details

Table 1 summarize the specific input of this launch vehicle basing on the equations 1 and 2.

Table 1. Launch vehicule propulsion details

Structural weight		Trajectory details	
Propellant	LOX /LCH4	Ideal Exhaust velocity(km/ s)	5.5 Km/S
First stage total mass (kg)	3500	Ideal velocity increment (km/ s)	9 Km/S
Second stage total mass (kg)	1500	Tank 1 propellant mass (kg)	2800
Payload mass (kg)	100	Tank 2propellant mass(kg)	850
Rocket Total mass (kg)	5100		

2.2. Mathematical formulation

And in order to define the thrust needed we use the following formula:

$$Fthrust = mdot\ C \tag{3}$$

Where, Fthrust is a rocket total force, mdot is the mass flow rate and C represents the effective exhaust velocity.

For equal specific impulse, in order to define the needed thrust, we use the following formula:

$$\Delta V = Isp\ g0\ \ ln\ (\mu 1.\ \mu 2) \tag{4}$$

$$\mu 1 = mintial\ /\ mpayload \tag{5}$$

$$\mu 2 = mstage2\ /\ mpayload \tag{6}$$

$$\gamma 1 = mpayload\ /\ minitial \tag{7}$$

$$\gamma\ 2 = mpayload\ /\ mstage2 \tag{8}$$

It needs to be verified with the following formula:

$$\mu required = exp\ (\Delta V\ /\ Isp\ g0) \tag{9}$$

Thrust to weight ratio, also called load factor is defined for the two operations as follow:

$$Thrust\ /\ weight = mdot\ C\ /\ mg0 \tag{10}$$

3. Methodology of analysis

Following the previous section and in order to reach the final design, the design variables listed before must be handled following the algorithm in figure 2:

For this platform, the first step to establish our model is to set the parametric design of the nozzle. The calculation of the effective velocity is considered as a resultant of the previous analysis, for a parallel step, we establish the propellant debit. From the equation 1 we can deduce the thrust of our rockets. From this last variable we can establish the significant ratios, the mass ratio and the payload ratio:

- The payload ratio is needed to verify if the model rocket can lift the required weight of the payload
- The mass ratio will be introduced to calculate the required mass ratio and if this condition is satisfied, following the equation 2 we establish the ideal increment of the rocket.

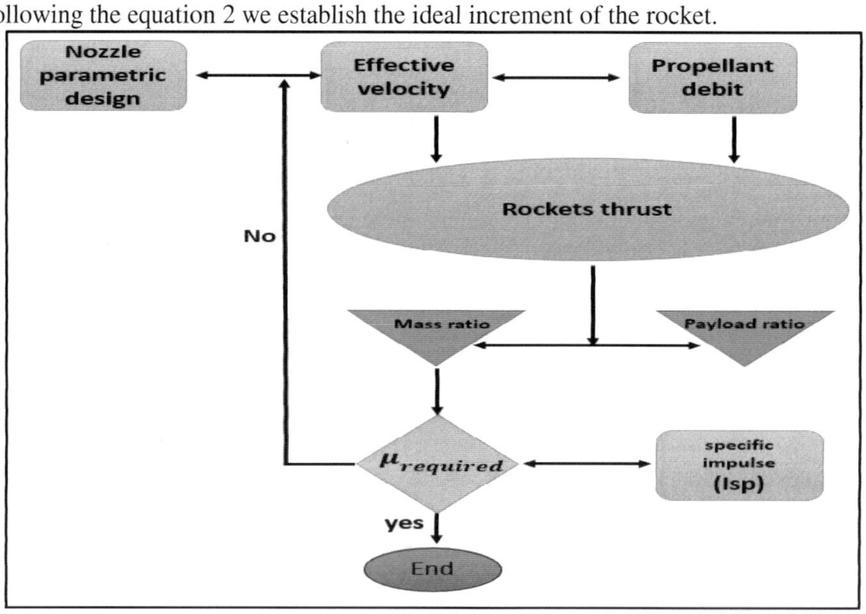

Figure 2. Algorithmic Synoptic scheme and design analysis

The primitive design set for this model is related to the parametric design of the nozzle; it determines the exhaust velocity following the equation3. In this section, we introduce the nozzle expansion ratio, which is a key design parameter, it represents a security factor for the propulsion system, as it is determined as follow:

$$C = A_t \ P / mdot \tag{11}$$

$$\varepsilon = A_e / A_t \tag{12}$$

ε is the nozzle expansion ratio and A_e is its exit area and A_t the engine throat area.

4. Assumptions and optimization procedure

In order to have a global view of the design set, we have to define the main parameters and this includes the main factors that ensure the security of the mission. After analyzing the different models, the specific missions and environment of the launch vehicles, we have opted to target the listed parameters:

A. Launch-vehicle propulsion subsystems must be designed for

- Thrust-to-weight ratio will be given to maximum performance in a way that the propulsive load is fully enough for the given lift-off mass
- Optimum ratio of nozzle expansion with respect to security aspect

B. By staging launch vehicles, we can:

- Reduce the total vehicle mass for a given payload and ΔV requirement, it is imperative to ensure the total needed propellant required for the whole mission first,
- Increase the total mass of the payload delivered to space
- Decrease the engine efficiency (Isp) required to deliver the payload to orbit

5. MATLAB-SIMULINK Platform presentation

Figure 3 show the Matlab SIMULINK platform, is divided to three main blocks, each of the three blocks is modeled to define a decisive parameter, we can mention:

The Simulink platform mainly of three blocks, which includes:

A. The nozzle parametric design: this block consist the calculation of the exhaust velocity, it was previously given as a key concept for our modeling, by introducing the defining factors the output will be the ideal exhaust velocity with interaction with the other blocks.

B. The propellant mass flow rate block: for this section, it is important to set the upper and the lower ranges flow, the maximum consumption by respect to table 1 that list the tank capacity of each stage of the launch vehicle.

C. The $\mu_{required}$ PID block: connected the two other blocks as a tracker, it is in aim to adjust the output results of other blocks and remain operative to ensure the required thrust.

Following the synopsis algorithmic design in figure 2, the verification of the required *Isp* of the propulsion module is given with similitude to the PID controller. this controller adjust the needed thrust by tracking the exhaust velocity which is the result of the nozzle parametric design following the formulation given by equation 11 and 12 and the expression of the mass flow rate. For this PID controller, it is important to set the suitable gains, for this, we refer to the expression of the mass and the payload ratio to standardize the variables.

IOP Conf. Series: Materials Science and Engineering **449** (2018) 012014 doi:10.1088/1757-899X/449/1/012014

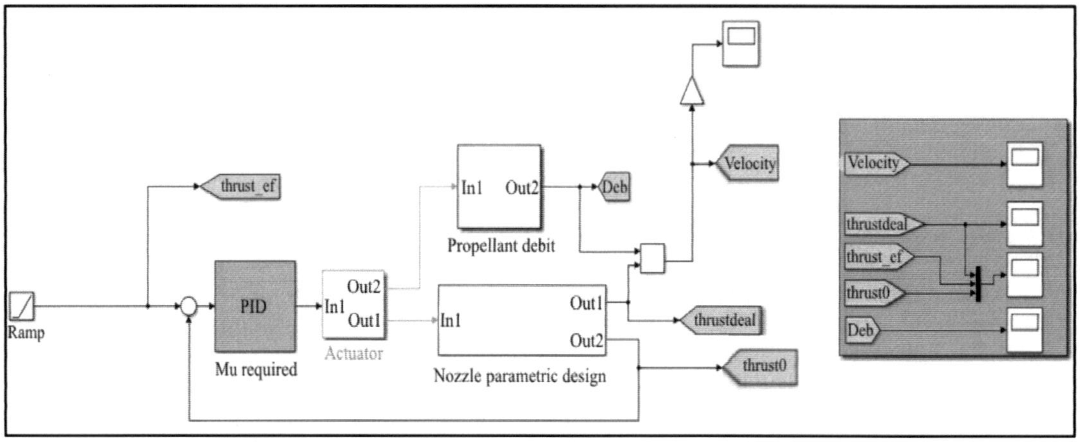

Figure 3. Small launch vehicle propulsion Simulink platform

The result of the simulation shows in figure 4 that the needed thrust for the full rocket is 110 KN force, the calculated thrust following the previous formulation was corrected by the adjustment of the mass flow rate of the propellant and the exhaust velocity. During this launch mission and in its first phase, the consumption of the propellant is ascendant to reach the maximum 18.33 Kg/s figure 5. The exhaust velocity resultant from the parametric design of the nozzle block is established to reach a peak of 6km/s as mentioned in figure 6.

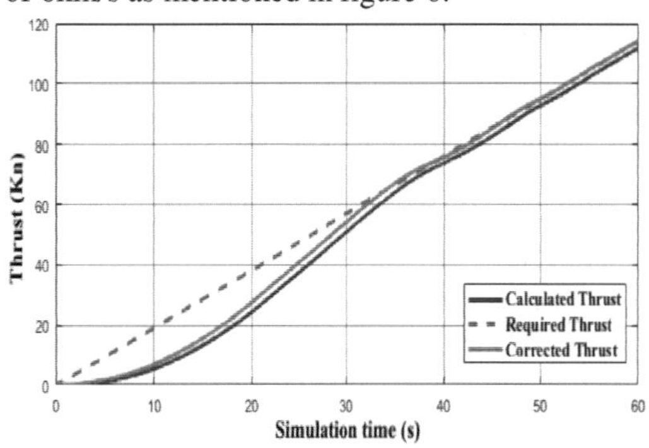

Figure 4. Simulation results (the adjustment of the thrust for the full rocket).

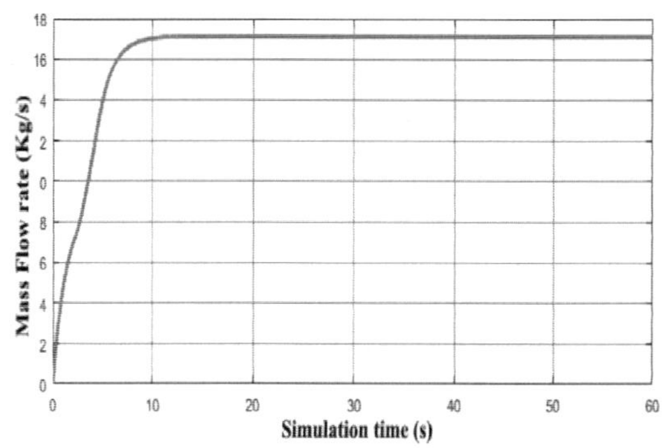

Figure 5. Mass flow rate of the propellant full rock.

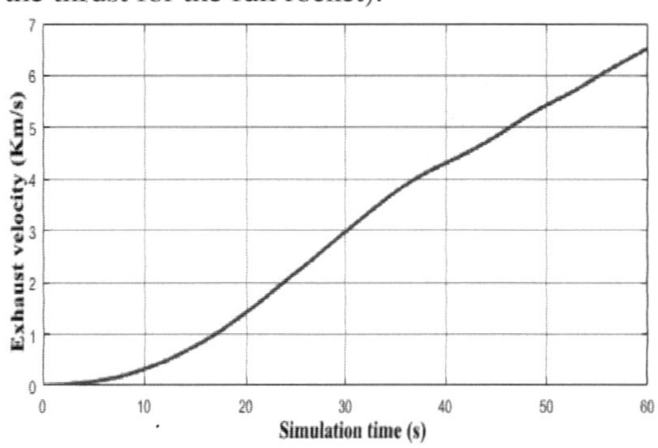

Figure 6. Exhaust velocity for the full rocket.

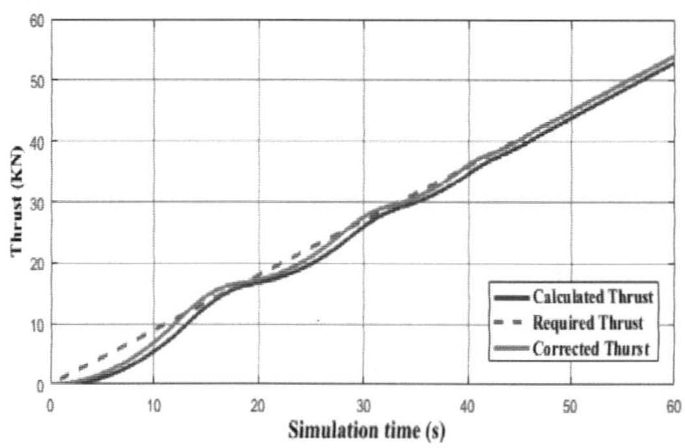

Figure 7. re-adjustment of the thrust after the separation of the first stage

After the separation of the first stage which is calculated by the consumption of the propellant fuel and by reaching the specified altitude, the second stage including the payload will carry on its mission. By the decrease of the mass of the rocket, the needed thrust will decrease, from figure 7. The needed thrust is given at 50 KN force and for this, from figure 8, the needed a consumption of 8.66 Kg/s propellant as a flow rate for the same exhaust velocity (figure 9).

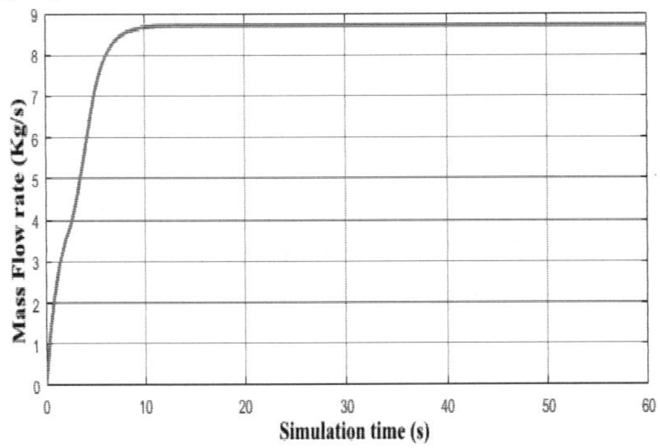

Figure 8. Mass flow rate of the propellant after the separation of the first stage.

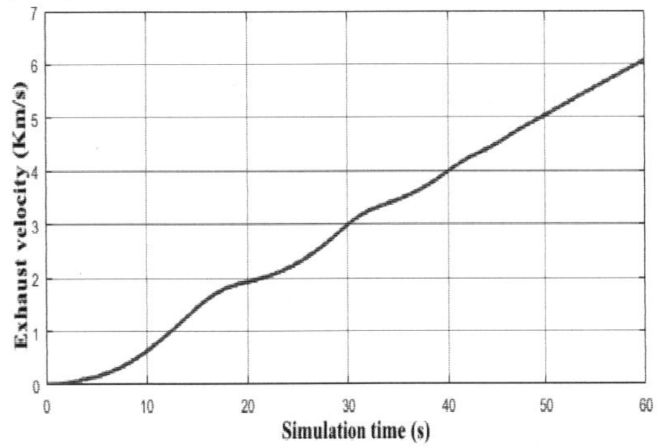

Figure 9. Exhaust velocity after the separation of the first stage.

6. Conclusion

This paper was in aim to understand the basics of space launch vehicles with a special focus on the propulsion system, by mathematical formulation we have determined the critical factors for a success of such mission. As we can notice from the design point of view, the targeted variables were analysed to ensure an optimum model. A great focus was given to the parametric design of the nozzle, many configurations were investigated and by interacting with the other design variable, we deduced the appropriate scenario.

The established MATLAB-SIMULINK platform is a numerical tool for analysis and visualization; we reproduced the theoretical model in order to track the mission. This platform gives a global view and offers the flexibility to switch the decisive factors in order to handle the design sets.

As a perspective for this study, is will be more important to carry on the research and the analysis of the launch operation in dynamic environment, and by enlarging the design variables set we can introduce new concept and handle more set-up configurations.

References

[1] D. Krejci and P. Lozano 2018 Space Propulsion Technology for Small Spacecraft *Proceedings of the IEEE.*

[2] B. Pamadi, P. Tartabini, and B. Starr 2004 Ascent, stage separation and glideback performance of a partially reusable small launch vehicle in *42nd AIAA Aerospace Sciences Meeting and Exhibit*, p. 876.

[3] E. Civek and M. K. Özgören 2017 Space Launch Vehicle Design with Simultaneous Optimization of Thrust Profile and Trajectory in *AIAA SPACE and Astronautics Forum and Exposition*, p. 5333.

[4] V. Hutchinson and J. Olds 2004 Estimation of launch vehicle propellant tank structural weight using simplified beam approximation in *40th AIAA/ASME/SAE/ASEE Joint Propulsion Conference and Exhibit*, p. 3661.

[5] G. Adheena, K. A. Sunny, A. Vasanth, and N. M. Kumar 2018 Influence of Combustion Chamber Pressure on Injector Performance *International Journal of Mechanical Engineering & Technology (IJMET)-Scopus Indexed,* vol. 9, pp. 810-820.

[6] R. C. Vemula, J. Gustavsson, and R. Kumar 2018 Rocket Nozzle Thrust and Flow Field Measurements Using Particle Image Velocimetry *New Space,* vol. 6, pp. 37-47.

[7] L. Brevault, M. Balesdent, N. Bérend, and R. Le Riche 2015 Multi-level hierarchical MDO formulation with functional coupling satisfaction under uncertainty, application to sounding rocket design in *11th World Congress on Structural and Multidisciplinary Optimization, Sydney*.

[8] R. Braun, R. Powell, R. Lepsch, D. Stanley, and I. Kroo 1995 Comparison of two multidisciplinary optimization strategies for launch-vehicle design *Journal of Spacecraft and Rockets,* vol. 32, pp. 404-410.

[9] A. M. Sallam and A. E.-S. Makled 2016 Two-Stage Launch Vehicle Trajectory Modeling for Low Earth Orbit Applications *World Academy of Science, Engineering and Technology, International Journal of Mechanical, Aerospace, Industrial, Mechatronic and Manufacturing Engineering,* vol. 11, pp. 86-93.

[10] T. Shimada, K. Kitagawa, M. Kanazaki, S. Yuasa, T. Sakurai, I. Nakagawa, *et al. 2016* Study of low-cost 100kg-satellite launcher using hybrid rocket engines *Session7 of Space Propulsion.*

[11] C. R. KOPPEL, "SYSTEMES PROPULSIFS A PROPERGOLS LIQUIDES PETITE PROPULSION SPATIALE."

[12] G. A. Hrinda2008 Structural optimization of conceptual aerospace vehicles *High Performance Structures and Materials IV,* vol. 97, p. 411.

[13] M. Cerf 2017 Optimal thrust level for orbit insertion *Acta Astronautica,* vol. 136, pp. 55-63.

IOP Conf. Series: Materials Science and Engineering **449** (2018) 012015 doi:10.1088/1757-899X/449/1/012015

Landing Performance Prediction of Commuter Aircraft Due to Wings Ice Accretion on High Terrain Operation

D I Pratama[1], W Nirbito[1,2] and M Adhitya[1]

[1] Department of Mechanical Engineering, Universitas Indonesia, Depok, West Java 16424, Indonesia

[2] Corresponding author: wahyu.nirbito@ui.ac.id

Abstract. Ice accretion on a wing is one of the aviation accident factors because it will interrupt the flow over an airfoil that will reduce its performance. The effect of wings leading edge ice accretion on the landing distance of commuter class aircraft in remote high terrain operation was obtained with the help of LEWICE code for ice shape predictions and XFLR5 for airfoil polar data prediction. The ice accretion process occurred within various icing atmospheric conditions based on Icing Design Envelopes 14 CFR part 25.1419, Appendix C. Two category of ice accretion, horn ice and streamwise ice, were obtained on the leading edge of the wing. The performance degradation caused by the ice accretion will increase aircraft landing speed that it will affect the landing distance.

1. Introduction

The development of commuter aircraft is important to fulfill the transportation needs for an archipelago country like Indonesia that have many semi-prepared airstrips and short runaway. The problem that will be encountered by the pilot on high terrain remote areas like Papua region in Indonesia is ice accretion. Ice accretion tends to occur at high altitude that has the air temperature below freezing point.

Ice accretion on a wing is one of the factors of aviation accident because it will interrupt the flow over the wing. Ice accretion on an aircraft airfoil is a condition when *supercooled liquid water* on atmosphere collided with the body of the aircraft and suddenly freeze on the impact. The Ice that accreted on the airfoil will affect the plane flight characteristic because it will increase drag and reduce lift. Longitudinal stability of the aircraft will be affected too because the lift from horizontal stabilizer will be reduced. Bragg, Broeren, and Blumenthal have categorized the ice accretion based on aerodynamic behavior into four categories. There are roughness, streamwise ice, horn ice, and spanwise-ridge ice[1].

The shape of ice accretion can be investigated through flight test, wind tunnel experiment, and numerical simulation. Flight test and wind tunnel experiment will determine the shape of ice accurately but usually too expensive and not practical. Therefore, numerical simulation is used to predict the shape of ice accretion because it is economical and can simulate the icing process and provide a relatively accurate evaluation of ice accretion[2]. The purpose of this research is to predict the landing distance changes that caused by the ice accretion on the wing.

2. Methodology

Ice accretion simulation and iced airfoil wind tunnel simulation are required to obtain the performance characteristics on the iced airfoil. The performance characteristics of iced airfoil then can be calculated to predict the aircraft landing distance affected by the ice accretion. LEWICE code is used to predict the geometry of ice accretion. XFLR5 is used to predict the airfoil performance characteristics. The following methods are used to predict the landing distance

2.1. Icing Atmospheric Condition and Flight Configuration Parameters Determination

For running an icing simulation with LEWICE code, atmosphere condition and flight configuration parameters are required. These parameters can be seen in Table 1 and Table 2. There are liquid water content (LWC), median volume diameter (MVD), static temperature, altitude, aircraft speed, and aircraft weight. Besides all of the parameters, the geometry of the airfoil is the main parameters. Icing Design Envelopes 14 CFR part 25.1419, Appendix C[3] was used to determine the atmospheric icing condition.

Table 1. Icing Atmospheric Condition Parameters

Temperature (C°)	MVD (μm)	LWC (g/m^3)
	15	0.74
-3	25	0.44
	40	0.14
	15	0.6
-10	25	0.31
	40	0.1
	15	0.3
-20	25	0.15
	40	0.06

Table 2. Icing Flight Configuration Parameters

Altitude (ft)	Weight (kg)	Airspeed (knots)	α (°)
8000	7031	153	3.04
10000	7031	157	2.32

2.2. Simulation Running

After all icing parameters and flight configuration obtained, the ice geometry coordinates can be obtained by LEWICE code[4]. The output iced airfoil geometry then become an input on XFLR5 for predicting the polar data of the iced airfoil with the same test condition with the previous wind tunnel test experiment that has 0.2 Mach Number and 2.9×10^6 of Reynolds Number condition. For the XFLR5 simulation, the airfoil should be on 30° flap for the landing configuration.

2.3. Data Correction

The polar data that were obtained by XFLR5 is still in the form of 2-dimensional airfoil polar. The 3-dimensional aircraft polar is required to achieve the landing distance. Therefore an approach is applied by assuming the degradation of airfoil performance on simulation will be the same as the degradation that will happen in the wind tunnel experiment. The following data in Table 3 is the previous aircraft wind tunnel experiment polar data on 30° flap configuration.

Table 3. Wind Tunnel Polar Data

Flap	$(C_L)_0$	$(C_L)_\alpha$	(C_L)max	(C_L)min	$(C_D)min$	k
30°	1.0839	0.0945	2.6379	0.3394	0.0763	0.0532

IOP Conf. Series: Materials Science and Engineering **449** (2018) 012015 doi:10.1088/1757-899X/449/1/012015

The following formula can be used to obtain the maximum aircraft coefficient of lift

$$(C_L)\,max\,iced = (C_L)\,max\,clean - (C_l)\,max\,clean - (C_l)\,max\,iced \quad (1)$$

To obtain the aircraft ground coefficient of lift and drag, the airfoil polar data at 2° is used because of the angle of incident. Aircraft ground coefficient of lift and drag value is assumed the same for each of ice category. Therefore these 4 following equation can be used to obtain the aircraft ground coefficient of lift and drag.

$$\Delta(C_L)_0 = (C_l)_2\,clean - (C_l)_2\,ice \quad (2)$$

$$(C_L)_0\,iced = (C_L)_0\,clean - \frac{1}{n}\sum_{i=1}^{n}\Delta(C_L)_0 \quad (3)$$

$$\Delta(C_D)_0 = (C_d)_2 clean - (C_d)_2 iced \quad (4)$$

$$(C_D)_0 iced = (C_D)_0\,clean - \frac{1}{n}\sum_{i=1}^{n}\Delta(C_D)_0 \quad (5)$$

When the aircraft is in a ground roll phase and the wing is near enough to the ground, there will be a ground effect that will change slightly the coefficient of lift and drag. An approach is applied by assuming these coefficient change at the previous wind tunnel experiment will be the same for the case in this study. The coefficient change due to ground effect from the previous wind tunnel experiment is +0.02696 for lift coefficient and -0.00466 for drag coefficient. Therefore these two following formula can be used

$$(C_L)_2\,iced\,\text{GE} = (C_L)_2\,iced + \Delta(C_L) wind\ tunnel\ ground\ effect \quad (6)$$

$$(C_D)_2\,iced\,\text{GE} = (C_D)_2 iced + \Delta(C_D) wind\ tunnel\ ground\ effect \quad (7)$$

2.4. *Landing Distance Calculation*

After all iced aircraft polar data is obtained, landing distance can be calculated in ILAGA airstrips in Papua that located at 7,500ft and only have 600m runway available. The total landing distance is the sum of the approach distance (s_a), flare distance (s_f) and ground roll distance (s_g). To obtain all of the distance required the following formula can be used

$$s_a = \frac{50-h_f}{tan\,\theta_a} \quad (8)$$

For the approach distance, the flare height (h_f) is constant at 10ft height and the approach angle is about (3.9° – 4.4°) depends on the aircraft Vstall.

$$s_f = R\ sin\,\theta_a \quad (9)$$

For the flare distance, the radius of flare (R) can be obtained by the following formula while assuming the value of load factor is 0.2

$$R = \frac{V_f}{0.2g} \quad (10)$$

For the ground roll distance, this formula was used and assuming the time of touchdown transition N = 2, $V_{TD} = 1.15\,V_{stall}$, maximum landing weight (6,940 kg), $\mu_r = 0.025$ (asphalt) and there is no braking force or reversed thrust[5].

$$s_g = N\,V_{TD} + \frac{W\,V_{TD}{}^2}{2g}\left[\frac{1}{T_{rev}+D+\mu_r(W-L)}\right]_{0.7\,V_{TD}} \quad (11)$$

3. Results and Discussion

3.1. *Ice Geometry*

The prediction of ice geometry case (-10121) and (-20123) that simulated using LEWICE code with atmospheric parameters from Icing Design Envelope 14 CFR Part 25, Appendix C can be seen in Figure 1. On this prediction, the shape of horn ice and streamwise ice were obtained and can be seen in Table 4.

Table 4. Ice Shape

Ice Case Code	*MVD* (μm)	*LWC* (g/m^3)	*T* (C°)	Altitude (ft)	Ice Shape
-3111	15	0.74	-3	10000	Horn
-3112	25	0.44	-3	1000	Horn
-3113	40	0.14	-3	1000	Horn
-3121	15	0.74	-3	8000	Horn
-3122	25	0.44	-3	8000	Horn
-3123	40	0.14	-3	8000	Horn
-10111	15	0.6	-10	10000	Horn
-10112	25	0.31	-10	1000	Horn
-10113	40	0.1	-10	1000	Steamwise
-10121	15	0.6	-10	8000	Horn
-10122	25	0.31	-10	8000	Horn
-10123	40	0.1	-10	8000	Steamwise
-20111	15	0.3	-20	10000	Steamwise
-20112	25	0.15	-20	1000	Steamwise
-20113	40	0.06	-20	1000	Steamwise
-20121	15	0.3	-20	8000	Steamwise
-20122	25	0.15	-20	8000	Steamwise
-20123	40	0.06	-20	8000	Steamwise

The horn ice is mainly formed at the higher temperature meanwhile, the streamwise ice is mainly formed at the lower temperature. On the higher temperature, droplets will freeze slower that it will have more time to spread from the stagnation point along the upper and lower surface of airfoil before it becomes frozen completely and shaped a horn geometry. On the lower temperature, droplets will freeze immediately that it will tend to follow the original shape of the airfoil. The geometry of ice may also change depends on the size of the droplets of water.

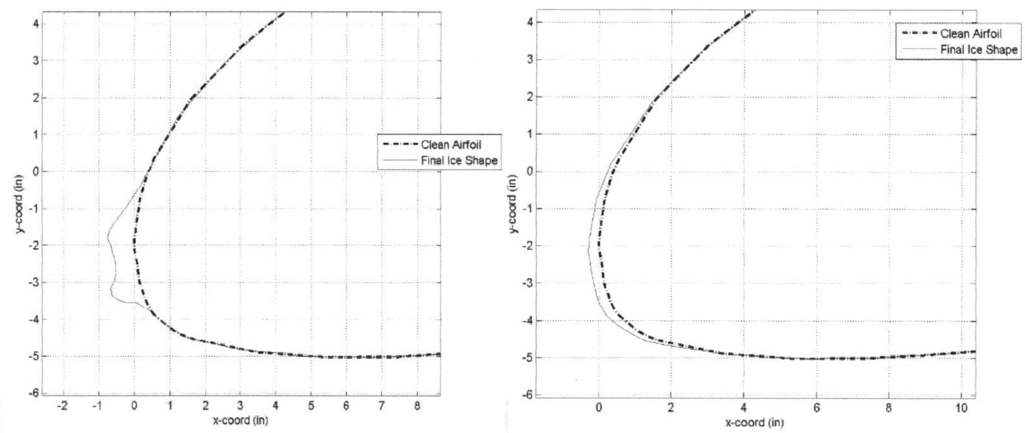

Figure 1. Horn Ice -10121 (left) and Streamwise Ice -20123 (right)

3.2. Airfoil Performance (XFLR5)

Airfoil performance degradation that caused by the ice accretion is needed for calculating the landing distance. $(C_L)max, (C_L)_{2^\circ}, dan \ (C_D)_{2^\circ}$ is used to predicting the landing distance. Airfoil performance data were obtained by XFLR5 and combined with previous wind tunnel experiment results. These data can be seen in Table 5.

Table 5. Iced Polar Data

Shape	Ice Case Code	XFLR5			Correction Calculation		
		$(C_L)max$	$(C_L)_{2°}$	$(C_D)_{2°}$	$(C_L)max$	$(C_L)_{2°}$	$(C_D)_{2°}$
Horn	-3121	2.0441	1.7314	0.07713	2.3698	0.97612	0.108632
	-3122	1.9915	1.6472	0.08022	2.3172	0.97612	0.108632
	-3123	2.1655	1.7072	0.07939	2.4912	0.97612	0.108632
	-10121	1.7565	1.6884	0.07773	2.0822	0.97612	0.108632
	-10122	1.8794	1.7359	0.07654	2.2051	0.97612	0.108632
	-3111	2.1259	1.7159	0.078	2.4516	0.97612	0.108632
	-3112	1.8314	1.6326	0.08134	2.1571	0.97612	0.108632
	-3113	2.1611	1.7406	0.07642	2.4868	0.97612	0.108632
	-10111	1.9504	1.71	0.0752	2.2761	0.97612	0.108632
	-10112	1.9983	1.7	0.076	2.324	0.97612	0.108632
Steamwise	-10123	2.2069	1.757	0.07558	2.5326	1.009563	0.107746
	-20121	2.17	1.7388	0.07577	2.4957	1.009563	0.107746
	-20122	2.1741	1.7247	0.07723	2.4998	1.009563	0.107746
	-20123	2.1607	1.7268	0.07692	2.4864	1.009563	0.107746
	-10113	2.2069	1.7544	0.07549	2.5326	1.009563	0.107746
	-20111	2.1838	1.726	0.07822	2.5095	1.009563	0.107746
	-20112	2.185	1.7022	0.08037	2.5107	1.009563	0.107746
	-20113	2.1787	1.745	0.07577	2.5044	1.009563	0.107746

The graphs shows that generally horn ice shape airfoil have lower coefficient of lift and higher coefficient of drag compared to the streamwise ice. It caused by the horn ice shape that have more random geometry that it will cause flow separation at lower angle of attack.

3.3. Landing Distance

Figure 2 shows that ice accretion will affect the landing distance. As the coefficient of lift decrease, aircraft have to increase the speed to maintain the lift force. The more speed increased, the landing distance needed of the aircraft to become completely stop will be longer. But if it compared to the condition where the ice accretion does not occur, some of the data shows that smaller degradation coefficient of lift has shorter landing distance.

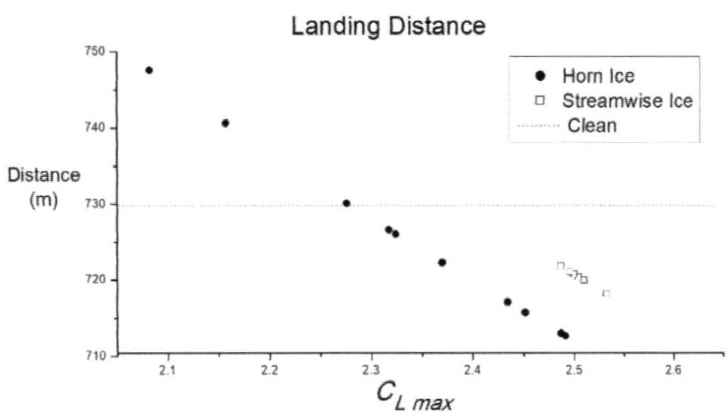

Figure 2. Landing Distance Change Due to Ice Accretion

This can happen because the increased drag force due to ice accretion still dominating the increased aircraft speed at the smaller degradation of the coefficient of lift. The landing distance varies from 712m – 747m for horn iced airfoil and 718m – 712m for the streamwise iced airfoil. Meanwhile, for the clean airfoil, the landing distance would be around 729m.

4. Conclusion

The maximum coefficient of lift horn ice shaped airfoil decrease within a range of 6.4% - 24.12% and the maximum coefficient of lift streamwise ice shaped airfoil decrease within a range of 4.64% - 6.637%. The most severe maximum coefficient of lift degradation occurred on (-10121) horn ice case. The results of landing distance calculation showed that the landing distance will have a deviation between -2.3% – 2.5% caused by horn ice and -1.6% – -1.1% caused by streamwise ice.

5. Acknowledgment

The authors would like to thank PITTA grant programme of the University of Indonesia for funding this research with grant number 2553/UN2.R3.1/HKP.05.00/2018 and PT. Dirgantara Indonesia (Indonesian Aerospace), for the facility, guidance, and chance to help the development of this commuter aircraft through this research.

Reference

[1] Bragg, M.B., A.P. Broeren, and L.A. Blumenthal, *Iced-airfoil aerodynamics.* Progress in Aerospace Sciences, 2005. **41**(5): p. 323-362.

[2] Cao, Y., et al., *Numerical simulation of ice accretions on an aircraft wing.* Aerospace Science and Technology, 2012. **23**(1): p. 296-304.

[3] Jeck, R.K., *Icing Design Envelopes (14 CFR Parts 25 and 29, Appenddix C) Converted to a Distance-Based Format.* 2002, FEDERAL AVIATION ADMINISTRATION TECHNICAL CENTER ATLANTIC CITY NJ.

[4] Wright, W., *User's manual for LEWICE version 3.2.* 2008.

[5] Anderson, J., *Aircraft performance and design, 1999.* International edition, WCB/McGraw-Hill, 1999.

IOP Conf. Series: Materials Science and Engineering **449** (2018) 012016 doi:10.1088/1757-899X/449/1/012016

Small Fault Detection for Satellite Attitude Control System Actuators with Stacked Autoencoder Network

L Li, Y M Gao, Z H Wu and X B Zhang

School of Space Information, Space Engineering University, Beijing 10416, China

lilei_0617@hotmail.com

Abstract. The fault detection method based on the analytical model can not completely deal with the model uncertainty, disturbance torque, measurement error and other disturbances, and it is difficult to detect small faults similar to disturbances. A detection method of small actuator fault based on stacked autoencoder(SAE) network is proposed. By learning historical data of system, SAE network can reconstruct the state data with stable errors. The relation between states of system will change when a fault occurs, and residual will change too. The variation trend of the residual can be used to detect the fault. Simulation results show that SAE network is more robust to disturbance compared with Elman neural network and nonlinear observer. Small faults under the disturbance can be detected by SAE network.

1. Introduction

The attitude control system is one of the key subsystems of the satellite. The function of the satellite load depends on the establishment of the correct satellite attitude. Once the attitude control system fails, it will cause very serious consequences [1]. Fault identification can provide an important basis for fault recovery and fault tolerance control of the system. When the fault causes could be identified accurately, it is pertinent to the fault response. Although the role of fault identification is very important, the research results are far less than fault detection and fault separation [2].

There are usually two kinds of fault identification methods: model-based method and knowledge based method. The model-based method pre-establishes a set of fault models, and uses the observer, parameter estimation and other methods to estimate the parameters of the fault model, and then determines the fault type and fault level. In Refs.[3-4], adaptive observer was used to realize the fault diagnosis of the nonlinear system and fault parameter estimate. In Ref.[5], the robust augmented observer was used to achieve a higher precision fault estimation than the traditional observer. This method could identify the sudden and slow faults of the actuator, but unable to identify the specific classification of the faults. In Ref.[6], the least squares parameter estimation method was used to realize the fault diagnosis of the satellite attitude control system, and fault-tolerant control was carried out by identifying the parameters. The knowledge-based method uses neural network, support vector machine and other methods to extract the fault mode through the analysis of historical data and prior knowledge, and then identify the fault by pattern matching. In Refs.[7-8], neural networks was used to identify and estimate fault by estimating fault parameters. In Ref.[9], neural network was used to improve the sensitivity of fault detection by learning historical data. All of the above methods only achieve component-level fault identification, and could not identify the causes of the fault more deeply. The model-based method has such problems: the fault model is difficult to establish, and the application scenario is limited; the accuracy of the identification is not high due to the influence of model accuracy, measurement noise and coupling relationship of different fault parameters. The knowledge-based method has such problems: the selection of feature quantities depends on human experience, which makes it difficult to find the characteristics of comprehensive response fault information; the fault mode extraction method could not adapt to complex systems and multi-mass data, which makes it difficult to extract the hidden patterns in the data.

In recent years, deep learning has developed rapidly, and has achieved good results in many recognition problems. Compared with traditional methods, it greatly improves the accuracy of recognition. Deep learning has better nonlinear expression ability, better generalization ability and deeper feature extraction ability than traditional machine learning methods. Therefore, deep learning is introduced into the field of fault diagnosis. In Refs.[10-13], different deep learning frameworks was used to realize fault diagnosis in the fields of gearboxes and robots, and achieved good results. However, there are few reports on the application of deep learning in the field of satellite fault diagnosis.

In this paper, the application of deep learning in satellite attitude control system fault detection is attempted, and a fault detection method based on SAE network is proposed. The method uses the SAE network to extract the relation between the states of satellite attitude control system by learning historical data, and then establishes the normal state model of the system. The predicted values of the model and the measured values of the system form a residual, and fault detection is carried out through the variation of residual.

2. Models

Consider the three-axis stabilized satellite attitude control system shown in figure 1, where the sensors are gyros and star sensors; the actuators are reaction flywheels; and the control algorithm uses a nonlinear quaternion feedback control law.

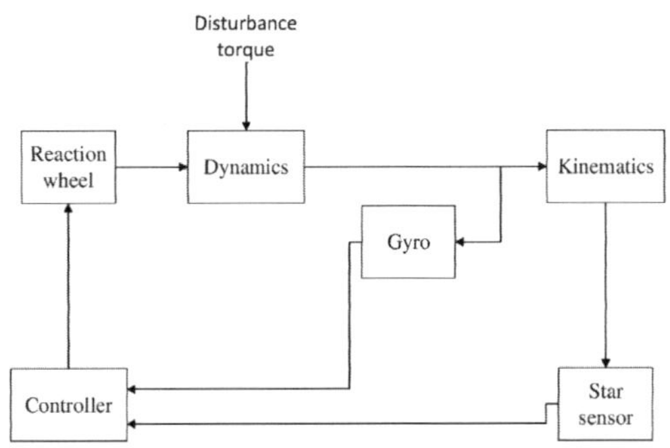

Figure 1. Satellite attitude control system

Satellite attitude control system can be divided into kinematics subsystems and dynamics subsystems [14]. The kinematics subsystem includes gyros, star sensors and kinematics, reflecting the relationship between satellite attitude angle and angular velocity; the dynamics subsystem includes reaction flywheels, gyros and dynamics, reflecting the relationship between the control torque and angular velocity.

The attitude kinematics equation described by quaternion is given by

$$\begin{cases} \dot{q}_0 = -\frac{1}{2}\bar{q}^T\omega \\ \dot{\bar{q}} = \frac{1}{2}\left(\bar{q}^\times + q_0 I_{3\times 3}\right)\omega \end{cases} \tag{1}$$

Here $\bar{q} = [q_1, q_2, q_3]^T$, $\omega = [\omega_x, \omega_y, \omega_z]^T$ is the angular velocity of the body of satellite with respect to the inertial frame expressed in the body-fixed reference frame. The skew-symmetric matrix is given by

$$\bar{q}^{\times} = \begin{bmatrix} 0 & -q_3 & q_2 \\ q_3 & 0 & -q_1 \\ -q_2 & q_1 & 0 \end{bmatrix} \tag{2}$$

Considering the satellite ontology as a rigid body, the following dynamics equation of the satellite can be obtained from the momentum moment equation of the rigid body motion [14]:

$$\begin{cases} I_x\dot{\omega}_x + \left(I_z - I_y\right)\omega_y\omega_z = M_x + T_{dx} \\ I_y\dot{\omega}_y + \left(I_x - I_z\right)\omega_z\omega_x = M_y + T_{dy} \\ I_z\dot{\omega}_z + \left(I_y - I_x\right)\omega_x\omega_y = M_z + T_{dz} \end{cases} \tag{3}$$

Here $I = diag\{I_x, I_y, I_z\}$ is the tensor of inertia in the body frame, and $M = \left[M_x, M_y, M_z\right]^{\mathrm{T}}$ is the control torque of the actuator, and $T_d = \left[T_{dx}, T_{dy}, T_{dz}\right]^{\mathrm{T}}$ is the external disturbance torque.

3. Methodologies

3.1. SAE network

The SAE network is a structure of deep learning, usually composed of a series of stacked autoencoder(AE). The SAE network can learn the internal features of the data and reconstruct the original data. By stacking multiple AEs, more abstract and more global features can be extracted. AE is a type of neural network that can be trained to copy input to output. The simplest AE is a three-layer feedforward neural network as shown in figure 2, which contains an input layer with v neurons, an implicit layer with h neurons, and an output layer with v neurons. The input layer is also called the visible layer, the hidden layer is called the encoder, and the output layer is called the decoder.

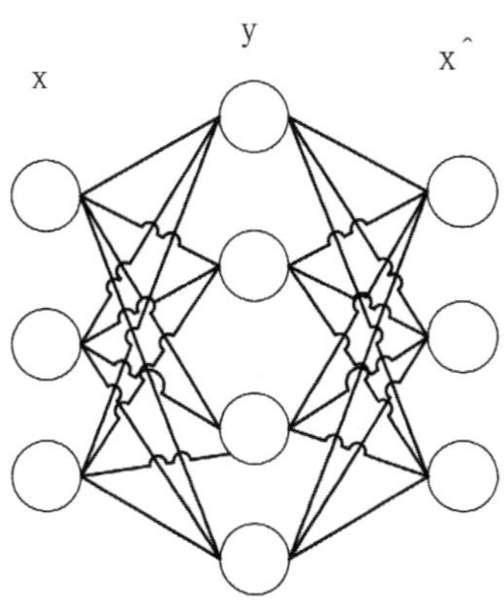

Figure 2. Structure of AE

The mathematical model of AE is shown in the equation (4)

$$\begin{cases} y = \sigma_a\left(W_a x + b_a\right) \\ \hat{x} = \sigma_s\left(W_s y + b_s\right) \end{cases} \tag{4}$$

Here x is the input data, y is the code obtained by the encoder, $\hat{x}$ is the reconstruction of input data, W_a are the weights between input layer and hidden layer, W_s are the weights between hidden layer and output layer, b_a and b_s are bias terms, σ_a and σ_s are activation functions.

In order to learn more robust features, random noise can be added to the input data. At this time, for the network, the input is noise-bearing data, and the expected output is noise-free data.

The learning goal of the network is to minimize the error of the input value x and the reconstructed value $\hat{x}$ as shown in equation (5).

$$E = \underset{W_a, W_s, b_a, b_s}{\arg\min}\left(c\left(x, \hat{x}\right)\right) \tag{5}$$

Here E is the optimization objective function, $c\left(x, \hat{x}\right)$ is the error of reconstruction.

3.2. Diagnosis decision

Residuals are often affected by uncertainties and disturbances, which leads to a decrease in fault detection capability. In this case, the evaluation function of the residual can be selected as a root mean square function, and its expression is as follows:

$$R_e = \left(\frac{1}{T}\int_{t_1}^{t_2} c^{\mathrm{T}}\left(x, \hat{x}\right) c\left(x, \hat{x}\right)\right)^{\frac{1}{2}} \tag{6}$$

Here R_e is the residual, $T = t_2 - t_1$ is the window of fault detection. The residual threshold is determined based on the mean and standard deviation of the residuals as equation (7).

$$J = u \pm 3\sigma \tag{7}$$

Here J is the threshold, u is the mean of residual, σ is the standard deviation of residual. The fault diagnosis logic can be given as follows:

$$\begin{cases} R_e > J \Rightarrow \text{fault} \\ R_e \leq J \Rightarrow \text{no fault} \end{cases} \tag{8}$$

4. Simulation

4.1. Parameters setting

The parameters of controller: $K_p = \left[40, 40, 40\right]^{\mathrm{T}}$ and $K_d = \left[1000, 1000, 1000\right]^{\mathrm{T}}$.

The inertia matrix of the satellite: $I = diag\left\{930, 800, 1070\right\} \mathrm{kg \cdot m^2}$.

The orbital angular velocity: $\omega_0 = 7.3 \times 10^{-5} rad/s$.

Initial attitude angle velocities: $\omega\left(0\right) = [0, 0, 7.3]^T \times 10^{-5} rad/s$.

Initial attitude: $Q_0 = \left[0.9786, -0.0704, -0.1798, 0.0704\right]^{\mathrm{T}}$.

Disturbance torque: $T_d = \left[A_x \sin \omega_d t, A_y \sin \omega_d t, A_z \sin \omega_d t\right]^T$, where $A_x = 1.4e{-}5N \cdot m$, $A_y = 1.5e{-}5N \cdot m$, $A_z = 1.6e{-}5N \cdot m$, $\omega_d = 0.02 rad/s$.

Gyro constant drift: $8.72e{-}5\mathrm{rad/s}$.

Noise is Gaussian white noise: $N\left(0, 10^{-7}\right)$.

An actuator fault in X-axis occurs at simulation time 500s, and the fault amplitude is $1e{-}3N \cdot m$.

The SAE network structure used for fault detection is as shown in figure 3. The net has 4 AEs, each containing 400, 200, 100, and 50 neurons. The initial parameters are random small values obeying Gaussian distribution, the initial learning rate is 1, and the input noise ratio is 0.5. The activation function is sigmoid function. Each layer training iteration epochs is 200, and the whole network fine tuning training iteration epochs is 20.

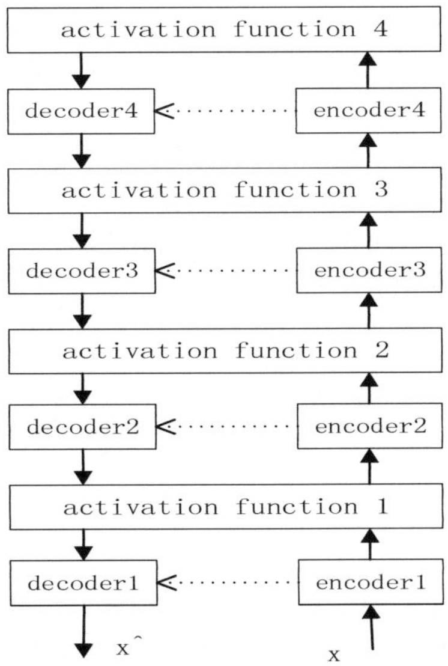

Figure 3. Structure of SAE network

4.2. Evaluation

The simulation results are as shown in figure 4-6. Figure 4 is the result using SAE network, figure 5 is the result using Elman neural network, figure 6 is the result using nonlinear observer.

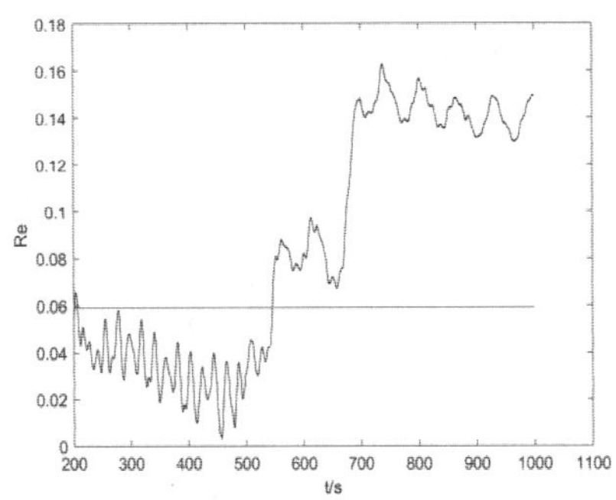

Figure 4. Result of SAE network

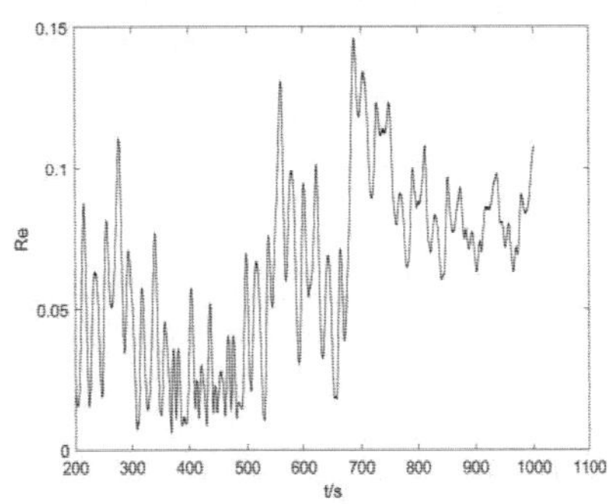

Figure 5. Result of Elman neural network

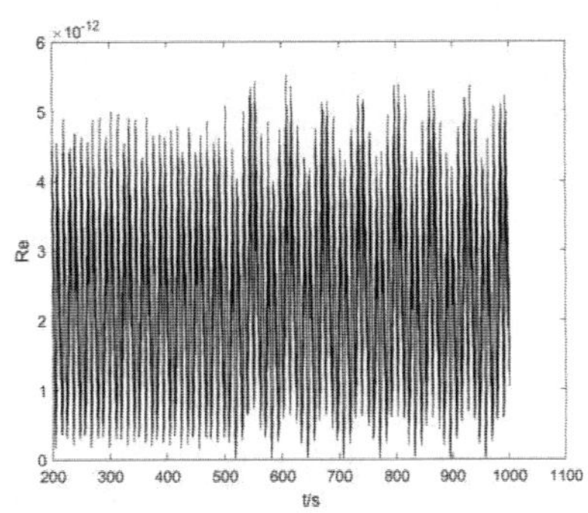

Figure 6. Result of nonlinear observer

System disturbances and noise can interfere with the residuals of fault detection. For small amplitude faults, the residuals produced by the Elman neural network and the nonlinear observer are completely obscured in the disturbance as shown in Figures 5 and 6. The SAE network is more robust to disturbances and noise due to its ability to learn deeper features between the states of system.

5. Conclusions

In this paper, a deep learning technique based on SAE network has been proposed to detection fault of actuator in satellite attitude control system. The method uses SAE network to reconstruct states of the system by learning historical data. Simulation results show that SAE network can detect small fault under disturbances and noise, while both Elamn neural network and nonlinear observer can not detect.

References

[1] ZHAO G, LI Z, LI Y J. Research on method of fault prediction for onboard spacecrafts based on times series event omen pattern mining[J]. Computer Meassurement & Control, 2009, 17(12): 2352-2354.
[2] SU L, SHANG C X, LIU W J. Survey on the technology of fault diagnosis for spacecraft attitude control system[J]. Journal of Changchun University of Science and Technology(Natural Science Edition), 2010, 33(4): 23-27
[3] CEN C H, WEI J L, JIANG R, et al. Accurate diagnosis on process fault parameters of nonlinear actuator based on adaptive observer and extended state observer[J]. Journal of Astronautics, 2011, 32(6): 1318-1326.
[4] HU Z G, ZHAO G R, LI F, et al. Fault diagnosis for nonlinear dynamical system based on adaptive unknown input observer[J]. Control and Decision, 2016, 31(5): 901-906(in Chinese).
[5] CHENG Y, WANG R X, XU M Q. Simultaneous state and actuator fault estimation for satellite attitude control systems[J]. Chinese Journal of Aeronautics, 2016, 29(3):714-721.
[6] Jiang T, Khorasani K, Tafazoli S. Parameter estimation-based fault detection, isolation and recovery for nonlinear satellite models[J]. IEEE Transactions on Control Systems Technology, 2008, 16(4):799-808.
[7] TALEBI H A, KHORASANI K. A neural network-based multiplicative actuator fault detection and isolation of nonlinear systems[J]. IEEE Transactions on Control Systems Technology, 2013, 21(3): 842-851.
[8] SHEN Q K, JIANG B, SHI P. Novel neural networks-based fault tolerant control scheme with fault alarm[J]. IEEE Transactions on Cybernetics, 2014, 44(11): 2190-2201.
[9] Cheng Y, Wang R, Xu M. A Combined Model-Based and Intelligent Method for Small Fault Detection and Isolation of Actuators[J]. IEEE Transactions on Industrial Electronics, 2016, 63(4):2403-2413.
[10] Erfani S M, Rajasegarar S, Karunasekera S, et al. High-dimensional and large-scale anomaly detection using a linear one-class SVM with deep learning[J]. Pattern Recognition, 2016, 58(C):121-134.
[11] Utkin L V, Zaborovskii V S, Popov S G. Detection of anomalous behavior in a robot system based on deep learning elements[7]. Automatic Control and Computer Sciences, 2016, 50(8): 726-733.
[12] Chen Z Q, Li C, Sanchez R V. Gearbox Fault Identification and Classification with Convolutional Neural Networks[J]. Shock & Vibration, 2015, 2015(2):1-10.
[13] TALEBI H A, KHORASANI K. A neural network-based multiplicative actuator fault detection and isolation of nonlinear systems[J]. IEEE Transactions on Control Systems Technology, 2013, 21(3): 842-851.
[14] WANG R X, CHENG Y, XU M Q. Analytical redundancy based fault diagnosis scheme for satellite attitude control systems[J]. Journal of the Franklin Institute, 2015, 352(5):1906-1931.

IOP Conf. Series: Materials Science and Engineering **449** (2018) 012017 doi:10.1088/1757-899X/449/1/012017

Numerical Simulation of Aeroacoustics of Hovering Helicopter Rotor

A N Kusyumov[1], S A Mikhailov[1], S A Kusyumov[1], K V Fayzullin[1] and G N Barakos[2]

[1]Department of Aerohydrodynamics, KNRTU-KAI, 10 K. Marx str., Kazan, 420111, Russia
[2]James Watt South Building, School of Engineering, University of Glasgow, G12 8QQ, Scotland

postbox7@mail.ru

Abstract. The specifics of the problem of estimating the noise of a hovering rotor, allows for some simplification of the Ffowcs Williams-Hawkins (FW-H) equation. Most published works dedicated to helicopter tonal noise estimates use the far-field formulation. This paper estimates the aeroacoustic emissions of a helicopter rotor in hover, for observers placed at different distances using the FW-H equation, including near-field and far-field terms. The blade pressure distribution is obtained from numerical simulations with the RANS equations. To demonstrate this approach, the near- and far-field contributions are analyzed for the model-scale UH-1H main helicopter rotor. For the numerical simulations, the HMB solver of Glasgow University and the ANSYS Fluent13 commercial solver are used. As the rotor blade behaviour is characterized by a complex motion in the computer program it is assumed that the blade is seen in as a rotating rigid body. The most commonly used mathematical model of the FW-H equation corresponds to the classical impermeable formulation. In this case, the source surface corresponds to the blade surface. Then, the acoustic pressure (based on the FW-H 1A formulation) is modified with empirical adjustments, based on the radiation Mach number. This was applied for the near- or far-field thickness noise depending on the rotor-observer distance.

1. Introduction

The prediction of aircraft noise falls in the area of Computational Aeroacoustics (CAA). CAA combines flow data obtained by Computational Fluid Dynamics (CFD) for a region near the aircraft surface, and wave propagation equations, to estimate the aircraft noise far away from their sources.

For a conventional helicopter, there are two main components contributing to the generation of near- and far-field noise, the main rotor and the tail rotor [1]. The helicopter main rotor generates tonal and broadband noise. The tonal noise in hover is dominant in a conical region directed 30 to 40 degrees downwards of the rotor plane, while broad-band noise radiates mostly out of the plane of the rotor. Most of works on helicopter rotor noise employ the Ffowcs Williams-Hawkings (FW-H) equation [2]. This equation separates the noise in thickness and loading noise components, and its general formulation allows for estimates of the acoustic emissions for observers in the near- and far-field of the rotor.

Both thickness and loading noise components can be presented as a sum of several terms, what are functions of the distance between the rotor and the observers. The nature of the dependence of noise components on the rotor-observer distance determines the near- and far-field terms. It allows for simplification of the FW-H equation for different observer locations. Most published works dedicated to helicopter tonal noise estimation use the far-field formulation (see, for example, references [3, 4]). Typically, the far-field formulation is applied to the Mach-scaled rotors. In the case of low blade tip Mach number the direct CFD simulation can be used for the near-field noise estimation, based on Reynolds-Averaged Navier-Stokes (RANS) equation (see reference [5]). Due to the limited space of wind tunnels, comparisons with experimental measurements require the complete formulation of the FW-H equation, including the near- and far-field terms.

This paper estimates the aeroacoustics of the UH-1H model-scale, helicopter rotor [6] in hover, for different observer distances. The acoustic pressure is determined via the FW-H equation. The far-field FW-H equation solution is compared to experimental reference data. And the contributions of the different FW-H equation components for different rotor-observer distances are analyzed. For the numerical modelling, the HMB solver of Glasgow University and ANSYS Fluent13 are used.

2. FW-H equation

The general retarded-time formulation of Farassat, for the observer aeroacoustic pressure p' is commonly referred to as Formulation 1A [2] and is given by the following expression:

$$p'(x,t) = p_T'(x,t) + p_L'(x,t), \tag{1}$$

were $p_T'(x,t)$ and $p_L'(x,t)$ are the thickness and load noise respectively. For an arbitrary rotor-observer distance the thickness and load noise components can be written as

$$p_T'(x,t) = p_{TN}'(x,t) + p_{TF}'(x,t), \qquad p_L'(x,t) = p_{LN}'(x,t) + p_{LF}'(x,t). \tag{2}$$

The subscripts N and F stand for the near- and far-field terms defined by

$$p_{TF}'(x,t) = \frac{1}{4\pi}\int_{f=0}\left[\frac{\rho_0 \dot{v}_n}{r|1=M_r|^2} + \frac{\rho_0 v_n \hat{r}_i \dot{M}_i}{r|1=M_r|^3}\right]_{ret} dS, \tag{3}$$

$$p_{TN}'(x,t) = \frac{1}{4\pi}\int_{f=0}\left[\frac{\rho_0 a_0 v_n (M_r - M^2)}{r^2|1=M_r|^3}\right]_{ret} dS, \tag{4}$$

$$p_{LF}'(x,t) = \frac{1}{4\pi a_0}\int_{f=0}\left[\frac{\dot{p}\cos(\theta)}{r|1\,sM_r|^2} + \frac{p\cos(\theta)\hat{r}_i \dot{M}_i}{r|1\,sM_r|^3}\right]_{ret} dS, \tag{5}$$

$$p_{LN}'(x,t) = \frac{1}{4\pi}\int_{f=0}\left[\frac{p(\cos(\theta) - M_i n_i)}{r^2|1\,sM_r|^2} + \frac{p\cos(\theta)(M_r - M^2)}{r^2|1\,sM_r|^3}\right]_{ret} dS. \tag{6}$$

Here p is the pressure at emission point; a_0 is the speed of sound; $M_r = M_i \hat{r}_i$ is the Mach number of source in radiation direction; $r = |\mathbf{x}-\mathbf{y}|$ is the distance between observer and source, $\mathbf{x}$ is the observer position vector, and $\mathbf{y}$ is the source position vector. The above expressions also include $\cos(\theta) = \mathbf{n} \cdot \hat{\mathbf{r}}$, with θ the local angle between the normal vector $\mathbf{n}$ to the emission surface, and the radiation direction $\hat{\mathbf{r}}$ at the time of emission. A dot over variables is used as a symbol of differentiation with respect to the time of the emission source. The subscript *ret* stands for the retarded time with the integration evaluated over the emission surface. For solution of the FW-H equation numerical (CFD) simulation results are used for the acoustic pressure at source.

The mathematical model (1)–(6), corresponds to the classical impermeable formulation of FW–H equations. In this case, the source surface $f = 0$ corresponds to the blade surface. This model, for high blade tip Mach numbers, usually gives under predicted values of the noise level. Nevertheless the classical FW–H formulation was used in resent publications for relatively low values of the blade tip Mach numbers (see reference [7], for example). Then, the acoustic pressure (based on the Formulation 1A) can be modified with empirical adjustments, based on the radiation Mach number M_r. In the present paper the radiation vector $\hat{\mathbf{r}}$ of the source – observer distance projected to shaft – observer direction is used instead of the source-observer distance. This was applied for the near- or far-field thickness noise depending on the rotor-observer distance.

The employed algorithm is similar to the algorithm, described in [3]. The first stage of FW-H solution algorithm is to divide the rotor blade surface in a number of panels. Integration over each panel is approximated using the value at its centroid.

The algorithm was implemented to the in-house H-FWH computer code.

3. CFD Modelling

The numerical simulations were conducted for 1/7 scaled UH-1H rotor; the geometry of the full-scale UH-1H rotor is presented in table 1. Since temporal and spatial periodicity is assumed for the flow, the computational grid was constructed for a single rotor blade (with appropriate periodic boundary conditions). For this two-bladed rotor, the computational domain is a half of a cylinder. The multi-block topology used in this paper, can be seen in Figure 1. C-type blocking is used around the blade. The structured and multi-block computational grid consisted of 88 blocks and 8.4 million cells. Along the aerofoil surface 218 points are located with concentration near the leading and trailing edges. Normal to the surface the first cell size is 10^{-5} of the blade's chord length and the cell aspect ratio is less than 1.2.

Table 1.UH-1H operational characteristics

Parameter	Value
Number of blades N	2
Rotor diameter $2R$, m	14.63
Rotor solidity σ	0.0464
Blade chord$c$, m	0.53
Blade airfoil	NACA 0012
Blade twist (root to tip), deg	-10.9
Max gross weight W, N	43000

The multi-block topology used in this work, can be seen in Figure 1. Around the blade a C–topology is used.

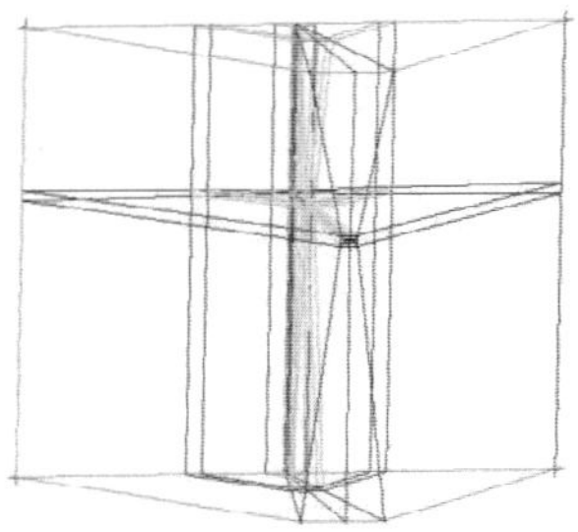

(a)

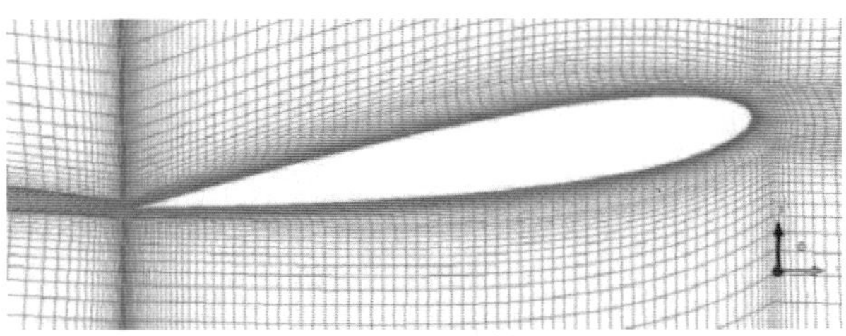

(b)

Figure 1. Multi-block topology (a) and mesh section (b).

The flow fields were numerically simulated using the 3D steady RANS equations.

For HMB the "source-sink" boundary conditions [8] were used at all surfaces of the computational domain apart from the symmetry plane. The "periodic" boundary condition provides the periodicity of the flow around the blade. The inflow (W_{in}) and outflow (W_{out}) velocities were obtained from momentum theory according to the expressions:

$$W_{in}(x,y,z_u) = -\frac{U}{8}\sqrt{C_T}\left(\frac{R}{L}\right)^2; W_{out} = -U\sqrt{C_T}. \quad (7)$$

Here U is the speed of the blade tip; L is the distance between the rotor centre and an arbitrary (x,y,z_u) point on the upper surface z_u=const of the computational domain. The thrust coefficient C_T is determined as

$$C_T = \frac{T}{\pi R^2 q_{tip}}, \quad (8)$$

where q_{tip} is the dynamic pressure at the blade tip and T is the rotor thrust, obtained after iterative computations. The rotor thrust coefficient C_{TW} (under the condition $T=W$) is first used as an initial guess and then the thrust coefficient value C_T is recomputed using successive approximations.

Similar boundary conditions with some differences at the bottom part of modeling domain were used for the HMB and Fluent solvers: in Fluent for the bottom part of computational domain the «pressure-outlet» boundary condition was assigned.

The flow conditions, and the values of the trust coefficient obtained with HMB and Fluent are presented in table 2. A collective pitch angle $\theta_{0.75}$ was chosen according to approximate trimming methods [9]. The values $\overline{W}_{in}$ of averaged on the upper part of the computational domain used for computations in both HMB and Fluent are shown in table 2. Some discrepancy between HMB and Fluent results can be explained, by the different type of boundary conditions at the upper and bottom parts of the computational domain: in Fluent $W_{in} = \overline{W}_{in} = const$, and for the bottom part of computational domain the «pressure-outlet» boundary condition was assigned.

Table 2. Conditions for computations

Parameter	Value
Collective pitch, $\theta_{0.75}$	7°
Tip Mach number, M_{tip}	0.8
Tip Reynolds number, Re	9.5×10^6
Thrust coefficient (for $T=W$), C_{TW}	0.0055
Inflow velocity (HMB), $\overline{W}_{in}$	0.201 m/s
Inflow velocity (Fluent), $\overline{W}_{in}$	0.2 m/s
Thrust coefficient (HMB), C_T	0.00577
Thrust coefficient (Fluent), C_T	0.0054

Simulations were conducted with the k-ω SST turbulence model. Results for the surface pressure coefficient distribution are show in Figure 2.

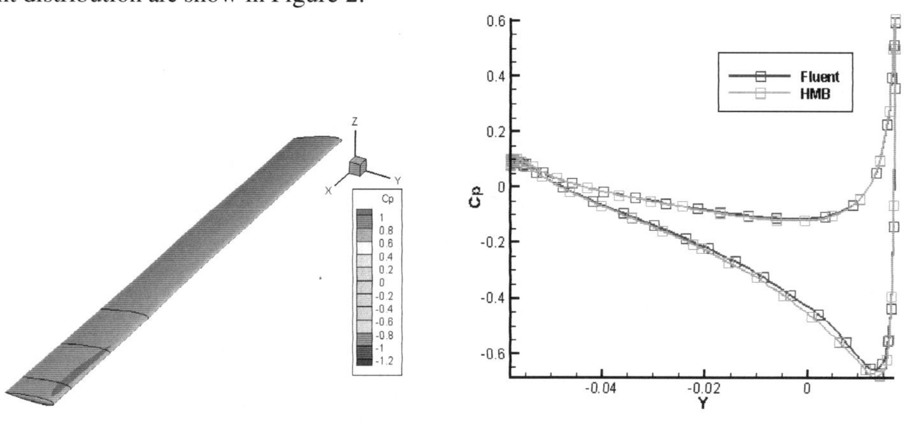

(a) (b)

Figure 2. Distribution of the pressure coefficient Cp: on the UH–1H rotor blade surface (a), and at the blade section $\overline{r} = r/R = 0.75$ (b).

The pressure distribution at the observer plane section is shown in Figure 3.

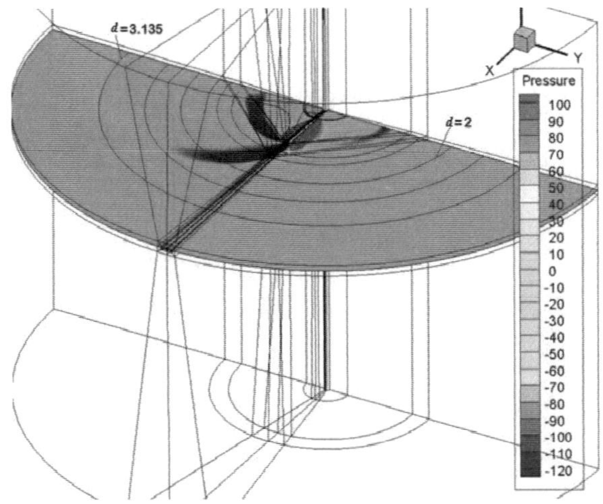

Figure 3. Absolute pressure distribution at the observer plane section.

From Figure 3 it follows that the obtained CFD simulation results for the near-field area (distance between the rotor center and the observer $d < 2R \approx 2\ m$) allows for comparison of the aeroacoustic signal at the CFD post processing stage (denoted as "direct" evaluation, without any further processing) to the FW-H equation solution results. At the distance $d > 2.5R$ comparison of the "direct" evaluation of the CFD results to the FW-H equation solution results can be not correct because of poor grid resolution in the far-field area.

4. FW-H equation solutions

Figure 4 presents the results of numerical simulation and experiments [6] for a far-field observer located at the rotor disc plane on the distance of $3R$=3.135 m (see Figure 3) away of the rotor.

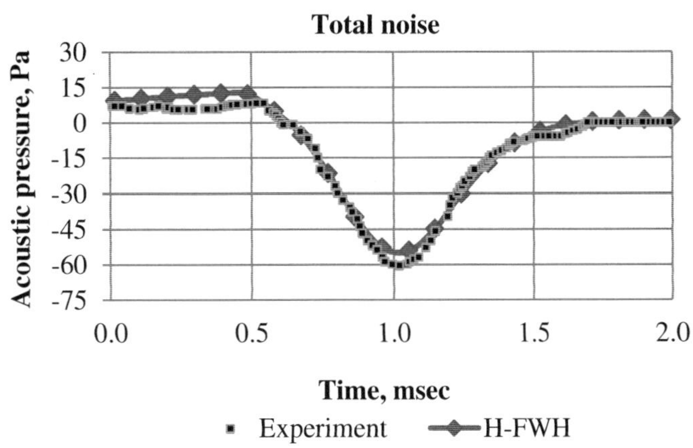

Figure 4. Comparison of CFD and experimental data for a far-field observer at 3R on the rotor disk plane.

The total noise predicted by the H-FWH code for an observer located in the rotor plane is in good agreement with the measurements. Comparison of the computed FW-H peak values to theoretical asymptotes

$$p'_{Fmin}(d) = p'_{min}(3R)\frac{3R}{d}, \qquad p'_{Nmin}(d) = p'_{min}(3R)\left(\frac{3R}{d}\right)^2 \qquad (9)$$

is presented in Figure 5.

The expressions (9) provide approximations of the limited peak values for observers located in the near- and far-field areas. The difference between the FW-H peak and the ′ values corresponds to the contribution of the near-field terms of the FW-H equation that were omitted in this analysis.

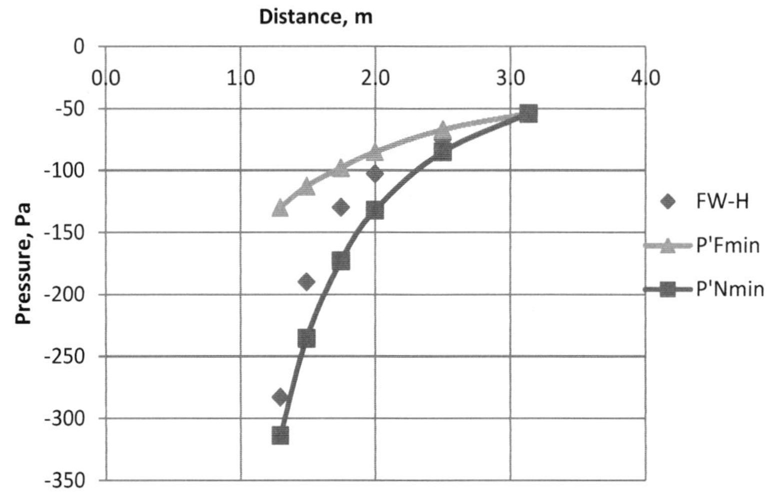

Figure 5. Comparison approximate and FW-H peak values.

From Figure 5 it follows that for short rotor-observer distances (*d* *R*) computation of the near-field terms of the system (1)–(6) is important.

5. Conclusions

Application of the in-house code for the FW–H (Ffowcs Williams – Hawkings) numerical solution was considered for near-field observers. The total rotor noise was calculated as a sum of the thickness and load noise including the near- and far-field components. To determine the near-field flow parameters for the UH–1H rotor in hover mode, the CFD code is used. Comparison of the computed FW-H peak values to theoretical asymptotes shows that computation of the near-field terms for short rotor-observer distances is important for accurate predictions of the observer noise.

Acknowledgements

This work was supported by the grant No. 9.1577.2017/4.6 of the Ministry of Education and Science of the Russian Federation.

References

[1] Edwards Band Cox Ch 2002 *NASA/CR-2002-211650.*
[2] Ffowcs Williams J E, Hawkings D L 1969 *Philosophical Transactions of the Royal Society* A **264** (1151) pp 321–343
[3] Brentner K S, Farassat F 2003 *Progress in Aerospace Sciences* **39** pp 83–120
[4] Song Wen-ping, Han Zhong-hua, Qiao Zhi-de 2006 *25th International Congress of the Aeronautical Sciences ICAS* pp 1–9
[5] Stepanov R, Pakhov V, Bozhenko A, Batrakov A, Garipova L, Kusyumov A, Mikhailov S, Barakos G *International Journal of Aeroacustics* 2017 vol 16 issue 6 pp 460–475
[6] Shmitz F H and Yu Y H 1983 *NASA Technical memorandum* 84390 p 105
[7] Sharma K, Brentner K *AHS 72nd Annual Forum* 2016 pp 1–14
[8] Mohd N A, Barakos G N, *RAeS Aerodynamics Conference* 2010 pp 1–19
[9] Seddon J *Basic helicopter aerodynamics* 1990 (BSP professional books) p 137

IOP Conf. Series: Materials Science and Engineering **449** (2018) 012018 doi:10.1088/1757-899X/449/1/012018

Rapid Decompression Technology for the Oxygen Supply System

Xian Zhang[1], Wei Leng[1], Ying Zhou[1], Ran Liu[1] and Junwei Wang[1]

[1] Beijing Institute of Spacecraft Environment Engineering, Beijing 100094, China

15652368545@163.com

Abstract. To ensure the safety of the pilot, the design scheme of a ground experiment platform for testing the performance of oxygen supply system under rapid decompression was presented. Detail design process of the subsystems such as vacuum system, rapid decompression system and temperature control system was introduced. Numerical simulations were carried out for the latter two subsystems. The results show that, the requirements both for the temperature uniformity of the experiment cabin and for the rapid decompression time can be achieved, based on the design of the present work which is very valuable for constructing the rapid decompression test platform of oxygen supply system.

1. Introduction

When the cabin is structurally damaged at high altitude, the air pressure in the cabin drops sharply. Rapid decompression, which occurs in less than a second or even tens of milliseconds, is often called explosion decompression [1]. Rapid decompression is extremely harmful [2-5]. In order to ensure the safety of pilots, when the cabin explodes and decompresses, the oxygen supply system [6-7] should first ensure that the expiratory valve is free for a short time to release the pressure on the lungs, and then quickly pressurize the body surface and mask to avoid the occurrence of explosive hypoxia [8]. In order to study the working performance of the oxygen supply system and ensure the life safety of the pilot, the ground simulation explosion decompression test of the oxygen supply system is needed.

2. Overview

A design technology of a rapid decompression test platform for oxygen supply system was introduced in the present work, which is mainly used for the comprehensive dynamic performance test of oxygen supply system by simulating the rapid decompression process. The test platform consists of vacuum system, rapid decompression system and temperature control system. Its main functions are as follows:

Environmental pressure: altitude > 25000m, control accuracy > 30m (0-8000m), + / -100m (8000-18000m)

Cabin temperature: 55 ℃ ~ room temperature, temperature uniformity: within 2 ℃

Rapid decompression: time for explosion decompression: less than 100ms; time for rapid decompression: 210ms-500ms.

3. Vacuum system

The main equipment of the vacuum system includes screw dry pump unit, vacuum valves and control valves, vacuum pressure measuring instruments, mufflers and other devices. The schematic diagram is shown in Figure 1.

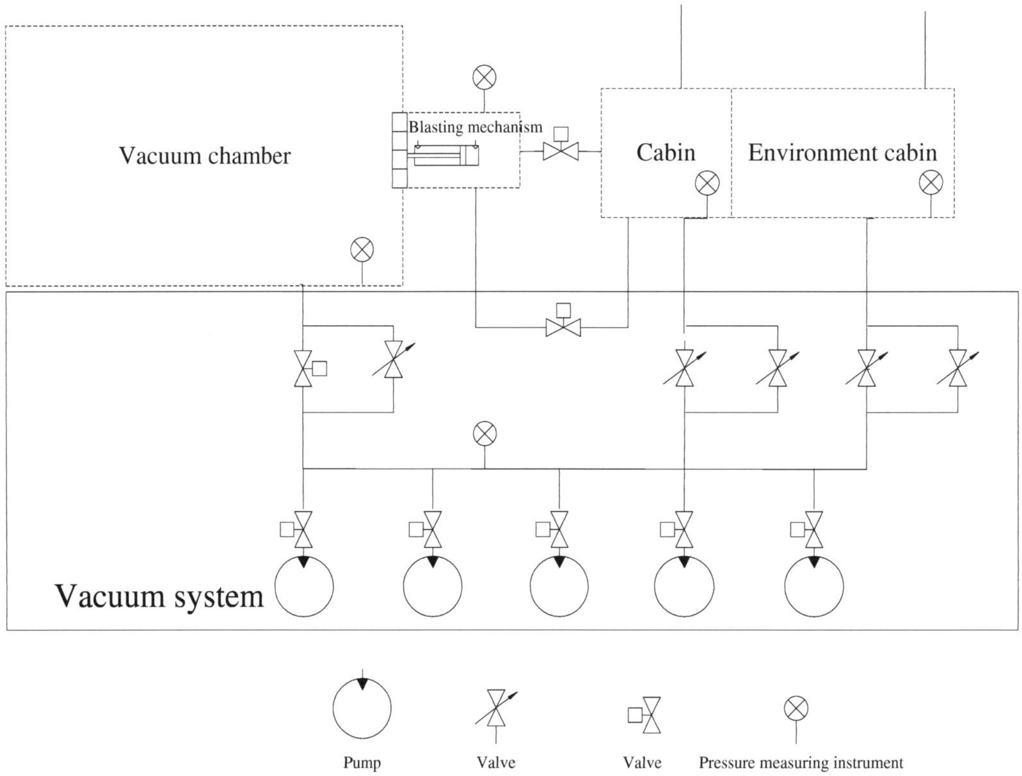

Figure 1. Schematic diagram of vacuum system

The pressure relief channel is set up between the cockpit and the vacuum reserve cabin, and the vacuum valve is set in the middle to ensure the pressure safety before the pressure relief. The vacuum system is equipped with a screw dry pump unit, which is used to pump air out of each compartment. The pump unit is connected with each compartment through regulating valves. Pressure measuring instruments are installed in each section to measure the pressure in real time. When the cabin pressure or lifting pressure rate needs to be maintained and controlled, the valve can be adjusted through the feedback system to control the flow rates of air extraction and air intake. Therefore, the constant maintenance and control of the final required pressure or lifting pressure rate can be realized.

The pumping speed required for the vacuum system to reach the set pressure is determined by the following equation(1):

$$P_g = P_0 + \frac{Q}{S_{eff}} + \frac{Q_1}{S_{eff}} \tag{1}$$

P_g represents the maximum vacuum that the cabin can reach, Pa. P0 represents the limit vacuum for the pump, Pa. Q_0 represents the air load of the vacuum chamber (including leakage Q_L and gas release of material surface Q_e), PaL/s after a certain period of time. Q_l represents process gas load of vacuum chamber. S_{eff} represents the effective pumping speed of the pump near the suction port of the vacuum chamber, L/s.

In the above equation(1), the ultimate vacuum of the vacuum cabin is always lower than that of the vacuum pumping unit, i.e. $P_0 << P_j$. Under the condition of a certain effective pumping speed of the pump near the suction port, the ultimate vacuum of the vacuum cabin under no-load condition is proportional to the leakage and air output of the vacuum cabin.

In the case of loading, assuming the exhaust flow in the cabin is 700L/min at room temperature and the exhaust pressure is 0.3Mpa. The leakage gas load of the vacuum container itself is negligible compared with the product load flow. At this point, the design height of the vacuum cabin is required to be 25000m, and the height is converted to the pressure of about 2.5×10^{3}Pa. According to the basic law of ideal gas and equation(1), the extraction speed required to maintain the dynamic vacuum degree of the vacuum system is:

$$S_{eff} = PS_l T_g / P_g T_l \tag{2}$$

The vacuum unit adopts the combination of roots pump and dry pump to pump air. Therefore, the pumping speed also needs to meet the requirements of vacuum reserve tank pumping time. The pumping speed can be calculated according to the following formula:

$$S_e = 23\frac{V}{t} K \times Lg\frac{P_1}{P_2} \tag{3}$$

T represents pumping time, h; S_e represents effective pumping speed of dry pump, m^3/h; *V* represents volume of the chamber, m^3; P_1 represents pressure at the beginning of pumping, Pa; P_2 represents the air pressure reached after time t, Pa. *K* represents correction factor, whose value is shown in Table 1.

Table 1. Value of correction factor K

Pressure/Pa	*K*
$10^5 \sim 10^4$	1
$10^4 \sim 10^3$	1.25
$10^3 \sim 10^2$	1.5
$10^2 \sim 10^1$	2
$10^1 \sim 10^0$	4

By calculating the requirement of pumping speed under the ultimate pressure and pumping time, ensure that the vacuum system can still meet the pressure control requirements under the maximum working condition of the product load, and the pumping time of each cabin can be achieved.

4. Rapid decompression system

The rapid decompression system is used to realize the rapid decompression and blasting decompression of the cabin. The glass blasting method is adopted in this scheme, which has the advantages of reasonable structure, easy operation and convenient for using. The system consists of tempered glass and mounting flange, blasting mechanism, opening adjustment valve, vacuum butterfly valve and cleaning mechanism. The schematic diagram of system composition is shown in Figure 2 below:

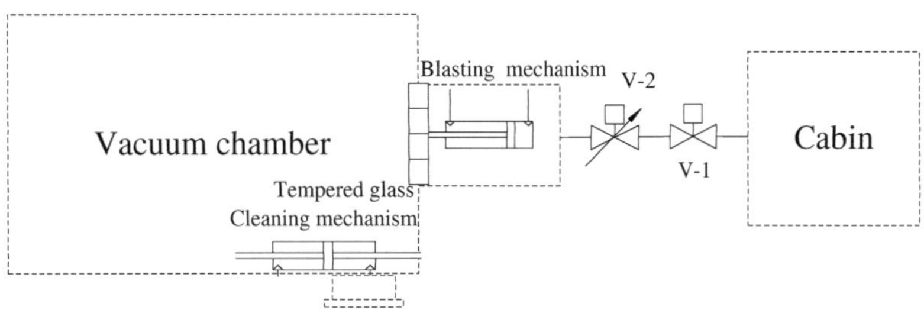

Figure 2. Schematic diagram of rapid decompression system

In order to realize rapid decompression, the blasting glass is set at the blasting decompression interface. The glass is installed on the side of the vacuum storage tank, and sealed by fluoro rubber ring. The blasting actuator is set up on the side of the cockpit of the blasting decompression interface. The actuator is driven by the cylinder and the tip collider is installed at the end of the cylinder pusher. When the blasting started, the crash needle is pushed by the cylinder to crush the tempered glass. According to the properties of the toughened glass, the whole glass will break up quickly. With the help of the huge pressure difference on different sides of the window, the glass will break to the side of the vacuum reserve. After the blasting is finished, the cylinder drives the impact needle back automatically.

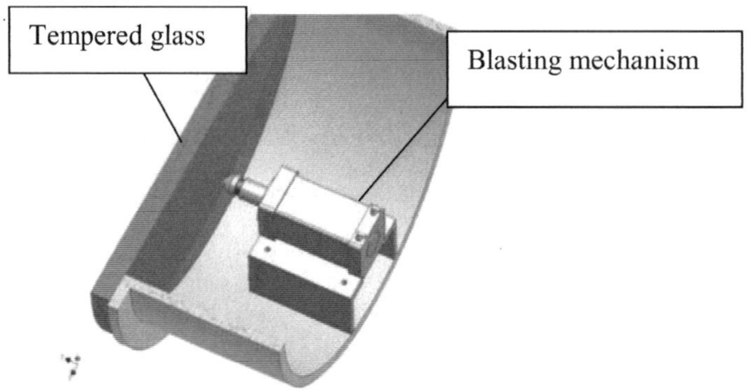

Figure 3. Break device and glass window

As shown in Figure 3, the pneumatic system of the blasting mechanism is mainly composed of air filtration combination, control valve and cylinder. Electromagnetic directional valve controls the direction of cylinder movement. Two-way throttle valve and one-way throttle valve regulate the cylinder speed.

The control of rapid decompression time is realized through the electric opening valve in the decompression channel, and the throat area during the decompression process of the pressure relief channel is controlled by adjusting the opening of the valve to adjust the decompression time. Fluent was used to simulate the dynamic process of rapid decompression of the cockpit and the vacuum reserve tank. The calculation results are shown in Figure 4.

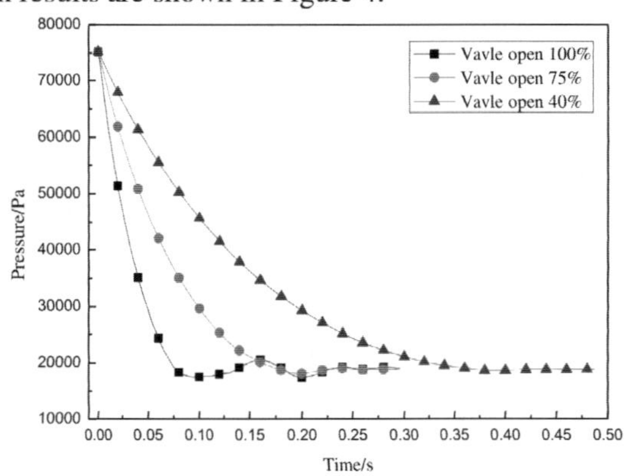

Figure 4. Pressure in environment cabin

5. Temperature control system

The temperature control system is mainly used to adjust the temperature in the environment test chamber, which consists of air conditioning unit and heat sink unit. The air conditioning unit consists of a fan, an electric heater, an evaporator and a refrigeration unit. The temperature control system is mainly used to adjust the temperature of the test space under normal pressure and low pressure. The heat sink unit is composed of heat sink and heat sink refrigeration unit, which is mainly used to adjust the temperature uniformity of test space under low pressure environment. The system flow diagram is shown in Figure 5 below.

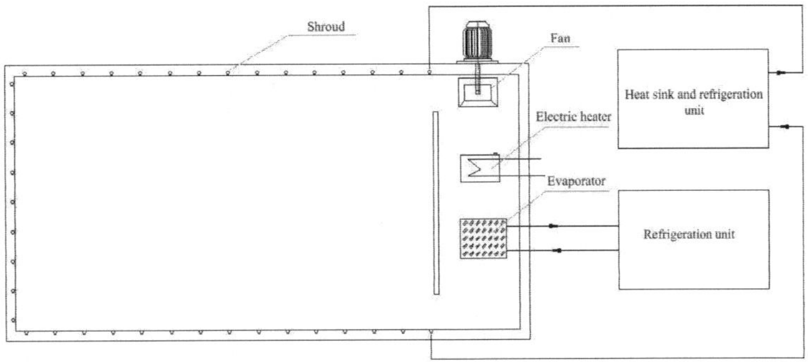

Figure 5. Schematic diagram of temperature control system

The system adopts the combination of active temperature control and passive temperature control to adjust the space temperature in the environment cabin. The active temperature control method is to use the air through convection to exchange heat with the test piece in the test box, transfer heat to evaporator or electric heater, and use the fan to complete the circulation heat exchange process. The passive temperature control method is to complete the heat transfer between the test piece and the heat sink by using heat sink to radiate heat in the test chamber space. The combined heat sink and air circulation temperature control method can effectively solve the problem of temperature uniformity in the test chamber under low pressure.

The design and calculation of temperature control system mainly include the power design of heating system and refrigeration system. The calculation method is as follows:

(1) Power calculation for heating system

In the cabin, its cooling load is mainly composed of inner tank load (Q_n), air load (Q_k) in the test chamber, heat dissipation of maintenance structure (Q_w), air conditioning unit load (Q_t) and load of test device (Q_f).Since the maximum temperature is normal temperature, the final heating power is determined by the power of temperature rise period. The heating power required in the heating stage is calculated by the following calculation formula:

$$Q_{heating} = Q_n + Q_k + Q_w + Q_t + Q_f \tag{4}$$

(2) Power calculation for refrigeration system

In the cabin, its thermal load includes inner tank load, air load in the test box, heat dissipation of maintenance structure, air temperature regulating unit load, load of test device, heat dissipation of motor and associated heat (Q'), and heat load of exhaust gas (Q_q). The refrigeration power required in the cooling phase and the refrigeration power required in the stabilization phase are calculated as follows:

$$\begin{cases} Q_{cooling} = Q_n + Q_k + Q_w + Q_t + Q_f + Q' \\ Q_{stabilization} = Q_w + Q' + Q_q \\ Q_{refrigeration} = \max\left\{Q_{cooling}, Q_{stabilization}\right\} \end{cases} \tag{5}$$

Fluent was used to simulate the temperature distributions of the environment chamber. The obtained temperature contours and velocity contours are shown in Figure 6.

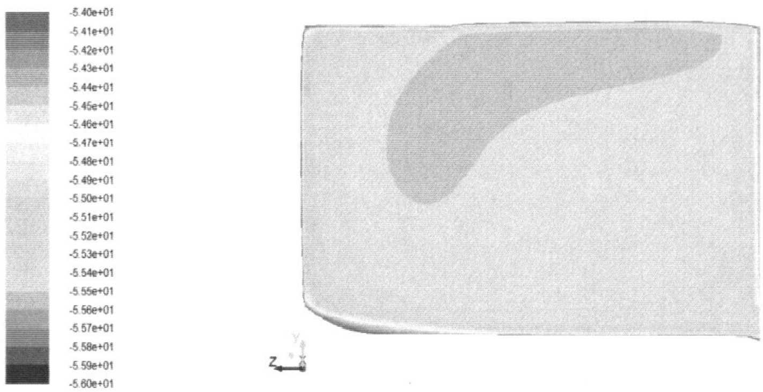

Figure 6 .Temperature distribution in environment cabin

6. Results

In order to ensure the safety of the pilot's life in the environment of rapid decompression, a design scheme for the rapid decompression test platform of the oxygen supply system was provided in this paper, the design methods of the vacuum system, the rapid decompression system and the temperature control system of the test platform were introduced in detail, and simulation verification for the rapid decompression and temperature control system were carried out. The main conclusions are as follows:

(1)A rapid decompression method of glass blasting was designed. Combined with the adjustment of the opening degree of the pressure relief valve, different rapid decompression time can be controlled.

(2) The temperature control method combining active temperature control and passive temperature control was proposed, and the temperature uniformity of the environment chamber was verified by numerical calculation.

In summary, by using the design advocated in this paper, the temperature uniformity of the test system and the requirements of rapid decompression time were realized, which provides a reference for the design of rapid decompression test of oxygen supply system in the future and has certain guiding significance.

References

[1] Ge RL,Kubo K,Kobayashi T,et al. Blunted hypoxic pulmonary vasoconstrictive response in the rodent Ochotona curzoniae(pika) at high altitude. Am J Phusiol,1998,274(5 Pt 2): H1792-H1799.

[2] Bass T, White LE,Wood RD,et al. Rapid decompression of congenital hydrocephalus associated with parenchymal hemorrhage. J Neyroimaging, 1995, 5:249-251.

[3] Fraser WD, Goodman LS, Ackles KN, et al. Cardiovascular responses with standard and extended bladder coverage G-suits during rapid decompression. Aviat Space Environ Med, 1994,65:209-213.

[4] Xiao HJ,He DY,Yuan XG. The physiologic effects of various oxygen levels following rapid decompression [J]. 1997, SAFE J. Vol. 27(2): 109.

[5] Lovett N P,Ashdown R. Development of the OBOGS and BRAG valve for the F-22 life support system[C]. Proceedings of SAFE the 36th Annual Symposium. 1998.

[6] 4 Buic F. Oxyhemoglobin saturation following rapid decompression to 18288 m preceded by diluted oxygen breathing [J]. Aviat Space Environ Med. 1991, Vol. 62(12):1119.

[7] Pokroy R, Barenboim E, Cartel D,et a1. Unilateral Optic Disc Swelling in a Fighter Pilot[J]. Aviat Space Environ Med. 2009, Vol. 80(10): 894-897.

[8] Fitzpatrick D T. Visual manifestations of neurologic decompression sickness [J]. Avlat Space Environ Med. 1994, Vol. 65(8): 736-738.

IOP Conf. Series: Materials Science and Engineering **449** (2018) 012019 doi:10.1088/1757-899X/449/1/012019

Numerical Study of the Effects of Fluid Conductance and the Capacity of Negatively Pressured Cabin to the Process of Explosive Decompression

Bing Bai[1, 2], Junwei Wang[1] and Ying Zhou[1]

[1] Beijing Institute of Spacecraft Environment Engineering, Beijing 100094, China

[2]cast511bai@163.com

Abstract. Explosive decompression is the rapid decompression process of aircraft in the sky caused by the damage of cabin which challenges the safety of people and equipment. The explosive decompression tester usually uses negative pressure reserve cabin technology. Numerical study was conducted to analyse the influence of the throttling and negative pressure reserve cabin volume on the explosive decompression process. The results show that as the throttle opening, the explosive decompression time decreases and the fluctuation amplitude of the two chambers after the pressure balance is enhanced; the volume of the negative pressure reserve cabin has little effects on the rate of explosive decompression and has no obvious influence on the fluctuation.

1. Introduction

At high altitudes, the air pressure is low. To ensure the safety of personnel and facility during the flight, the aircraft cabin will be artificially pressurized. Explosive decompression refers to the rapid pressure loss in the cabin caused by the direct contact of the air in and out of the cabin in the high air, which causes damage to personnel and equipment in the cabin.[1, 2] Explosive decompression and its countermeasures are the challenges that aviation industry has to face. To research the explosive decompression process and its influences on the physiology and equipment, the simulator of this process is in need [1], which decompresses rapidly. The test cabin needs to decompress from 75.2kPa to 18.8kPa in 15s [3] , which can hardly be realized via vacuum pumps. The common method is to use negative pressure reserve cabin technology, which uses a bigger cabin with lower pressure connecting the test cabin to realize the rapid decompression [3-5]. The parameters such as the connection type and the volume of each cabin have significant influence on explosive depression process, on which researches were done around the world.

Higins established a mathematic model to describe the decompressing process of exposing to vacuum and conducted numerical study. [6-8] Sun conducted experimental and theoretical research to the pressure change of test cabin in explosive decompression and derived a formula of decompression time. The results show that the decompression time is proportional to the volume of the cabin and inversely proportional to the effective area of the decompression blast hole [9]. Gao combined the fluid conductance equation with the gas thermodynamics theory, simplified the calculation with an isothermal process, and derived mathematical equations for the airflow balance between the cabins. The instantaneous pressure difference between the cabins is proportional to the initial pressure difference and the cabin volume, has an exponential relationship with the duration and an inverse relationship with the flow conductance of the connecting pipeline. [10]

Dynamic numerical simulation will be conducted to study the influence of throttling and the capacity of negatively pressured reserve cabin and provide guidance for the design of this type of equipment.

2. The influence of throttling on depression process

2.1. Mold and numerical method

A simulator with a test chamber volume V_A of 7.2 m^3 was numerical studied. The initial pressure p_{A0} of the chamber is 75.2 kPa, and the balance pressure p is 18.8 kPa. The negative pressure reserve chamber volume V_B is 400m^3. The two chambers are connected by a $\varphi 1\text{m} \times 2\text{m}$ pipe, as shown in Figure 1.

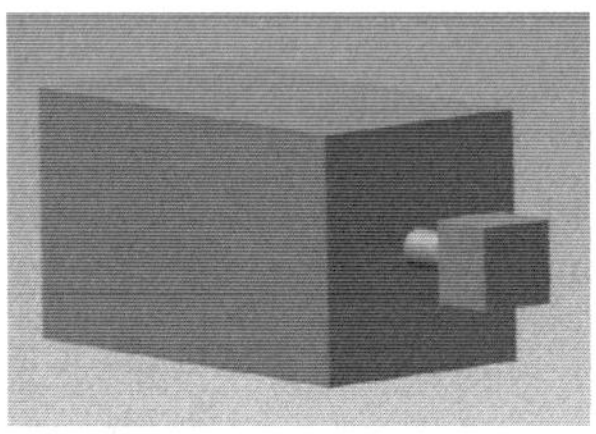

Figure 1. The simulator

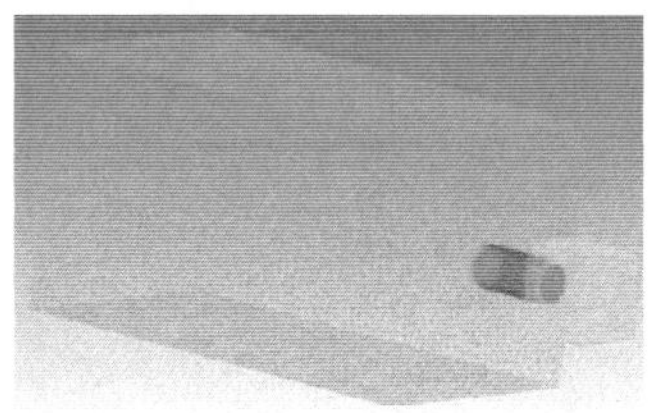

Figure 2. The meshing of the simulator

A throttle valve simplified to a orifice plate structure is placed at one end of the pipe. It changes flow conductance with its diameter. Three valves with diameters of 900mm, 630mm and 450mm were studied.

According to [10],

$$p_{A0}V_A + p_{B0}V_B = p(V_A + V_B) \quad (1)$$

The p_{B0} was calculated to be 17.8 kPa. The simulator was meshed with unstructured meshes, as shown in Figure 2. The initial pressure of the two chambers, p_{A0} and p_{B0}, were set by patch. Unsteady calculations of the decompression process were performedusing Realizable $k-\varepsilon$ model.

2.2. Calculation results and analysis

The total pressure distributions at different times were obtained by calculation. The total pressure contours and of the three testers with the three throttle valve diameters of 900mm, 630mm and 450mm at 0.01s and 0.02s are shown in Figure 3, and the static pressure contours are in Figure 4.

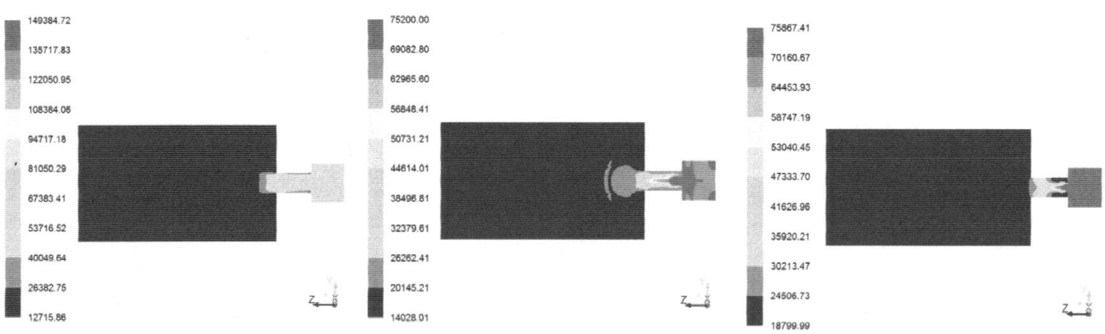

a) Total pressure contours of the system with 900/630/450mm valve at 0.01s

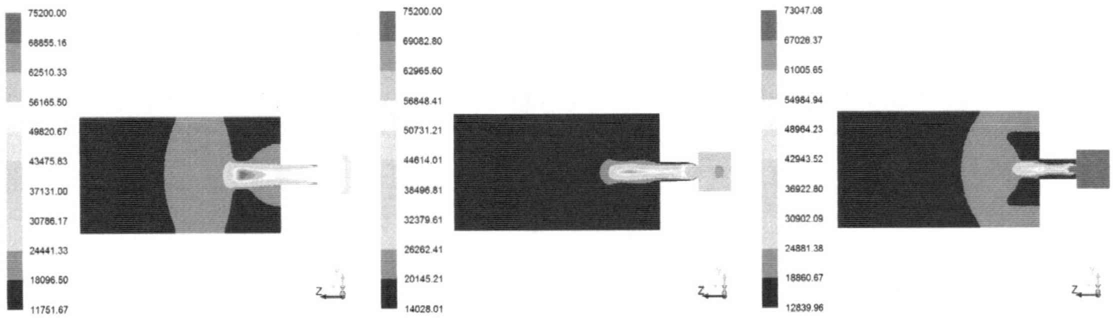

b) Total pressure contours of the system with 900/630/450mm valve at 0.02s

Figure 3. Total pressure contours of different valve and time

As can be seen from Figure 3, the high pressure air inside the test chamber quickly flows to the negative pressure reserve chamber during the decompression process. It can be found that as the valve closes the decompression becomes significantly slower due to the flow resistance increases.

Comparing Figure 3 with Figure 4, it can be seen that there is a distinct low pressure zone at the rear of the throttle valve. After the airflow flows into the negative pressure reserve tank, eddy currents are also formed, resulting in a low pressure zone. It can be drawn that the kinetic energy loss of the system is mainly concentrated in the cabin loss caused by the throttle valve and the jet.

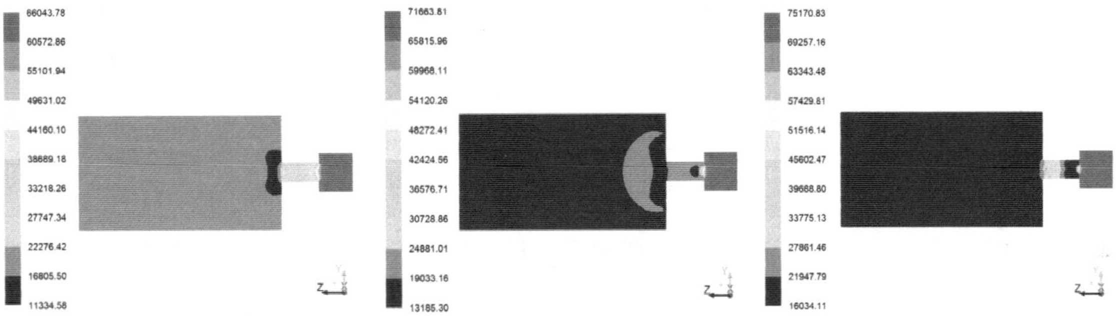

a) Static pressure contours of the system with 900/630/450mm valve at 0.01s

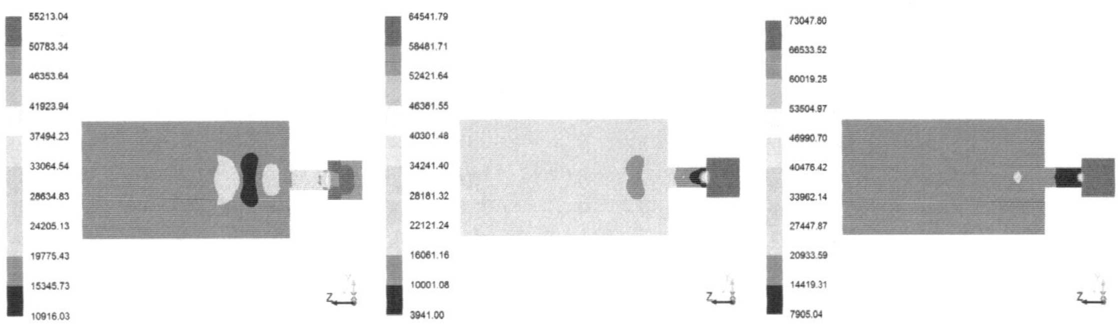

b) Static pressure contours of the system with 900/630/450mm valve at 0.02s

Figure 4. Static pressure contours of different valve and time

There is a distinct low pressure zone near the throttle valve. This is due to the formation of the vortex when air flow into the negative pressure reserve tank, mix with the low pressure gas and decelerates, consuming its kinetic energy. As the throttle valve closes, the vortex at the rear of the throttle valve is enhanced, the conductance is reduced, and the flow loss is increased.

For the three throttle valves, the calculation was done until the decompression process was completed and the fluctuation was reduced to ±500 Pa. The development curve of the static pressure of the test chamber over time in the three decompression processes is shown in Figure 5.

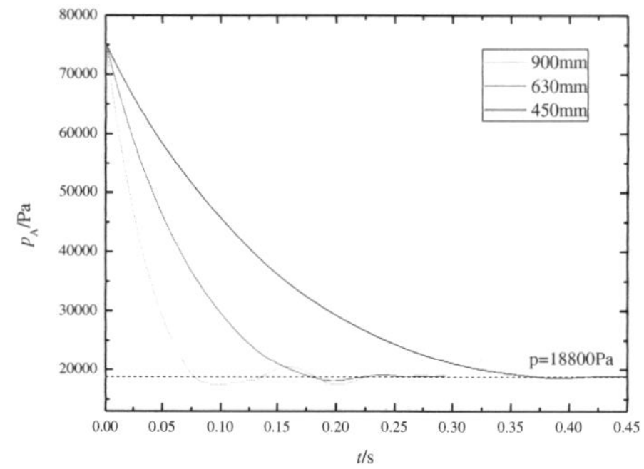

Figure 5. Development curves of static pressure of test chamber

As can be seen from Figure 5, the test chamber's static pressure drop rate increases significantly as the throttle valve opens. This is because the flow conductance is increased as the throttle valve is increased. Then the flow path resistance is reduced, and the gas flow is increased under the same conditions, which makes the decompression time reducing. The calculation results are in agreement with the theoretical values.

It's obvious that pressure fluctuations of different amplitudes will occur after the pressures of the two chambers get equal. This is caused by the airflow impact. When the pressures of the two chambers get equal, the airflow rate is not zero, so the flow will continue, resulting in the gas pressure in the test chamber continuing to decrease, causing the lower pressure than the negative pressure reserve tank. Then gas backflow and wave formation. Fluctuations gradually decrease and disappear with the loss of gas mechanical energy. Comparing the three curves, the fluctuation intensity increases when the throttle valve becomes larger, and the fluctuation phenomenon is basically not observed when the throttle valve diameter is 450 mm. This is because the flow conductance of the connecting pipe is reduced as the throttle valve closes. And, the mechanical energy loss during the decompression process is increased, the kinetic energy of the gas is reduced after the pressure balance of the two chambers, and the fluctuation is reduced.

3. The influences of negative pressure reserve tank volume

It can be drawn from the derivation of the literature [10] that the negative pressure reserve volume V_B has a significant impact on the decompression time. When design the test system, the test chamber volume V_A, the test chamber initial pressure p_{A0} and the balance pressure p are all specified by the design requirements. It can be seen from the formula (1) that the negative pressure reserve volume V_B also has an influence on the initial two-chamber pressure difference Δp_0. In order to guide the design of the test equipment, the influences of the volume of the negative pressure reserve tank on the blasting decompression process were explored.

Four volumes of the negative pressure reserve tank were selected as 200m^3, 400m^3, 800m^3 and 1200m^3. In order to ensure that the same pressure is reached after decompression, the initial pressure of the reserve tank under each volume is calculated according to equation (1) as shown in Table 1.

Table 1. Initial pressure of negative pressure reserve tank

V_B/m^3	200	400	800	1200
P_{B0}/Pa	16770	17785	18292	18462

To save computational resources, a 900 mm diameter throttle valve was uniformly selected. The pressure curves of each test chamber during decompression are shown in Figure 6.It can be seen from Figure 6 that the volume of the negative pressure reserve tank has little effect on the rate of blast decompression during the decompression phase, which is negligible. There is an impact in the fluctuation phase, but no obvious law.

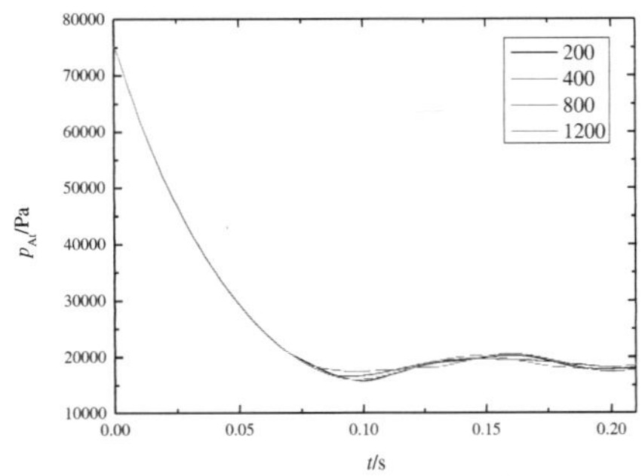

Figure 6. Static pressure curves of test chambers with different negative pressure reserve tank

As shown in Figure 3, the affected area of the jet in the tank is smaller than the volume of the tank, so the volume of the negative pressure reserve tank will not significantly affect the decompression process.

4. Conclusion

Unsteady simulations of the explosive decompression process were carried out. The working conditions of different flow conductance and different negative pressure reserve tanks are studied. The influence of these parameters is explored. The conclusions are as follows.

(1) As the throttle valve opens, the flow conductance increases, the gas flow increases under the same conditions, so the explosive decompression time is reduced;

(2) During the explosive decompression process, pressure fluctuation of different amplitudes will occur when the pressures of the two chambers get equal;

(3) As the throttle valve closing, the flow conductance of the connecting pipe is reduced, and the mechanical energy loss during the decompression process is increased. After the pressure of the two chambers gets balanced, the kinetic energy of the gas is reduced, and then the fluctuation is reduced;

(4) The volume of the negative pressure reserve tank has little effect on the rate of blast decompression during the decompression phase. It has a influence during the fluctuation phase, but there is no obvious law.

Reference

[1] A G. Mathematical Model Research of Human Exposure to Vacuum[J]. Journal of the British Interplanetary Society. 2008, 28(2): 148-152.

[2] M R E. Rapid(Explosive) Decompression Emergencies in Pressure-Suited Subjects[R]. NASA, 1969.

[3] Xingrong S. Research on Rapid Decompression Test Technology for Airborne Equipment [J]. Environmental Technology. 2011, 33(2): 18-22.

[4] Dongchen Y, Huajun X, Bin Z, et al. Development of Experiment Chambers of Complex Low-pressure Environment [J]. Chinese Medical Equipment Journal. 2011, 32(12): 6-12.

[5] Demtrad. On the Decompression of a Punctured Pressu-rized Cabin in Vacuum Flight[J]. Jet Propulsion. 2006, 24(2): 112-123.

[6] M E. Rapid Decompression Emergencies in Pressure Suited Subjects[J]. Air Medical Journal. 2007, 28(1): 82-91.

[7] Pagani A, Carrera E. Gasdynamics of Rapid and Explosive Decompressions of Pressurized Aircraft Including Active Venting[J]. Advances in Aircraft and Spacecraft Science. 2016, 3(1):

39-77.
[8] Higins. Simulation Model Research of Vacuum Exposure[J].
[9] Sun B, Xiao H, Yuan X. Effect of Cabin Decompression Time on Parameters of Emergency Oxygen Supply[J]. China Safety Science Journal. 2002, 2(5): 1-7.
[10] Gao H, Liu M, Wang J. Novel Technique to Rapidly Pump Spacecraft Cabin with Negatively Pressured Cabin and Its Theoretical Basis[J]. Chinese Journal of Vacuum Science and Technology. 2013, 33(12): 1191-1198.

IOP Conf. Series: Materials Science and Engineering **449** (2018) 012020 doi:10.1088/1757-899X/449/1/012020

A Rapid Estimation for Interplanetary Low-Thrust Trajectories Using Support Vector Regression

L Xu[1], H Shang[1,2] and X Qin[1]

[1]Beijing Institute of Technology, 100081 Beijing, People's Republic of China

[2]E-mail: shanghb@bit.edu.cn

Abstract. During the preliminary phase of space mission planning and design, a large quantity of trajectory optimization problems have to be solved. Obtaining the optimal solutions of low-thrust trajectory is computationally challenging since the optimization problems usually involve an iterative numerical algorithm and the complicated numerical integration of the equations of motion. It is necessary to develop the rapid trajectory estimation methods for low-thrust transfer. In this paper, a new method based on machine learning has been proposed to estimate the optimal interplanetary low-thrust trajectory. The minimum-propellant low-thrust trajectory is optimized by using the hybrid optimization algorithm, which would provide the high-quality training samples for machine learning. Support vector regression is adopted to construct and train the estimation model. Numerical simulations demonstrate that the proposed estimation method and the percentage errors of random test samples are all lower than 5%. This application of machine learning method can accomplish very efficient low-thrust interplanetary trajectory evaluation and it is therefore suitable to extend the design flexibility in the practical exploration mission.

1. Introduction

Application of low-thrust propulsion system is effective and prospective for interplanetary exploration. Benefiting from its high specific impulse, low-thrust propulsion can decrease the fuel consumption significantly [1]. Some accomplished missions have profited from that, such as Hayabusa, Dawn, etc. The optimization of low-thrust trajectory has always been an intensive research area. Various techniques have been developed which are mainly categorized as indirect and direct methods [2]. Both methods have their strengths and shortcomings but they all require the time-consuming iteration. However, in the preliminary mission design, there are usually a lot of low-thrust trajectories need to be evaluated, which is important for the appropriate mission planning. For instance, in the selection of target asteroids, the fuel consumption is a key factor to determine the feasibility of missions. Therefore, the new methods for evaluating optimal low-thrust trajectory effectively are very desirable.

In this paper, a novel estimation method for the optimal interplanetary low-thrust trajectory is developed based on machine learning. Considering the minimum-propellant transfer, support vector regression (SVR) is chosen to train the prediction model of velocity increment [3]. Through the rational data processing, the SVR-based model can predict the continuous output values efficiently for the given inputs.

This paper is organized as follows. Section 2 describes the evaluation problem of optimal low-thrust trajectory. In Sec. 3, low-thrust optimization problem is solved firstly by using numerical algorithms, which provides the reliable database. Then the prediction model of trajectory parameters is constructed and trained based on orbital dynamics and SVR. The numerical results are shown in Sec. 4 to verify and analyse the accuracy of proposed method.

2. Problem statement

Efficient estimation for optimal trajectories is of good importance in the preliminary mission design and tend to be solved by the trajectory optimization methods. Some of the well-known software, such as MALTO and VARITOP, are all medium- or high-fidelity [4]. These optimization methods solve low-thrust optimization problem by the numerical iteration, leading to a complicated and time-consuming computation.

The low-fidelity optimization tools to quickly obtain optimal trajectory are lacking. Only a kind of new methods based on curve-fitting is developed for interplanetary circular-coplanar low-thrust transfers. Kluever given the analytic functions for computing mass-optimized trajectory using solar electric propulsion [5]. This method is derived by fitting the database derived from the low-thrust coplanar and circular transfers between Earth orbit and a target orbit. It is assumed that the transfer starts with a powered arc, followed by a coast arc, and finally ends with another powered arc. Then, the analytic function of velocity increment for optimal low-thrust transfers is fitting as

$$\Delta V = A\bar{a}_T^B \text{ km/s} \tag{1}$$

where A and B are the fit coefficients. $\bar{a}_T$ is the average thrust acceleration.

The numerical results proved that the analytic function of Eq. (1) is extremely efficient to predict velocity increment by comparing with the solutions from optimization algorithms. However, this method is only available for low-thrust circular-coplanar transfers. In terms of the general low-thrust transfer, the new technique must be developed to evaluate optimal trajectory.

3. Optimal trajectory evaluation based on support vector regression

In this section, an evaluation method for propellant-optimal interplanetary low-thrust trajectory is developed to compute trajectory parameters. The general low-thrust transfer from Earth is considered to generate the database and support vector regression is adopted to train the evaluation model.

3.1. Generation of optimal low-thrust trajectory database

Considering the minimum-propellant low-thrust transfer from Earth, the trajectory optimization problem is a nonlinear optimal control problem. In this paper, the Sims-Flanagan transcription is adopted to solve this problem [6]. This method discretizes the whole trajectory into N equal-size time steps. The continuous thrust of each step can be approximated by applying a bounded velocity impulse at the center of each time step. It assumes that the spacecraft moves in an unperturbed Keplerian orbit between applied thrust impulses. Figure 1 shows the diagram of a low-thrust trajectory using the Sims-Flanagan transcription. The maximal impulse $\Delta V_{\max}$ that the spacecraft can produce is determined as:

$$\Delta V_{\max} = \frac{DT_{\max}\Delta t}{m} \tag{2}$$

where $T_{\max}$ is the maximal thrust magnitude that spacecraft can produce, D is the thruster duty cycle, Δt is the flight time of each time steps.

Taking the ephemeris-free low-thrust transfer into consideration, the decision variables of trajectory optimization problem are

$$X = \left[E_0\ E_f\ t_f\ m_f\ \Delta v_{ix}\ \Delta v_{iy}\ \Delta v_{iz}\right]^{\mathrm{T}} \quad (1 \le i \le N) \tag{3}$$

where E_0 and E_f are the eccentric anomaly of the initial and final orbit, respectively. t_f is the transfer time. m_f is the final mass of the spacecraft. And $\Delta v_{ix}, \Delta v_{iy}, \Delta v_{iz}$ are the three components to express the i th discontinuous impulse at center of time step.

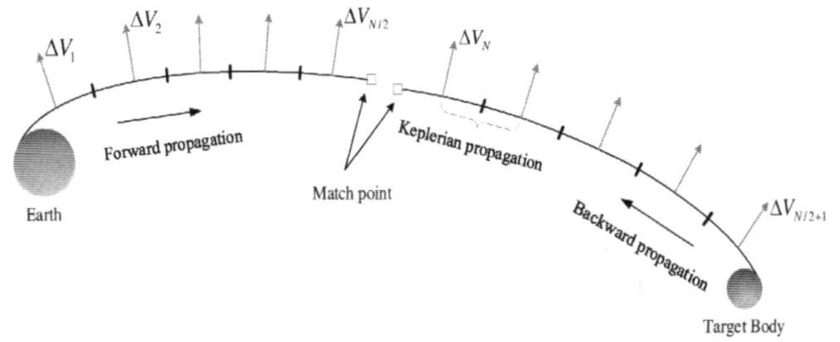

Figure 1. Schematic of low-thrust trajectory using the Sims-Flanagan transcription.

In order to reduce the sensitivity of the constraints to the decision variables, the whole trajectory is divided into two equal-time halves. The first half begins at the launch body and is propagated forward in time. So does the second half. Therefore, the state constraints of match point are:

$$\Psi(X) = z_F(X) - z_B(X) = \left[\Delta x \; \Delta y \; \Delta z \; \Delta v_x \; \Delta v_y \; \Delta v_z \; \Delta m\right] = 0 \quad (4)$$

where $\Delta x, \Delta y, \Delta z$ and $\Delta v_x, \Delta v_y, \Delta v_z$ are the three components of position and velocity vector difference in match points, Δm is the difference of spacecraft mass in two match points.

The hybrid optimization algorithm combining DE and SNOPT is adopted to solve optimization model. In detail, DE is used as a global optimizer to produce good potential solutions and then SQP as a local optimizer to fine tune the solution starting with the initial guess from DE. Particularly, the penalty function is used to handle the equality constraints.

3.2. Construction of evaluation model

In this section, the evaluation model to predict velocity increment for optimal low-thrust transfer is built and a popular data-driven machine learning method is adopted to train the prediction model.

To evaluate the minimum-propellant trajectory, the velocity increment ΔV for low-thrust transfer is always chosen as objective to be predicted. ΔV is determined theoretically and empirically by the propulsive parameters and the state vector of target body. Considering the ephemeris-free transfer, the target orbit can be described mainly by the orbital elements including semi-major axis a, eccentricity e and inclination i. The transfer time is also chosen as inputs, in other words, the optimal low-thrust transfer with the fixed time is evaluated. Then the prediction model is expressed as

$$f : R^7 \to R \;\Rightarrow\; [a, e, i, T_{\max}, I_{sp}, m_0, t_f]^T \mapsto \Delta V \quad (5)$$

Support Vector Regression (SVR) is used to train the prediction model [3]. It is assumed that the training samples set $U = \{(x_i, y_i) | i = 1, 2, \ldots, N\}$ contains the inputs x_i and the corresponding output y_i where N is the number of samples. The basic idea of regression is to map the input x_i to y_i with a function $f(x)$. In the nonlinear problem, the mapping function $\Phi(x)$ is used to simplify the problem. Then the regression function form is presented by

$$f(x) = \langle \omega, \Phi(x) \rangle + b \quad (6)$$

where $\langle \,,\, \rangle$ denotes the dot product. ω, b represent the weight coefficient and constant respectively. Theoretically, the function $f(x)$ fit the training samples as flat as possible by seeking a small ω. One way to satisfy this is to minimize $\|\omega\|^2 = \langle \omega, \omega \rangle$. There are two slack variables ξ_i, ξ_i^* introduced to deal with other infeasible constraints. With a ε for the prediction accuracy, the problem is formulated as

Minimize
$$\frac{1}{2}\|\omega\|^2 + C\sum_{i=1}^{l}(\xi_i, \xi_i^*) \quad (7)$$

Subject to
$$\begin{cases} y_i - \langle \omega, x_i \rangle - b \le \varepsilon + \xi_i \\ \langle \omega, x_i \rangle + b - y_i \le \varepsilon + \xi_i^* \\ \xi_i, \xi_i^* \ge 0 \end{cases} \quad (8)$$

where the constant C denotes the penalty factor. ξ_i, ξ_i^* represent the difference between the predicted and true values. This problem is solved more easily by the transformation of the dual formulation. The Lagrangian function is shown as follows

$$\begin{aligned} L := \frac{1}{2}\|\omega\|^2 + C\sum_{i=1}^{l}(\xi_i + \xi_i^*) - \sum_{i=1}^{l}(\eta_i\xi_i + \eta_i^*\xi_i^*) \\ -\sum_{i=1}^{l}\alpha_i(\varepsilon + \xi_i - y_i + \langle\omega, x_i\rangle + b) \\ -\sum_{i=1}^{l}\alpha_i^*(\varepsilon + \xi_i^* + y_i - \langle\omega, x_i\rangle - b) \end{aligned} \tag{9}$$

where $\eta_i, \eta_i^*, \alpha_i, \alpha_i^*$ are the Lagrange multipliers. This function in Eq. (9) has a saddle point with respect to the dual variables at the solution which is substituted by inferring the partial derivatives of L with respect to the primal variables. Then the dual optimization problem is written as

$$\text{Minimize} \quad \begin{aligned} \frac{1}{2}\sum_{i,j=1}^{l}(\alpha_i^* - \alpha_i)(\alpha_j^* - \alpha_j)K(x_i, y_i) \\ +\varepsilon\sum_{i=1}^{l}(\alpha_i^* + \alpha_i) - \sum_{i=1}^{l}y_i(\alpha_i^* - \alpha_i) \end{aligned} \tag{10}$$

$$\text{Subject to} \quad \begin{cases} \sum_{i=1}^{l}(\alpha_i^* - \alpha_i) = 0 \\ 0 \le \alpha_i, \alpha_i^* \le C \end{cases} \tag{11}$$

where $K(x_i, y_i)$ is the kernel function. The non-zero α_i and α_i^* are also the support vectors. By solving the dual optimization problem, the weight coefficient ω is computed

$$\omega = \sum_{i=1}^{l}(\alpha_i^* - \alpha_i)x_i \tag{12}$$

Besides, b can be determined by exploiting the Karush-Kuhn-Tucker (KKT) conditions.

$$\begin{aligned} b = y_j - \langle\omega, \Phi(x)\rangle - \varepsilon \quad \text{for } 0 \le \alpha_i \le C \\ b = y_j - \langle\omega, \Phi(x)\rangle + \varepsilon \quad \text{for } 0 \le \alpha_i^* \le C \end{aligned} \tag{13}$$

Then the mapping function can be presented in the support vector expansion as

$$f(x) = \sum_{i=1}^{l}(\alpha_i - \alpha_i^*)K(x_i, x) + b \tag{14}$$

According to Eq. (14), the determination of kernel function is the key component in SVR method. The commonly used kernel functions include linear kernels, polynomial kernels and Radial Basis Function (RBF). Their corresponding formulations are

$$\begin{aligned} K_{\text{linear}}(x_1, x_2) = x_1'x_2 \\ K_{\text{ploynomial}}(x_1, x_2) = (1 + x_1'x_2)^p \\ K_{\text{RBF}}(x_1, x_2) = \exp(-\|x_1 - x_2\|^2) \end{aligned} \tag{15}$$

Each kernel function has distinct characteristic because of its special and inherent formulation for computing the predictors which should be compared as the problem. Finally, to determine the suitable training sample set, the Mean Square Error (MSE) is chosen as objective to present the prediction accuracy. The lower MSE means the more accurate predicted values and the number of training samples can be determined by limiting an acceptable MSE.

4. Numerical results

To verify the efficiency of proposed method, the evaluation model is constructed to compute trajectory parameters. Considering low-thrust transfer from Earth's orbit with zero excess velocity, the model in Eq. (5) is trained based on SVR to predict the total velocity increment. To guarantee the performance of prediction, the seven input parameters are generated following a uniform distribution in the input domain. Without loss of generality, the range for each parameter is set as table 1.

Table 1. Ranges of seven input parameters.

Input parameters	ranges
Semi-major axis a (AU)	2.1~2.5
Orbital eccentricity e	0~0.3
Orbital inclination i (deg)	0~20
Maximal thrust magnitude T_{max} (mN)	300~500
Thrust specific impulse I_{sp} (s)	2500~3500
Initial mass of spacecraft m_0 (kg)	1000~2000
Flight time t_f (days)	1000~1500

For the inputs from table 1, the optimal low-thrust trajectory database used for SVR is obtained by using the Sims-Flanagan transcription and DE-SNOPT. Then, SVR is adopted to train the prediction model. The prediction model mainly depends on the determination of kernel function and training samples set. Choosing 200 test samples randomly, figure 2 shows the relationship between MSE and training samples set size for three kernels in Eq. (15). The degree of polynomial kernel is set to 3.

As shown in figure 2, it is obvious that polynomial kernel function performs better for this prediction problem. In addition, with the increase of training samples set size, MSE of test samples decreases and tend to be stable after fluctuation for all three kernels. According to this selection criterion, the number of training samples for polynomial kernel function are chosen as 800.

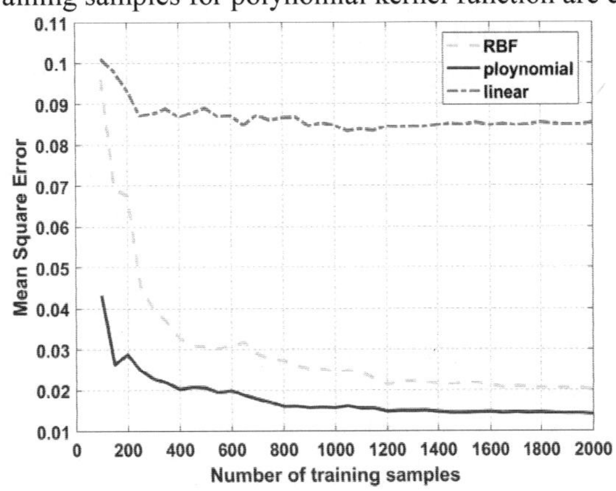

Figure 2. Relationship between MSE and number of training samples for three kernels.

To show the prediction deviation more clearly, the percentage error (PE) between the predicted outputs and optimal outputs of test samples is chosen for the prediction of SVR-based model. For 200 random test samples, the PE results of polynomial kernel are exhibited by histograms in figure 3.

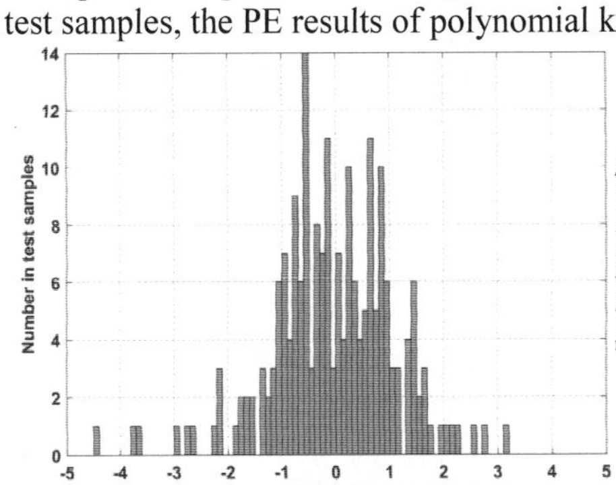

Figure 3. PE of polynomial kernels.

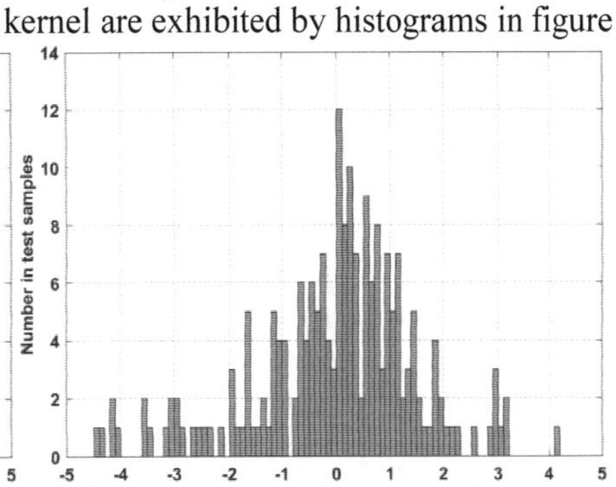

Figure 4. PE of RBF kernels.

It can be seen that the maximal PE is less than ± 5%. The SVR-based model is capable of predicting the velocity increment for interplanetary low-thrust transfer with sufficient precision. Comparatively, figure 4 shows that of RBF kernel function. The distribution of PE in figure 4 is more dispersive and the number of PE over ±3% is more than that in figure 3. Note that the generation of one optimal low-thrust trajectory by using hybrid optimization algorithm requires about 2 minutes, which means obtaining the 200 test samples takes about 6.67 hours. While the SVR-based prediction model can accomplish the computation like the analytical function.

By using the prediction model from polynomial kernel for low-thrust optimal transfer, we set the asteroid mission to Vesta for evaluating optimal velocity increment ΔV. Table 2 lists the predicted ΔV to Vesta for various inputs. Furthermore, it also shows the global optimal ΔV calculated by numerical optimization algorithm. As shown in table 2, the PE is varied from -0.59%~1.54%. The corresponding deviation of ΔV is less than 157 m/s. Therefore, the evaluation method based on SVR is really efficient in the preliminary low-thrust mission design.

Table 2. Prediction accuracy of propulsion systems for low-thrust transfers to Vesta.

T_{max}, mN	I_{sp}, s	m_0, kg	t_f, days	Optimal ΔV, km/s	predicted ΔV, km/s	PE of ΔV, %
300	2500	2000	1450	10.574	10.702	1.21
350	3500	1000	1400	10.187	10.127	-0.59
400	3000	1000	1050	10.245	10.234	-0.11
450	2500	1500	1200	10.182	10.226	0.43
500	3500	1000	1400	10.183	10.340	1.54
500	2500	1500	1200	10.182	10.156	-0.26

5. Conclusion

In this paper, a novel approach is developed for evaluating interplanetary optimal low-thrust trajectory quickly and efficiently. Considering the minimum-propellant low-thrust transfer, the SVR-based model can predict optimal velocity increment accurately. Comparing to the numerical optimal solution, the percentage errors of 200 random test samples are all less than 5%. Thus, the proposed SVR-based models are powerful tools for evaluation of optimal low-thrust trajectory, especially in the case that a large number of candidate trajectories are considered. The introduction of machine learning also provides an advantage to reveal the relationship of trajectory parameters for low-thrust transfer, which makes it well suited for performing preliminary interplanetary low-thrust mission design.

References

[1] Yarnoz D G, Jehn R and Croon M 2006 Interplanetary navigation along the low-thrust trajectory of Bepi-Colombo *Acta Astronautica*. **59** 284

[2] Yam C H, Lorenzo D D and Izzo D 2011 Low-thrust trajectory design as a constrained global optimization problem *Journal of Aerospace Engineering*. **225** 1243

[3] Kim D W, Seo S and De Silva C W 2009 Use of support vector regression in stable trajectory generation for walking Humanoid robots. *ETRI Journal,31,5(2009-10-05)*. **31** 565

[4] Kos L D, Polsgrove T and Hopkins R 2006 Overview of the development for a suite of low-thrust trajectory analysis tools *AIAA/AAS Astrodynamics Specialist Conference and Exhibit (AIAA)*

[5] Kluever C A 2015 Efficient computation of optimal interplanetary trajectories using solar electric propulsion *Journal of Guidance Control & Dynamics*. **38** 821

[6] Sims J A, and Flanagan S N 1999 Preliminary design of low-thrust interplanetary missions *AAS/AIAA Astrodynamics Specialist Conference*, American Astronautical Soc. 338

Evaluating Gravity-Assist Range Set Based on Supervised Machine Learning

K Zhang[1], H Shang[1,2], Q Chen[1] and X Qin [1]

[1]Beijing Institute of Technology, 100081 Beijing, People's Republic of China

[2]E-mail: shanghb@bit.edu.cn

Abstract. The dynamics of gravity-assist (GA) trajectories contain strong nonlinearity, which makes the traditional methods for impulse transfer range set (RS) are intractable to deal with the gravity-assist RS. This paper develops a novel method to evaluate the gravity-assist RS based on regression methods in supervised machine learning (SML) field. The performances of three powerful regression methods with several common kernel functions are assessed. The Gaussian Processes Regression (GPR) method with Matérn 3/2 kernel is selected because of the minimum mean squared error (1.11×10^{-3} km^2/s^2). The predicting model based on GPR is constructed to make prediction form the orbital elements of destination orbits to the total velocity increment of corresponding optimal GA trajectories. The percentage error of predicting model is no more than 2%. Millions pairs of sample points are generated by the trained predicting model. The points with specified value of total velocity increment are extracted, of which the envelope constitutes the gravity-assist RS. Both of Venus GA and Mars GA trajectories are considered in this paper.

1. Introduction

The reachable domain of spacecraft designates the set of positions that the spacecraft can reach with given initial orbit and specified fuel constraint. Reachable domain is a useful tool for mission planning and collisions risk assessment between two spacecraft [1]. Existing researches on reachable domain are based on the description of the positional parameters in Cartesian coordinate [2] [3] or the Keplerian orbital elements [4]. Reachable domain is also known as range set (RS) in the researches descripted by Keplerian orbital elements. These researches were usually based on impulse transfer, in which appropriate mathematics tools could be used to derive the relationship between the boundary accessible points and the velocity increment. However, it is difficult to calculate the gravity-assist RS because the GA models are more complex comparing with impulse transfer models [5] [6]. For GA problem, the traditional methods in the field of RS are intractable to obtain the relationship discussed above.

Regression methods in supervised machine learning (SML) aim to predict unsuspected data accurately and efficiently by detecting implicit relationships in training set [7]. These methods can substitute for the traditional methods, and dig out the relationship between the orbital elements of destination orbit and the total velocity increment ΔV_{opt} of corresponding optimal GA trajectory. Inspired by this idea, a novel method based on SML is proposed to evaluate the gravity-assist RS numerically in this study. First, the feature and target of predicting model based on SML are determined. A hybrid optimization combining Differential Evolution (DE) and SNOPT is employed to solve the trajectory-optimization problem based on GA model with deep space maneuver (DSM).

Later, the performances of three powerful SML regression methods with several common choices of kernels are assessed. Finally, millions pairs of sample points are generated by the trained predicting model thanks to the high efficiency of SML methods. The points with specified value of ΔV_{opt} are extracted, and the envelope of these extracted sample points constitutes the gravity-assist RS. Both of Venus GA and Mars GA trajectories are considered in this paper.

2. Problem Formulation

The regression methods in SML field are able to construct the predicting model from the features to target based on the training set that contains numbers of pairs of feature values and corresponding target values. In this paper, we aim to evaluate the gravity-assist RS with the description of Keplerian orbital elements. Considering ephemeris-free model, the mapping from orbital elements $[a,e,i,\Omega,\omega]$ to the ΔV_{opt} of corresponding optimal GA trajectory is in strong demand. Therefore $\boldsymbol{x}=[a,e,i,\Omega,\omega]$ is selected as the feature of predicting model, and $y=\Delta V_{opt}$ is regard as the target. The basic idea of regression is to find the mapping $f:\boldsymbol{x}\rightarrow y$ based on the training set, expressed by $\mathcal{D}=\{\boldsymbol{x}_i, y_i \mid i=1,2,...,n\}$. The orbital elements in training set are randomly sampled in the region $\boldsymbol{U}$, defined as

$$\boldsymbol{U}=\{(a,e,i,\Omega,\omega) \mid 1.7\le a\le 3.5AU, 0\le e\le 0.5, 0\le i\le 30\deg, 0\le\Omega\le 360\deg, 0\le\omega\le 360\deg\} \quad (1)$$

Actually, this region covers more than 90% the current known main-belt asteroids. The calculation of gravity-assist RS within this region is significant for the target selection in practical mission.

In this paper, the single GA trajectory is considered, and the candidate GA bodies contain Venus and Mars. The gravity-assist model with DSM is introduced to design the GA trajectories. The objective function of GA trajectory with DSM is given by

$$\Delta V_{GA}=\|\Delta\boldsymbol{V}_D\|+\|\Delta\boldsymbol{V}_{DSM1}\|+\|\Delta\boldsymbol{V}_{DSM2}\|+\|\Delta\boldsymbol{V}_I\| \quad (2)$$

where $\Delta\boldsymbol{V}_D$ and $\Delta\boldsymbol{V}_I$ denote two impulses applied at departure from a circular Earth parking orbit (200km height) and the injection to the destination orbit. $\Delta\boldsymbol{V}_{DSM1}$ and $\Delta\boldsymbol{V}_{DSM2}$ represent two deep space maneuvers applied in the legs before and after GA flyby respectively. Hybrid of DE [8] and SNOPT is employed to solve the optimal GA trajectory with minimum ΔV_{GA}. DE executes a global search and provides a high-quality initial guess for SNOPT. SNOPT performs precise local search based on the solution from DE. The minimum ΔV_{GA} for different destination orbits are obtained, which are regarded as corresponding ΔV_{opt} in subsequent sections.

3. Predicting Model Building and Range Set Calculating

3.1. Performance assessment for supervised machine learning regression methods

In this section, the performances of three powerful SML regression methods are assessed to determine the final predicting model, including Gaussian Processes Regression (GPR), Support Vector Regression (SVR) and Kernel Ridge Regression (KRR). Considering that these methods are kernel based, the selections of kernel functions are also considered in the performance assessment, including the Squared Exponential (SE), the Matérn class (M3/2 or M5/2), the Rational Quadratic (RQ), the Polynomial (Poly) kernel, the Radial Basis Function (RBF) and the Sigmoid (Sig). A training set $\mathcal{D}$ with 2000 random orbital elements data in $\boldsymbol{U}$ and the corresponding ΔV_{opt} is constructed to train the predicting models based on different SML regression methods. Define $\mathcal{C}=\{(\boldsymbol{x}_i^*, y_i^*) \mid i=1,2,...,100\}$ as the test set, which is generated in the same way as training set $\mathcal{D}$. The test set $\mathcal{C}$ is used to measure the accuracy of predicting models by calculating Mean Squared Error (MSE) or Percentage Error (PE) between the prediction output and the true value. Taking Mars gravity-assist case as an example, the MSE distribution is shown in figure 1.

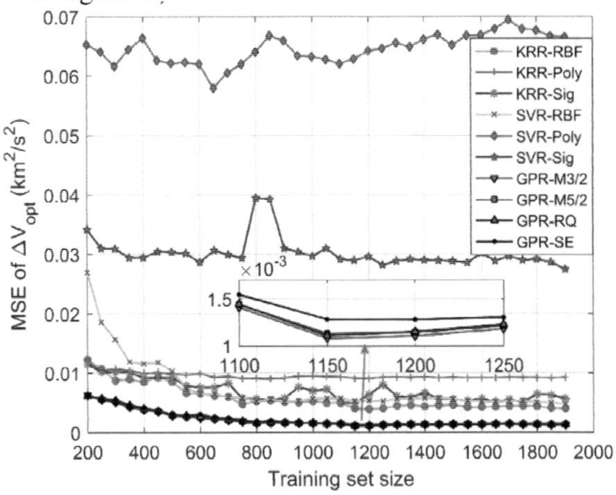

Figure 1. MSE distribution of different SML regression methods with different kernels

In general, the MSEs of SML regression models decline as the training set size grows, and the MSE curves tend to converge after the training set contains about 1400 training pairs. Among all the SML regression models in figure 1, GPR models show the overwhelming superiority with faster convergence rate and lower stable value of MSE. In addition, the GPR models are relatively insensitive to the selection of kernel functions, while the performance of other models shows significant disparity with different kernels. As shown in the enlarged view of GPR models, the Matérn 3/2 kernel has a slight advantage with 1.11×10^{-3} km^2/s^2 minimal MSE at 1150 training set size. In conclusion, GPR model with Matérn 3/2 kernel function is selected to generate the gravity-assist RS, and the training set size is chosen as 1150 in subsequent sections of this study. The PE distribution of predicting model based GPR with Matérn 3/2 kernel is shown as figure 2. It can be seen that the maximum PE is no more than 2%, and the vast majority of PEs are less then ±1%. According to the MSE and PE of predicting model, it can be concluded that the accuracy of the predicting model is satisfactory.

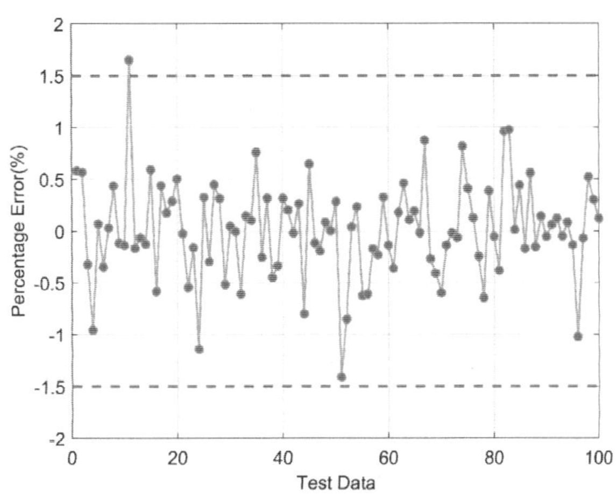

Figure 2. PE distribution of predicting model

3.2. Evaluating the gravity-assist range set

Thanks to the high-accuracy GPR predicting model constructed in this paper, the ΔV_{opt} of the optimal

GA trajectories can be easily obtained given the orbital elements inputs. A large-scale dataset contains millions pairs of orbital elements and corresponding ΔV_{opt} is generated using the predicting model. These orbital elements are sampled uniformly in the region ***U*** with the sampling frequency expressed by 50×50×50×25×25, of which the multipliers correspond to each element of $[a,e,i,\Omega,\omega]$. These 78,125,000 sampling points are used to detect the gravity-assist RS by extracting the points with specified value of ΔV_{opt}. The slack variable ΔV_{slack} =±0.01km/s is introduced because it is difficult to find the sample points with ΔV_{opt} equal to a given value exactly without any fractional error. For example, it means that ΔV_{opt} of 6.99~7.01km/s will be regarded as 7km/s in this study. The envelope of the extracted sampling points can be regarded as the corresponding RS via optimal gravity-assist trajectories with specified Total Velocity Increment Limit (TVIL) for missions. Finally, the RS via Venus GA trajectories and Mars GA trajectories are shown in figure 3 and 4 respectively.

As we can see in figure 3, only a small region in ***U*** is reachable when the TVIL for missions is set to 6km/s, and it tends to be completely unreachable given lower TVIL via Venus GA trajectory. The RS extends out obviously with the increase of TVIL. The pattern for Mars GA trajectory is analogous, while the minimum TVIL that make the region ***U*** partly reachable is lower than 4km/s (see figure 4). This reflects the advantage of Mars GA trajectory in reducing total velocity increment for missions. In general, the destination orbits with lower *a* and *i* are easier to reach via both Venus GA trajectories and Mars GA trajectories.

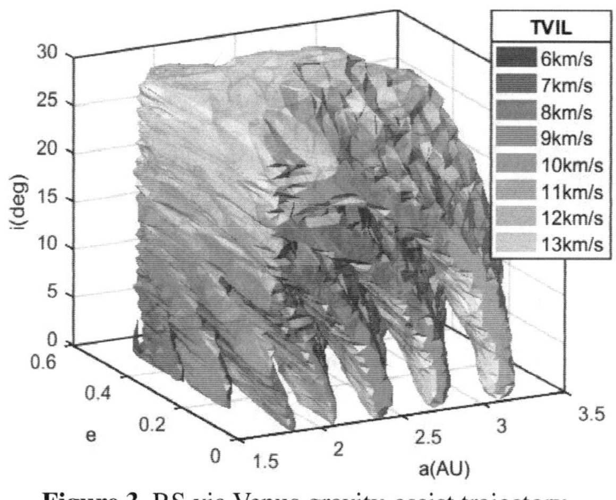

Figure 3. RS via Venus gravity-assist trajectory

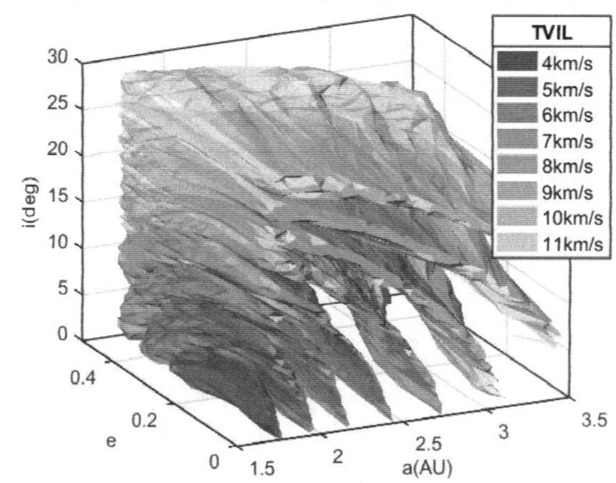

Figure 4. RS via Mars gravity-assist trajectory

4. Conclusions

In this paper, a novel method based on SML is proposed to evaluate the gravity-assist RS. The SML regression methods are introduced to dig out the implied relationship between the orbital elements $[a,e,i,\Omega,\omega]$ and the corresponding ΔV_{opt}. The property of several powerful SML methods with different kernel functions are assessed in this study, and the GPR model with Matérn 3/2 kernel is finally selected because it realizes minimum MSE (1.11×10^{-3} km^2/s^2) with fewer training data (1150). The orbital elements in training set $\mathcal{D}$ are randomly sampled in a significant region $\boldsymbol{U}$, which covers the majority of the main-belt asteroids. The corresponding ΔV_{opt} are calculated by solving a NLP problem based on the gravity-assist model with DSM. The hybrid of DE and SNOPT is employed to solve the optimal ΔV_{opt}. Using the trained predicting model, millions pairs of orbital elements and corresponding ΔV_{opt} are generated. These sample points are used to detect the gravity-assist RS by extracting the points with specified value of ΔV_{opt}. The envelope of the extracted sampling points with same ΔV_{opt} constitutes the corresponding range set. The RSs of different TVILs via both Venus GA trajectories and Mars GA trajectories are evaluated using proposed method.

References

[1] Xue D, Li J, Baoyin H and Jiang F 2010 Reachable domain for spacecraft with a single impulse *Journal of Guidance, Control, and Dynamics* **33** 934

[2] Nguyen X, Elmer G, Robert M, Donglong S and Ping L 1995 Reachable domain for interception at hyperbolic speeds *Acta Astronautica* **35** 1-8

[3] Wen C, Zhao Y and Shi P 2014 Precise Determination of reachable domain for spacecraft with single impulse *Journal of Guidance, Control, and Dynamics* **37** 1767-79

[4] Marcus J, Daniel J and Erwin R 2014 On-orbit operational range computation using gauss's variational equations with J2 perturbations *Journal of Guidance, Control, and Dynamics* **37** 608-22

[5] Neal D, Cynthia X and David F 1984 Optimum two-impulse transfer for preliminary interplanetary trajectory design *Journal of Guidance, Control, and Dynamics* **7** 458-61

[6] Qiao D, Cui H and Cui P 2006 Evaluating accessibility of near-earth asteroids via earth gravity sssists *Journal of Guidance, Control, and Dynamics* **29** 502-05

[7] Kevvin P 2012 *Machine learning, a probabilistic perspective* (Cambridge: MIT Press) p492

[8] Aaron D, Craig A and Michael L 2007 Interplanetary mission design using differential evolution *Journal of Spacecraft and Rockets* **44** 1060

IOP Conf. Series: Materials Science and Engineering **449** (2018) 012022 doi:10.1088/1757-899X/449/1/012022

Research on Influence of Geometric Adjustment on Performance of Variable Cycle Engine

Binglong Zhang, Shuiming Liu and He Liu

School of Energy and Power Engineering, Beihang University, Beijing, 100191, China

zgbglg@buaa.edu.cn

Abstract. Based on conventional turbofan engine, this paper adding mode selection valve, the front variable area bypass injectors and the rear variable area bypass injectors to build a mathematical model of the variable cycle engine. Then the influence of adjustable components on engine performance is simulated by computation program. The simulation results show that in the same work mode, the front variable area bypass injectors and the rear variable area bypass injectors have the same effect on engine performance, but different from the mode selection valve; Both front variable area bypass injectors and rear variable area bypass injectors have different effects on engine performance in different modes.

1. Introduction

For conventional aero engines, it can only maintain a good working condition under certain conditions. The performance will also drop due to deviation from the specified conditions. The Variable Cycle Engine (VCE) can change the thermodynamic cycle by changing the geometric position, shape and size of some components. The engine has a turbofan cycle with low fuel consumption during subsonic flight and a high unit thrust turbojet cycle for supersonic flight [1]. This performance advantage makes the variable cycle engine become one of the main directions of aviation research.

Research work on variable cycle engines has been in progress since around 1960. In 1976, GE used the Rear Variable Area Bypass Injector (RVABI) structure on the YJ101 engine [2]. In 1978, GE conducted a test of a variable cycle engine with a Forward Variable Area Bypass Injector (FVABI) to verify its feasibility [3, 4].

In recent years, research work on variable cycle engines has focused on the structure and overall performance of components [5, 6]. However, there is a lack of research on the effects of geometric adjustments such as FVABI, RVABI and MSV on the performance of variable cycle engines.

In this paper, the mathematical model of the variable cycle engine is built and the simulation calculation program is programmed to analyze the influence of components such as MSV, FVABI and RVABI on the performance of the variable cycle engine, and provide theoretical support for the subsequent research on the key adjustment components of the variable cycle engine.

2. Research object

The research object of this paper is the double external culvert cycle engine. The basic structure is shown in Figure 1. The main adjustable components affecting engine performance include MSV, FVABI and RVABI. The channel where FVABI is located is called the first outer culvert, and the channel regulated by MSV is called the second outer culvert. This engine has two modes and its working principle is as follows.

2.1. Single outer culvert mode

The engine operates in a small bypass ratio mode during high power conditions such as climb, acceleration and supersonic cruise (Figure 1 below). In this mode, the MSV is turned off, that is, the second outer culvert is closed. In order to allow more airflow into the core duct, the FVABI is turned down. Adjust the RVABI to the proper position to ensure the required bypass ratio and back pressure while achieving high thrust.

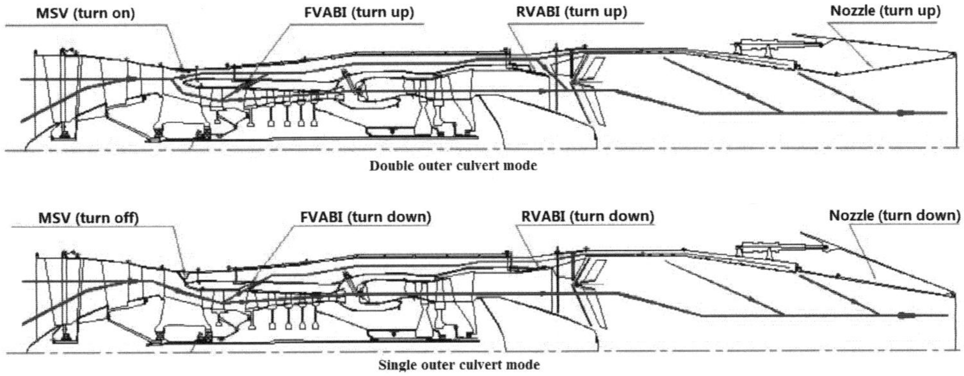

Figure 1. Variable cycle engine structure.

2.2. Double outer culvert mode

The engine operates in a large bypass ratio mode for take-off and subsonic cruise conditions (Figure 1 above). In this mode, the MSV is opened, and the first outer culvert and the second outer culvert are all open. By increasing the FVABI, the air flow into the core duct is reduced, and the air flow entering the outer duct is increased. The RVABI is increased, and the culvert airflow is injected into the exhaust gas, so that the engine thrust is further increased and the noise is reduced.

3. Research method

Compared to conventional turbofan engines, variable cycle engines add some adjustable components that bring new influences to the joint work. Therefore, in this calculation, the judgment parameters such as low-pressure rotor speed, high-pressure rotor speed, fan pressure ratio function, core drive stage fan pressure ratio function, compressor pressure ratio function, high-pressure turbine flow function, low-pressure turbine flow function, etc. are added. The following seven common working equations are selected based on engine airflow mass flow balance, pressure balance, physical speed equalization, and power balance relationship.

1. High pressure turbine flow balance

$$ERR(1) = \left(W_5 - W_{5g}\right)W_5^{-1} \tag{1}$$

ERR represents the residual, W_5 represents the high pressure turbine inlet converted flow, and W_{5g} represents the high pressure turbine inlet converted flow calculated from the high pressure turbine characteristic curve.

2. High pressure shaft power balance

$$ERR(2) = \left(N_C + N_D - N_{TH}\right)\left(N_C + N_D\right)^{-1} \tag{2}$$

N_C represents compressor power, N_D represents the core drive stage fan power, and NTH represents high pressure turbine power.

3. Low pressure turbine flow balance

$$ERR(3) = \left(W_{55} - W_{55g}\right)W_{55}^{-1} \tag{3}$$

W_{55} represents the low pressure turbine inlet converted flow, and W_{55g} represents the low pressure turbine inlet converted flow calculated from the low pressure turbine characteristic curve.

4. Low pressure shaft power balance

$$ERR(4) = \left(N_F - N_{TL}\right)N_F^{-1} \tag{4}$$

N_F represents fan power, and N_{TL} represents low pressure turbine power.

5. Mixing chamber static pressure balance

$$ERR(5) = \left(PS_{55} - PS_{25}\right)PS_{55}^{-1} \tag{5}$$

PS_{55} represents the static pressure of the core duct outlet, PS_{25} represents the static pressure at the exit of the outer duct.

6. Nozzle mass balance

$$ERR(6) = \left(W_9 - W_7\right)W_9^{-1} \tag{6}$$

W_9 represents the conversion flow of the nozzle, and W_7 represents the converted flow rate of the afterburner outlet.

7. CDFS mass balance

$$ERR(7) = \left(W_2 - W_{21} - W_{13}\right)W_2^{-1} \tag{7}$$

W_2 represents the fan inlet conversion flow, W_{21} represents the core drive stage fan import conversion flow, and W_{13} represents the outer culvert conversion flow.

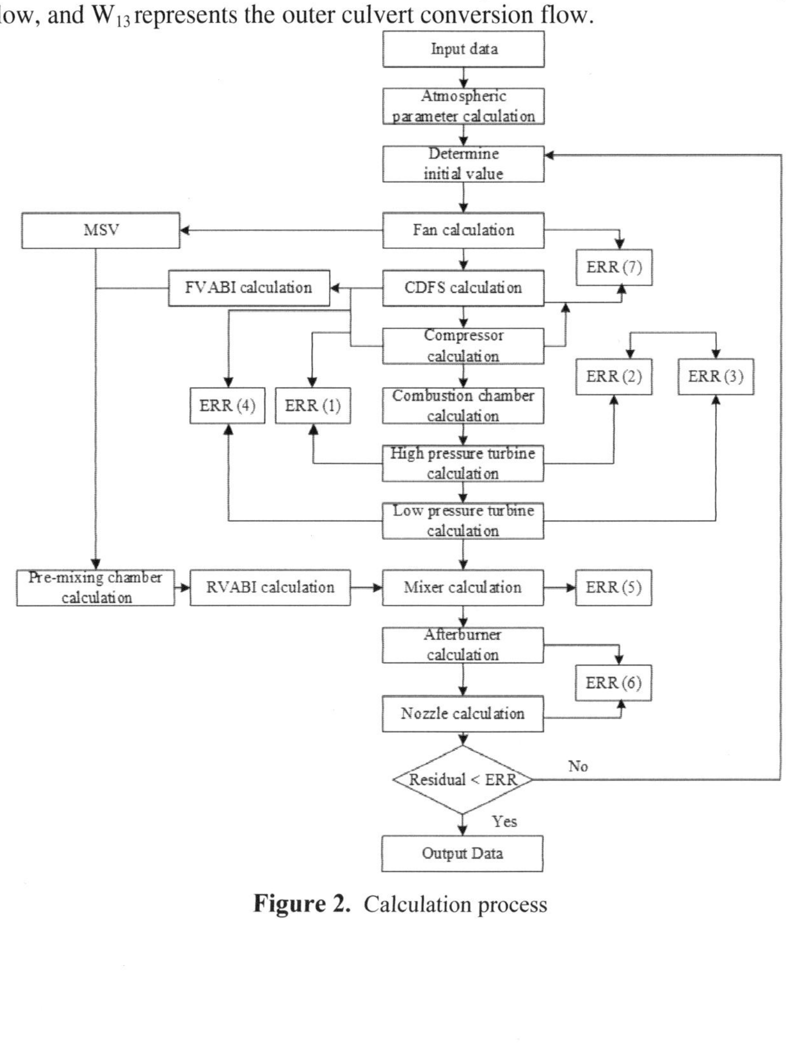

Figure 2. Calculation process

Thus, the mathematical model of the variable cycle engine is built and the simulation calculation program is written. The program calculation process is shown in Figure 2.

4. Research result

The design point parameters and control laws in the reference [7] are used for calculation. The simulation results of this paper are compared with the results in the reference, as shown in Table 1.

Table 1. Comparison of calculation results

H=0km, Ma=0.3	Reference [7]	This paper	Error
Fuel consumption rate	0.8315	0.8439	1.5%
Thrust	10597.25	10916.63	3.0%

The calculation results in this paper are basically consistent with the references [7-10]. The deviation of the calculation results is mainly due to the efficiency of the selected components and the Mach number of the design points of each section in this paper is inconsistent with the literature. Overall, the results illustrate the correctness of the simulation calculation program and can be used to further study the performance of the variable cycle engine.

The performance of the variable cycle engine is affected by adjusting the area of FVABI, RVABI, and MSV. The area of FVABI is represented by A22, the area of RVABI is represented by A25, and the area of MSV is represented by A13, in this paper, the change in area is expressed by the relative area, and the relative area is the actual area divided by the basic area. And the engine working state is H=0 km, Ma=0.3. In the single culvert mode, the control law is that the temperature before the turbine is equal to the constant, and the control law in the double culvert mode is that the relative physical speed of the low pressure rotor is equal to a constant.

4.1. Single outer culvert mode

In the single culvert mode, the MSV is off and the CDFS angle is 0°. In addition to the part to be studied can be changed, other parts are at the reference point.

4.1.1 FVABI geometry adjustment (A22)

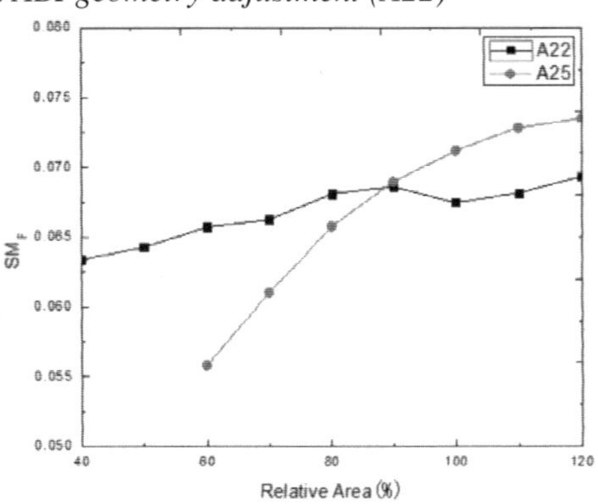

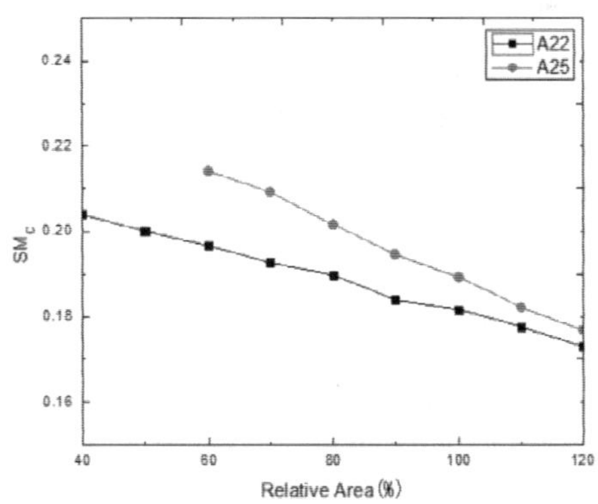

Figure 3. The fan surge margin **Figure 4.** The compressor surge margin

It can be seen from Figure 3 and Figure 4 that if other geometric components are unchanged, decreasing A22 will result in a decrease in the surge margin of the fan, and the surge margin of the compressor is slightly increased. Similarly, Figures 5, 6 and 7 show that as A22 decreases, the thrust and fuel consumption rate will increase, but the bypass ratio will decrease.

In the process of reducing A22, the external culvert flow will be reduced and the total pressure loss will be further increased, causing the operating point of the fan to move up, resulting in a decrease in the surge margin. Conversely, the air entering the core channel will increase, resulting in a slight increase in the surge margin of the compressor. That is to say, in this process, the bypass ratio will become smaller. This will result in an increase in unit thrust and fuel consumption, but the latter will vary greatly, so the total thrust and fuel consumption rate will increase.

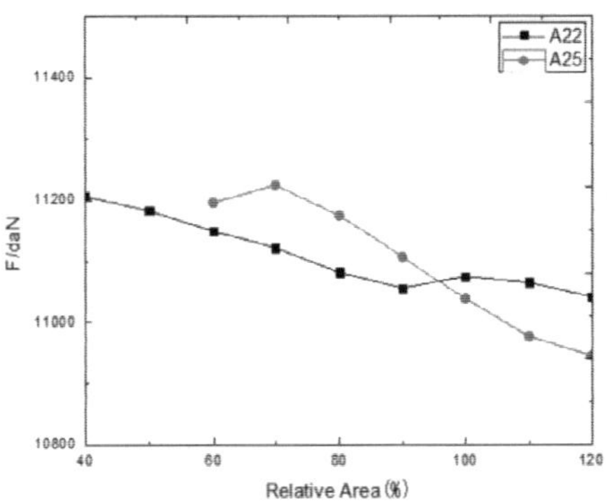

Figure 5. The engine thrust

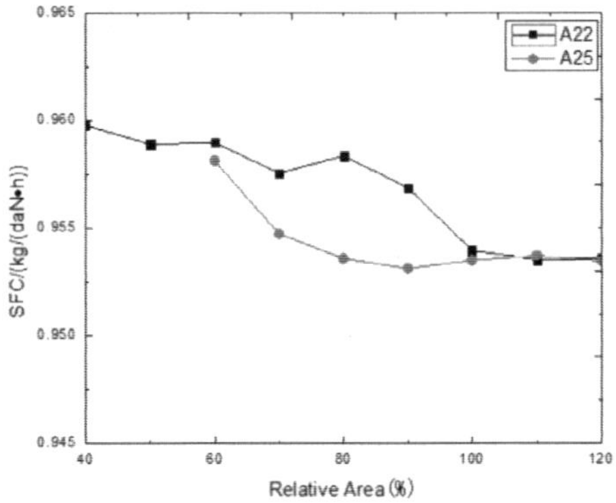

Figure 6. The fuel consumption rate

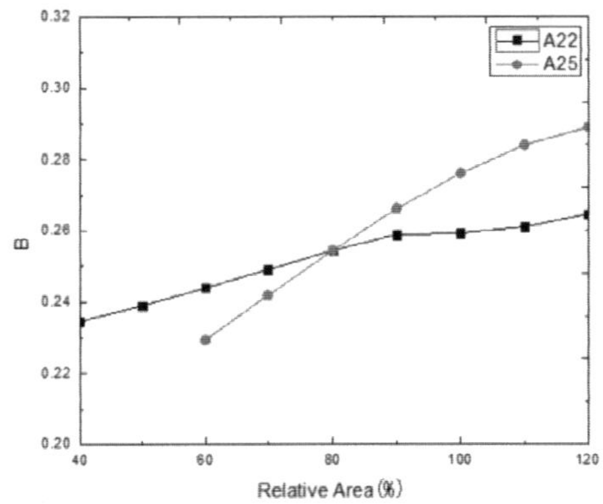

Figure 7. The bypass ratio

4.1.2 RVABI geometry adjustment (A25). As can be seen from the figures shown above, the variation trend of surge margin, bypass ratio, thrust and fuel consumption rate with A25 is consistent with that of A22. That is to say, the effect of the A25 on engine performance is similar to that of the A22.

When the A25 is reduced, the low pressure turbine drop ratio increases, but the high pressure turbine drop ratio does not change. Therefore, in order to ensure the regulation law, the low pressure turbine output power will increase, so that the low pressure rotor speed increases, causing the compressor surge margin to decrease. The decrease in A25 causes an increase in the air flowing into the core duct, causing the bypass ratio to decrease. As mentioned above, this will increase the thrust and fuel consumption rate.

4.2. Double outer culvert mode

In the single culvert mode, the MSV is open and the CDFS angle is 15°. In addition to the part to be studied can be changed, other parts are at the reference point.

4.2.1 FVABI geometry adjustment (A22). Figure 8 to Figure 12 show the effect of A22 on engine performance in double outer culvert mode. With the increase of A22, the fan surge margin does not change much, showing an increasing trend, and the compressor surge margin is reduced, the engine thrust is increased, but the fuel consumption rate and the bypass ratio are reduced.

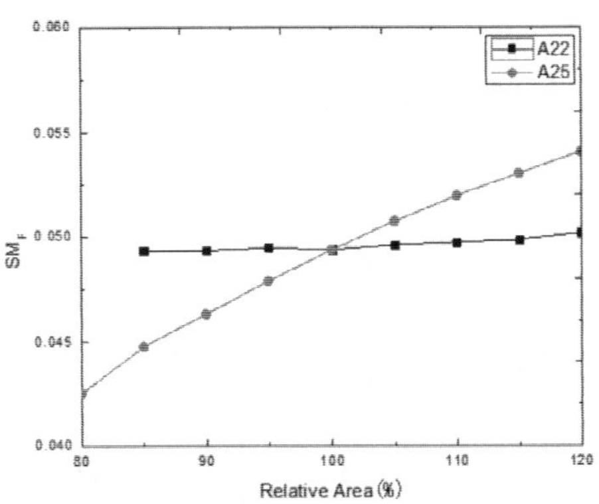

Figure 8. The fan surge margin

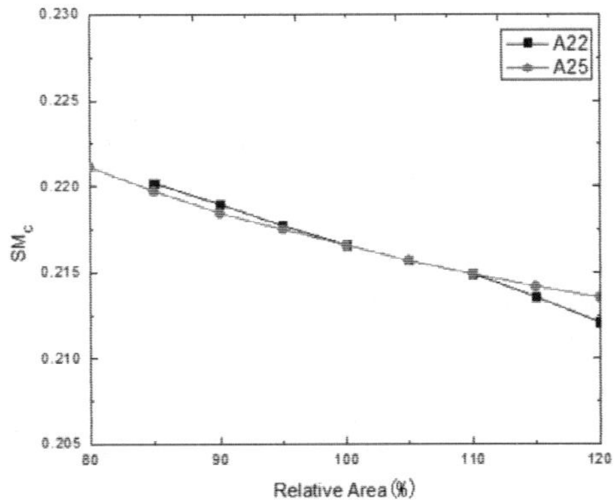

Figure 9. The compressor surge margin

When A22 is increased, the static pressure of the airflow entering the outer culvert is reduced, so that the flow rate of the second outer culvert is decreased. Conversely, the flow rate entering the compressor is increased, so the bypass ratio is decreased and the thrust is increased. At the same time, the fan moves down under the influence of back pressure, so the fan surge margin increases. The control law causes the high-speed rotor speed to increase, which reduces the compressor surge margin.

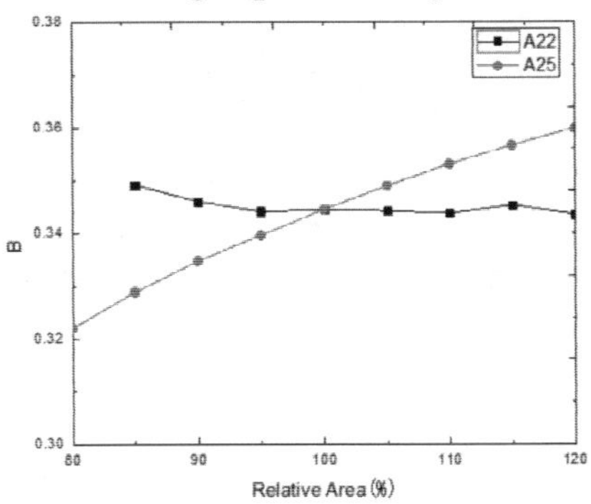

Figure 10. The engine thrust

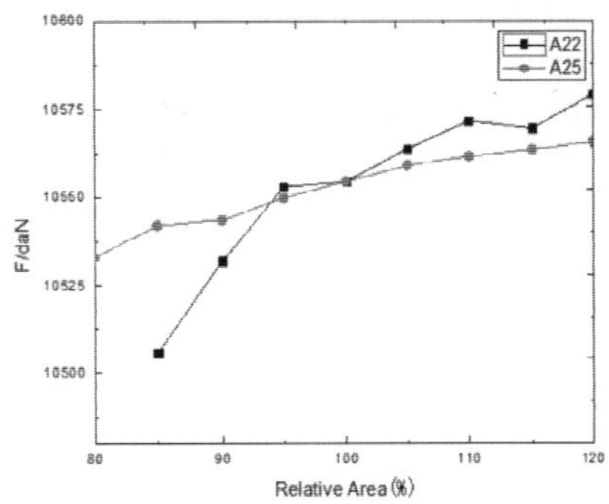

Figure 11. The fuel consumption rate

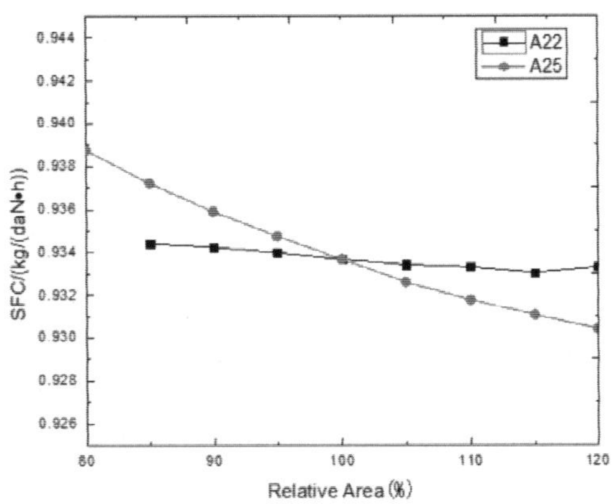

Figure 12. The bypass ratio

4.2.2 RVABI geometry adjustment (A25). As can be seen from the figures in the previous section, in the double outer culvert mode, when A25 is increased, the fan surge margin, thrust, and bypass ratio increase, but the compressor surge margin and fuel consumption rate decrease.

When A25 is increased, the low pressure turbine drop ratio is reduced, and the high speed rotor is increased in speed under the control of the regulation law, resulting in a decrease in the compressor surge margin. In addition, the working point of the fan moves down along the equal speed line, which means that the surge margin of the fan increases. At this time, the air flow of the outer duct increases, resulting in an increase in the duct ratio.

4.3. MSV geometry adjustment (A13)

The change process of MSV is also the switching process of the engine work modes. In this section, FVABI, RVABI and other components are at the reference point, and only the area change of MSV is studied.

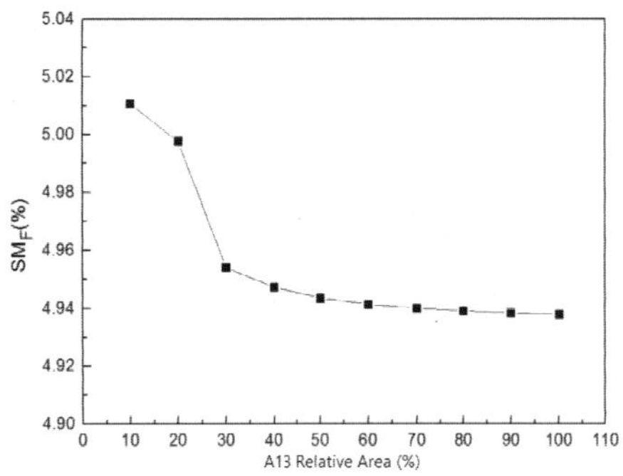

Figure 13. The fan surge margin

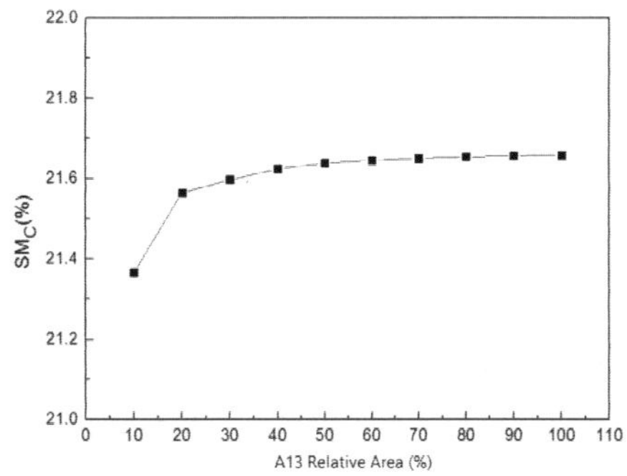

Figure 14. The compressor surge margin

As can be seen from Figure 13 to Figure 17, as the degree of closure increases, the flow area decreases, the fan surge margin and thrust increase, while the compressor surge margin, the fuel consumption ratio and the bypass ratio show a downward trend. When the area of MSV is gradually reduced to less than 30%, it can be seen that the fan surge margin and thrust rise rapidly, the compressor surge margin and the bypass ratio also decrease rapidly.

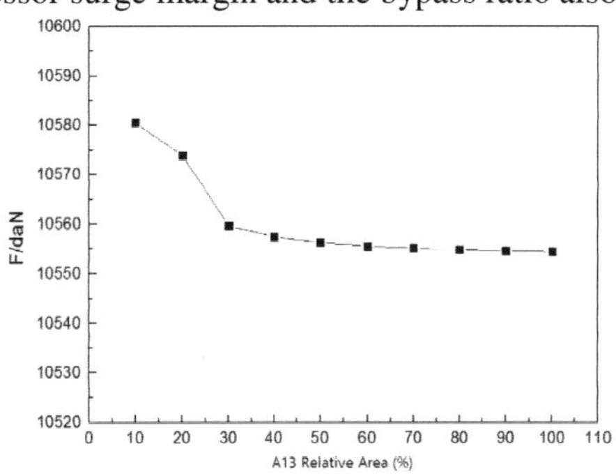

Figure 15. The engine thrust

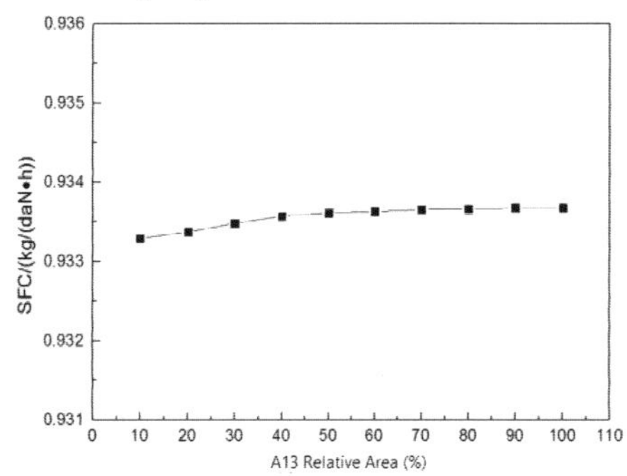

Figure 16. The fuel consumption rate

The process of reducing A13, that is, the engine is gradually approaching from the double culvert mode to the single culvert mode. At this time, the flow rate into the outer culvert is reduced, the flow rate into the CDFS is increased, and the high-speed rotor speed is increased, resulting in a decrease in the compressor surge margin. At the same time, due to the back pressure, the boost ratio of the fan decreases, and the operating point moves down, so the surge margin increases. The decrease in the flow rate of the outer culvert reduces the bypass ratio, and the magnitude of the increase in thrust becomes larger, resulting in a decrease in the fuel consumption rate.

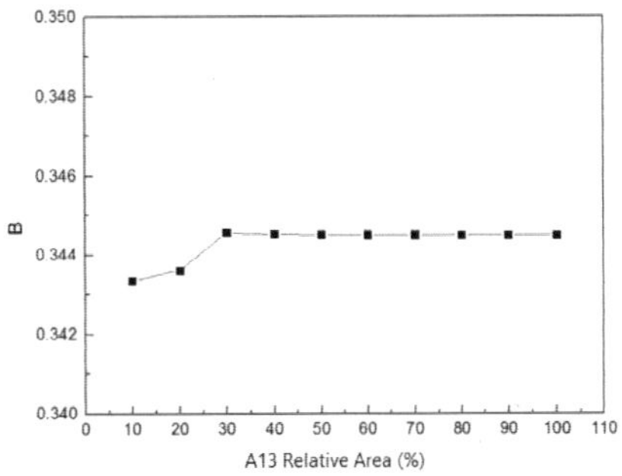

Figure 17. The bypass ratio

5. Conclusion

In this paper, the relationship between the adjustable components and the variable cycle engine performance is simulated. The following conclusions are obtained:

1. Based on the conventional turbofan engine, components such as MSV, CDFS, FVABI and RVABI are added to form a model of a double culvert variable cycle engine. The engine model and simulation calculation system are established, then the effectiveness of the model and method is verified by simulation of the reference point.

2. The effects of geometric adjustment components on engine performance under the two working modes are studied. The simulation results show that in the same work mode, the FVABI and the RVABI have the same effect on engine performance, but different from the MSV. Both FVABI and RVABI have different effects on engine performance in different modes.

References

[1] Johnson J E. Variable Cycle Engine Developments at General Electric-1955-1995. AIAA
[2] Vdoviak J W and Ebacher J A. VCE test bed engine for supersonic cruise research. NASA CP-2108
[3] French M W and Allen G L. 1981. NASA VCE test bed engine aerodynamic performance characteristics and test results. AIAA 81-1594
[4] Allan R D. 1979. General Electric Company variable cycle technology demonstrator program. AIAA 79-1311
[5] Tang Hailong and Zhang Jin. 1999 A study of object- oriented approach for aeroengine performance simulation. *Journal of Aerospace Power.* v14(4):421-424
[6] Allan R D and Joy W. 1975. Advanced supersonic propulsion study: phases III and IV-Final report .NASA CR-135236,
[7] Tang Hailong. 2006. Object-Oriented Aeroengine Performance Simulation System and its Application. Beihang university
[8] Hong Z, Wang Z, Liu Z, Zhang X and Xiang G. 2014. Variable geometry characteristics research of double bypass variable cycle engine. 35(8):2126-2135.
[9] Hong Z, Wang Z X, Liu Z W and Zhang M Y. 2016. Impact of variable area bypass injector on variable cycle engine performance. Journal of Aerospace Power. 31(12)
[10] Sullivan T J and Parker D E. 1979. Design study and performance analysis of a high-speed multistage variable-geometry fan for a variable cycle engine. NASA CR-159545

IOP Conf. Series: Materials Science and Engineering **449** (2018) 012023 doi:10.1088/1757-899X/449/1/012023

The Method of Field Simulation for Ultra-cryogenic Environment on Lunar

Ding Wenjing, Shan Weiwei and Xu Jinghao

Beijing Institute of Spacecraft Environment Engineering, Beijing, China

wjding@163.com

Abstract. Based on the requirement of ultra-cryogenic test for lunar roving vehicle and its component, theoretical analysis was performed in this paper. Gaseous helium cold box was designed and numerical simulation was conducted to optimize the design of cold box. After manufacturing, it was put into use and finished the task successfully. The Test performance of cold box shows that temperature uniformity of the cold box's during the test meets the requirement. The temperature of the cold box can be cooled down to 30K±1K, and the lowest temperature of test specimen is down to -205℃.

1. Introduction

The temperature of lunar ground changes with day and night period, and deepest temperature can reach to -180℃. As far as lunar roving vehicle concerned, the radiant heat comes from solar radiation, lunar reflect radiation and infrared radiation from lunar surface. Lunar roving vehicle receives the radiation during the daylight. However, it receives very little heat radiation during the night. What's more, specific heat of the lunar surface is very small. Hence, when it comes to the night, lunar roving vehicle has to survive in the environment with the temperature of -180℃.

Based on the above analysis, the environment is a big challenge which encountered by lunar roving vehicle on the lunar[1~4]. Lunar roving vehicle must work successfully whenever day and night. Hence, thermal vacuum test must be conducted to testify the performance of lunar roving vehicle.

According to the hypothesis, the temperature of test specimen during thermal vacuum test should maintain in the range from -200℃~+140℃.

Shrouds cooled by liquid nitrogen are designed to simulate cold black environment in thermal vacuum facility which can make the temperature of test specimen down to -100℃~-160℃. Hence, traditional nitrogen shroud can not meet the requirement of thermal vacuum test for lunar roving vehicle and its parts. The new test facility to conduct deep cryogenic test must be developed to solve the tough problem.

2. Simulation method for cryogenic environment of lunar surface

Deep cryogenic test should take new cold source to simulate the environment of lunar surface. Gaseous helium cryogenic system is the new cold source instead of liquid nitrogen system. The principle of vacuum thermal test is shown in Figure 1. It consists of vacuum chamber, liquid nitrogen shroud, gaseous helium cold box, vacuum pumping system, nitrogen system, and gaseous helium cryogenic system. Gaseous helium shrouds and gaseous helium cryogenic system are designed to achieve deep cryogenic environment.

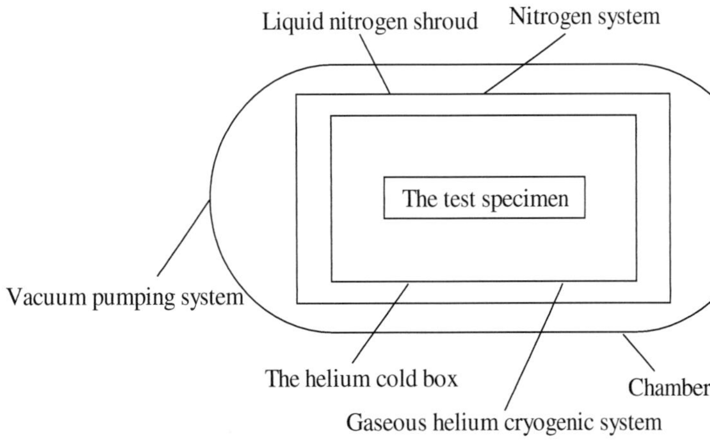

Figure 1. Subsystem of cryogenic simulation chamber for moon

3. The study of the helium cold box

3.1. The structure of the helium cold box
The helium cold box is the key equipment of the cryogenic simulation system. The structure of the cold box is shown in Figure 2.

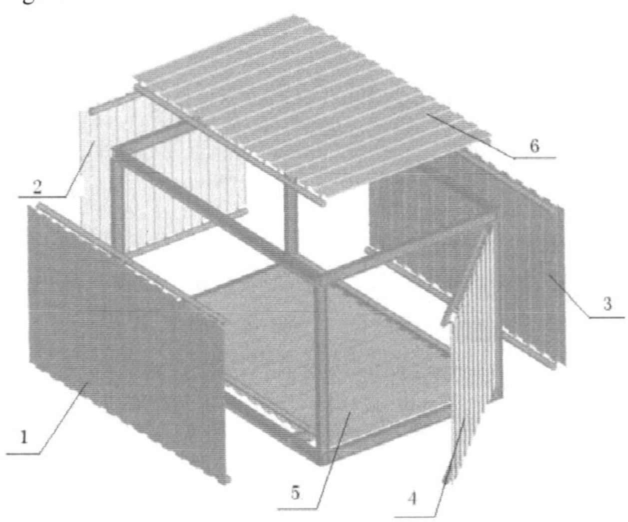

Figure 2. The structure of the cold box
1~6. 1#~6# Cold Shroud

3.2. Gas helium cold box numerical computation
The software called Flowmaster is performed to establish the model and simulate the heat transfer of gaseous helium filling in the cold box[5]. The temperature distribution curve can be achieved.

The fluid structure of the cold box is shown in Figure 3 and Figure 4.

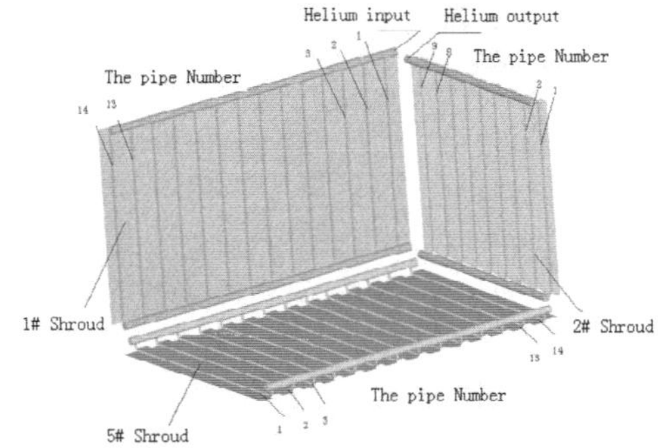

Figure 3. The pipe number of 1, 2, 5 Shroud

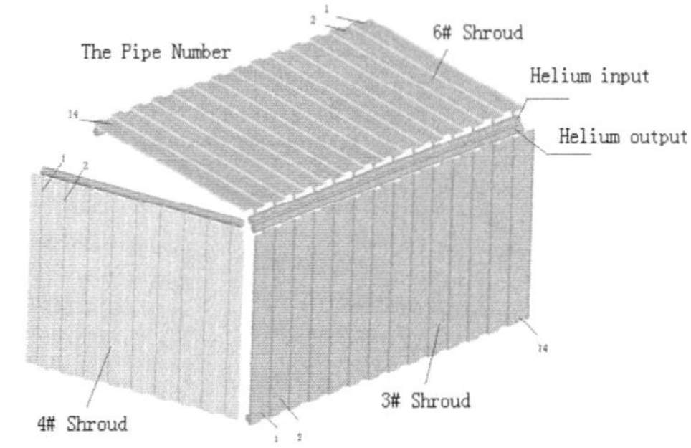

Figure 4. The pipe number of 3, 4, 6 Shroud

According to the technical specification of deep cryogenic system, the boundary of the model is as follows:

a. Working fluid of the cold box is gaseous helium;
b. The volumetric flux is 900 Nm^3/h;
c. Working pressure of gaseous helium is 0.15MPa;
d. Inlet temperature of gaseous helium is 28.7K.

Numerical model of the cold box is shown in Figure 5.

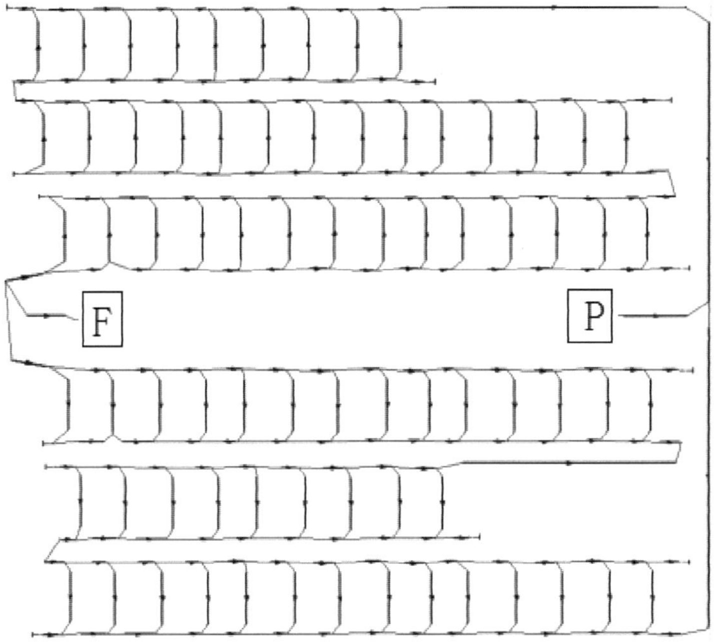

Figure 5. The numerical model of the cold box

Temperature distribution of branch tube in each cryopanel is shown in Figure 6, and the accurate result is shown in table 1.

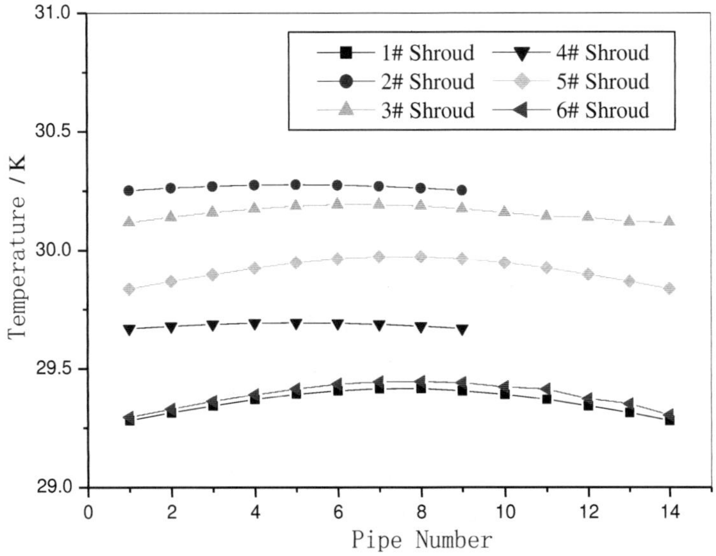

Figure 6. The distributing of the helium cold box temperature

Table 1. The temperature of the helium cold box /k

Pipe Number	1# Shroud	2# Shroud	3# Shroud	4# Shroud	5# Shroud	6# Shroud
1	29.28	30.25	30.12	29.67	29.84	29.30
2	29.31	30.26	30.14	29.68	29.87	29.33
3	29.34	30.27	30.16	29.69	29.90	29.36
4	29.37	30.28	30.18	29.69	29.92	29.39
5	29.39	30.28	30.19	29.69	29.95	29.42
6	29.41	30.27	30.19	29.69	29.96	29.44
7	29.42	30.27	30.19	29.69	29.97	29.44
8	29.42	30.26	30.19	29.68	29.97	29.45
9	29.41	30.25	30.18	29.67	29.96	29.44
10	29.39	--	30.16	--	29.95	29.42
11	29.37	--	30.14	--	29.92	29.41
12	29.34	--	30.14	--	29.90	29.37
13	29.31	--	30.12	--	29.87	29.35
14	29.28	--	30.12	--	29.84	29.30

From the numerical computation, the total pressure loss of the cold box is 0.01bar, and the maximum temperature gradient 1K. Hence, above results can meet the requirement of technical specification.

4. The experimental test result and discussion

IOP Conf. Series: Materials Science and Engineering **449** (2018) 012023 doi:10.1088/1757-899X/449/1/012023

4.1. The test system

The specimen was put into the helium cold box, the position of the helium cold box in the vacuum chamber is shown in Figure 7. The material of the test specimen is aluminum, of which the size is 500mm ×400mm ×5mm.The specimen emissivity is 0.07. The heat transfer between the specimen and the helium cold box only depends on the radiation.

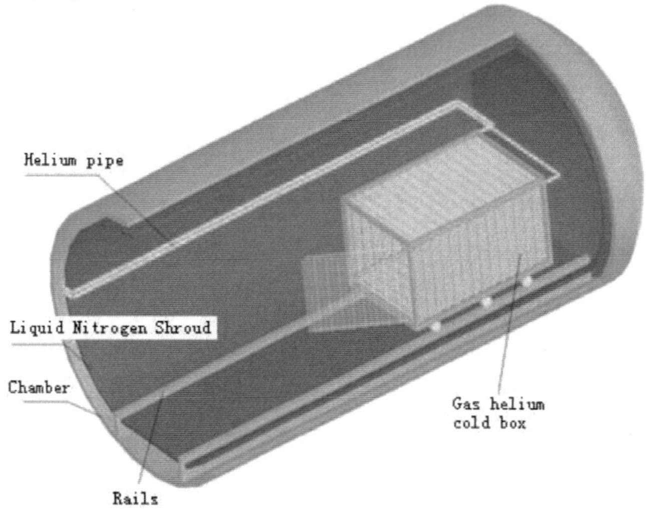

Figure 7. The position of the helium cold box in the vacuum chamber

4.2. The result and discussion

Figure 8 shows the curve of temperature versus time for the helium cold box. Figure 9 shows the specimen temperature versus experiment test time.

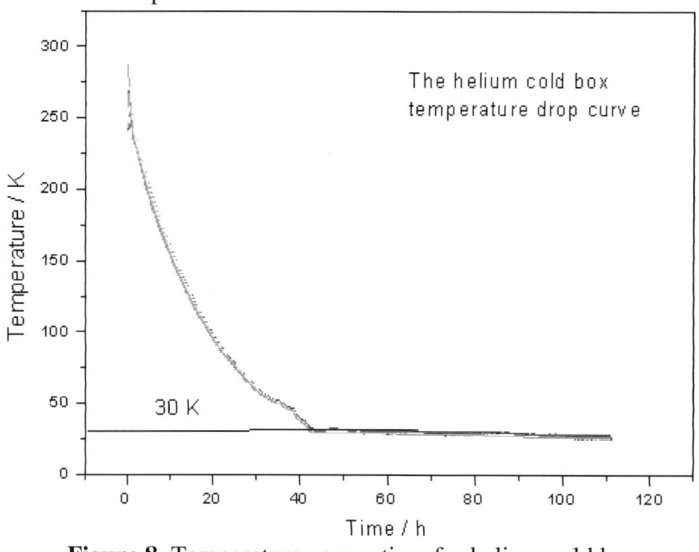

Figure 8. Temperature versus time for helium cold box

IOP Conf. Series: Materials Science and Engineering **449** (2018) 012023 doi:10.1088/1757-899X/449/1/012023

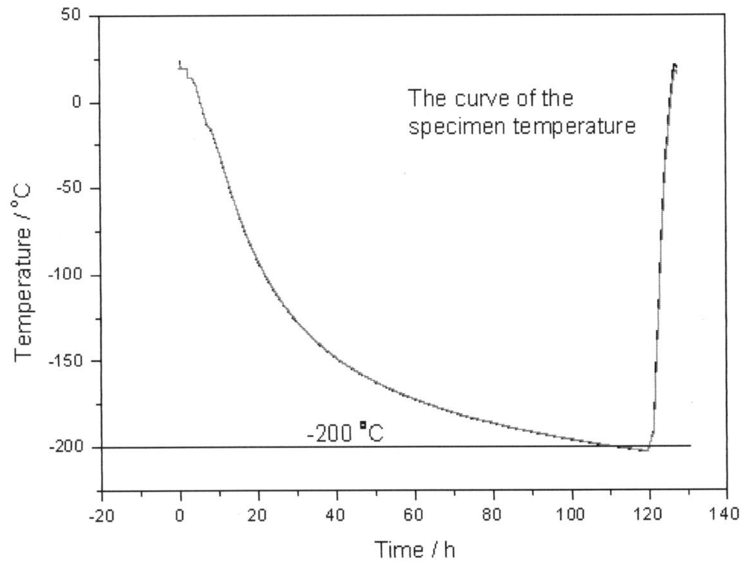

Figure9. The curve of specimen temperature

From Figure 8, it can be seen that the lowest helium cold box temperature is 30K±1K, it takes about 42hrs from 300K to 30K. Figure 9 shows the relationship of the specimen temperature versus the test process, the specimen temperature can reach -205℃, it takes about 116hrs from 27℃ to -205℃.

From above calculation and analysis, we can see that the temperature of the test specimen temperature can meet the requirement of the deep space exploration.

5. Conclusion

The gaseous helium cryogenic system can be taken as the cold source of the cryogenic environment simulated for deep space exploration. The lowest temperature of cold box is 30K±1K, and the specimen temperature can reach -205℃. The test certifies that the cold box can meet the needs of cryogenic environment test's requirement for the deep space detection.

Reference

[1] Seminara J L. Lunar surface simulation [J]. AIAA PAPER 67-866

[2] Jing Jiarong, Zhang Lixin. The simulation technology of the non-earth orbit spacecraft thermal environment in deep space. Spacecraft Environment Engineering. 2008 25(5)

[3] Ye Peijian Peng Jing, Deep Space Exploration and Its Prospect in China. Spacecraft Environment Engineering. 2006 8(10)

[4] Lwslie L Karafiath. Friction between solids and simulated lunar soils in ultrahigh vacuum and its significance for the design of lunar roving vehicles[J]

[5] Tao Wenquan. Heat transfer. (1998).

Experimental Research on the Helicopter Sliding Stability after Ditching

Chen Yinghua[1], Wu Ximing[2] and Yuan Libin[1]

[1]The China Helicopter of Research and Development Institute, Jingdezhen, 333001, China;
[2]Aviation Industry Corporation of China, Beijing, 100022, China;

Corresponding author:cyhj2336@163.com

Abstract. As the application of the helicopter increases rapidly, more and more attention is paid to the safety. Thus the survivability and crashworthiness has been an important assessment items for the helicopter design and flight certification. In this paper, a newly designed ditching experiment with a helicopter model scaled 1:8 was conducted to attain the dynamic response of loads and especially the posture. Parameters such as weight, bladder pressure, lateral wind and ditching velocity were set in different groups to investigate the effect on the sliding stability. Based on the result, comparison and analysis was carried out and some conclusions were given. Meanwhile, the method and process of the experiment was presented in detail and proved an effective and available reference for the helicopter flight certification.

1. Introduction

Ditching is an emergency measure when the helicopter is forced to land on water as the last possibility for protecting the passenger safety, which is divided into planed and un-planed cases. According to statistics [1], at least 98 events of helicopter ditching occurred between 1971 and 1992 in the worldwide, which sadly leaded to 338 deaths. In order to improve crashworthiness and safety, a regularity FAR29.801[2] that aim at conducting the rotor aircraft design was promoted by FAA in 1977.In which , the process and methods for the certification of the helicopter ditching such as numerical simulation and model test were prescribed. Similarly in China, the CCAR29[3] was published and refined for several times, in which some requirements were extra added for the stability of slide, balance and floatation after ditching.

For the experimental investigation of helicopter ditching, it has been started and developed for a long time. In comparison with the full-size model, a Scaled model was normally applied to decrease the expensive cost, which primarily based on the Froude similarity principle. In America, Bell-22 helicopter[4] with air bladder was scaly modeled and a series of ditching tests were performed to study the effect on the loads and sliding stability acted by float system. Based on which, a standard ditching process was established, but without taking consider of the effect by some parameters. H-46 helicopter[5] was modeled by a full-scale and scaled as 1:8 model respectively, and then different number and location of float bladders were set to analysis the effect on ditching and stability in different sea conditions. The Waist and England landing and special aircraft company[6] focused on the ditching load and float characteristic, taking the 2/3 lift of rotor into consideration with a scaled model.

In China, the first ditching test was recorded when Z8 helicopter was modeled and dropped into water from a designed height to attain the curve of ditching overload and static characteristic in 1982, however, the influence of float bladder was ignored. In 2008,a helicopter model[7] scaled 1:5 with float bladders was tested to statistic and analysis the ditching overload and pressure in different velocity and posture. In 2016, the civil helicopter AC313 [8] was modeled to investigate the dynamic response in different sea conditions and initial postures. The process and steps of experiment was introduced in detail and the results was analyzed and compare with Russia's .Based on which, the experimental ability is proved and technologic references were provided for further study. After that, the same model[9] was employed and the floatation experiment was carried out to study the lateral stability. A mathematic method of helicopter lateral stability calculation was established for the first time, considering the influence by the fuel liquid level correction. it was proved that the calculation was in good coincide with test results. But the method was not considering the coupling effect of longitudinal motion. At present, the research on the sliding stability of helicopter is just started and rarely involved in the field so far.

In this paper, a ditching test of a civil helicopter model scaled 1:8 with float bladders was newly designed and carried out. Based on which, the posture and sliding stability after ditching with different parameters such as weight, lateral wind, ditching velocity and bladder pressure were investigated. In conclusion, some advice and consult were given for the sliding stability design, and meanwhile the test process was proved effective and available for the certification of helicopter ditching.

2. Test facility

The facility used in the ditching test included a towing tank, a guide track, wave maker and measure equipment, some parameters of facility are shown as following in table 1.

- Size of the towing tank: length 510m, width 6.5m, while the depth of the tank and water is 6.8m and 5.0m respectively.
- Performance of the track: range of velocity is 0.1m/s～22m/s while the accuracy is more than 0.2%.
- Property of wave maker: range of regular wavelength is 0.5m～15m, range of regular wave height is 0.03m～0.5m; all irregular waves with large wavelength were available.
- Methods of measurement: an automatic data collection and analysis system.

Table 1. Measure equipment and justification list

No.	Name	Type	Range	Accuracy
1	High speed track	GT02	0.1~22m/s	0.2%
2	Inertial measurement unit	IMU610H	±90°	Dynamic accuracy<0.3°
3	PCB acceleration sensor	LW7646	±10g	0.5%
4	Pressure sensor	XPM10A1	2bar	0.25%FS
5	Micro data collection system	DRS104	±10VDC	0.5%
6	Temperature measure system	SWP-C80	-10~+55℃	0.1%
7	Digital electronic scale	DR-150	150kg	0.02kg
8	Platform scale	TGT-100	30Kg	0.02kg
9	Digital angle ruler	DXL360S	±90°	0.01°
10	Mass and inertia measure system	ZGT-200	100kg.m2	1%
11	Wave probe	NHWM-4	500mm	0.5%

In order to record the vertical overload in the cockpit, overloads of the center of gravity in three directions, and the pressure distribution at the bottom, a accelerometer and a inertia measurement unit were located in the inner of model and pressure sensors were located outside the bottom of the model. More details were shown as following:

a) Overload sensor

An acceleration sensor was installed near the header (300mm away from the header) to measure the overload in the cockpit.

b) Inertia measurement unit

The inertia measurement unit was located at the center of gravity (793mm away from the header) to record the overloads in three directions.

c) Pressure sensor

Ten sensors were used to measure the pressure at the bottom of the model, which are shown in figure 1.

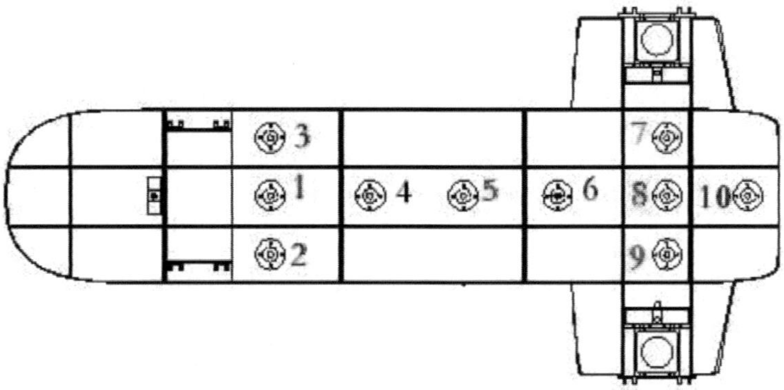

Figure 1. Locations of the pressure sensors at the bottom of model

3. Model and the principle of similarity

According to the References, it is strongly recommended to satisfy a similar principle between the test model and real structure. Mostly, the Froude similarity principle is applied, in which three similarity conditions should be followed.

Geometry similarity: Scale of all geometric sizes between the model and real helicopter is set exactly same.

Kinematics similarity: It is also called the time similarity. As required, the velocity and acceleration vectors of the two corresponding points on the model and the helicopter should be unanimously scaled, which can be also recognized as some combinations of time scale and geometry scale.

Dynamic similarity: The force and moment vectors of the two points at the corresponding location should be scaled the same.

The parameters such as length, Velocity, weight and inertia of both the modal and real helicopter are listed in table 2.

Table 2. The scale of parameters between model and real helicopter

Parameters	Real value	Scale	Model value
Length	L	λ	λ L
Inertia	I	λ^5	λ^5 I
Weight	m	λ^3	λ^3 m
Time	t	$\sqrt{\lambda}$	$\sqrt{\lambda}$ t
Velocity	V	$\sqrt{\lambda}$	$\sqrt{\lambda}$ V

Linear Acceleration	a	1	a
Pressure	P	λ	λ P

The Civil helicopter AC313 was employed for the test and the model scale was designed to 1:8 according to the size of towing tank and available counter weight, and the weight of the model was manufactured to 25.4kg finally, which is shown in figure 2.

Figure 2. The model of the ditching test

4. The supporter installation

The supporter was used for test model installation and posture adjustment. To be installed, a pipe instrument was firstly fixed on the measurement bridge platform of the track with the bolts, and then the angle adjustment set was installed at the bottom of the pipe. After that, the top of the supporter was locked together with the pipe in the connecting holes by some bolts. In order to check the liability of supporting system, a pretest that aim at simulating the motivations of model by lifting or dropping the measure bridge will be carried out, while a solid whose weight is slightly bigger than the model was suspended by a electric magnetic release after energization.

5. The test article installation

Before installed on the bracket, the model was adjusted to be horizontal on a platform. Adjusted the pitch angle, roll angle, and yaw angle to the right value, the model was connected to the bracket to make sure that the two plugs on the bracket match the receptacle on the top of model precisely. The ring on the top of model was inserted into the hook of the release and locked tightly after checking. When installation, a digital angle ruler was applied to set the roll angle and pitch angle of the modal bottom while the yaw angle was set by a mark line in the angle controller. The adjustment was preceded after installation until the angles satisfied the range as required.

6. Test in advance

To depart from the dock freely, the model was elevated to a suitable height by the measurement bridge and then the track was pushed out of the dock and aboard beside after the wave termination. The model was released to the lowest position (visual check) and then lifted to the certain height according to the designed vertical velocity. The electric magnetic release was then opened and the free fall and ditching was observed to check if any resistance exists. Before the formal test, it is recommended to perform the test for several times to make sure the stability of towing system. The installation of the system is shown as Figure 3.

IOP Conf. Series: Materials Science and Engineering **449** (2018) 012024 doi:10.1088/1757-899X/449/1/012024

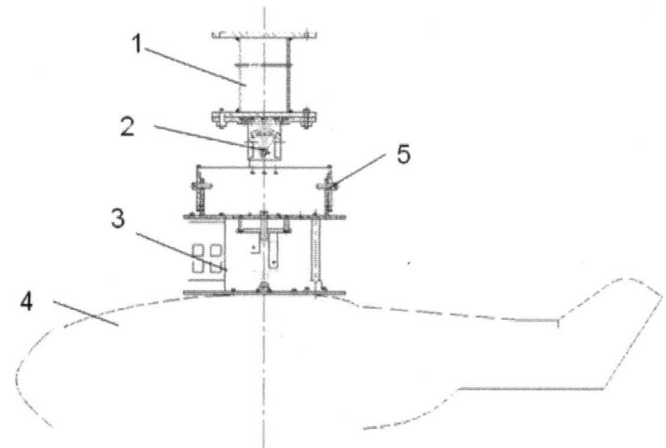

1-track;2-posture controller;3-bracket;4-helicopter model;5-measure bridge;

Figure 3. Installation of towing system

7. Test cases

In order to compare and analysis the effect, 4 groups of cases were designed with 4 parameters including the vertical velocity, weight, lateral wind and bladder pressure. In each group only one parameter was changed while others kept the same. Details about the test settings are shown in table3 to table 6 as following.

Table 3. Test settings with different vertical velocities (group 1#)

No.	Weight (kg)	β (°)	α (°)	γ (°)	V (m/s)	Vz (m/s)	W (m/s)	P_b (kPa)
1						0.40		
2	25.4	0	8	0	5.48	0.53	0	30
3						0.71		

Table 4. Test settings with different velocities of lateral wind (group 2#)

No.	Weight (kg)	β (°)	α (°)	γ (°)	V (m/s)	Vz (m/s)	W (m/s)	P_b (kPa)
1							3.50	
2	25.4	0	8	0	5.48	0.53	4.50	30
3							5.50	

Table 5. Test settings with different weights of model (group 3#)

No.	Weight (kg)	β (°)	α (°)	γ (°)	V (m/s)	Vz (m/s)	W (m/s)	P_b (kPa)
1	25.4							
2	30.4	0	8	0	5.48	0.53	0	30
3	35.4							

Table 6. Test settings with different pressure of air bladder (group 4#)

No.	Weight (kg)	β (°)	α (°)	γ (°)	V (m/s)	Vz (m/s)	W (m/s)	P_b (kPa)
1								20
2	25.4	0	8	0	5.48	0.53	0	30
3								40

8. Test process

Ten steps were included in the process of ditching test.

1)Installation

The bracket was fixed on the platform of the high-speed track and measure bridge, and then fastened to right position.

The pressure of the four bladders were checked and adjusted according to testing requirement (30±2kPa). The model was weighted and installed if the weight was in the range of tolerance. If not, the model would be dried to make the weight to fulfill the range before installed. The angles were adjusted to the value as required and the model was fixed on the bracket at last.

2)The posture adjustment

Angles were measured and finally set to the required value by a digital angle ruler if necessary.

The height of towing position was calculated according to the speed required, multiple the button to adjust the position to ensure the height of model out of water.

3)Camera system

Power on the camera and keep it standby.

4) Final check before test

- Check the model if the installation is fastened and liable.
- Check the camera
- Check the track system and make a safety check around and bottom of the track to make sure that track system run normally and there is no dangerous shelters or other irrelevant personnel.

5) Start the track

If it was okay after checking, an order was given to the track driver to start the track by the test director.

6) Start the micro data collector and camera system

When the track was speeding up, the camera and the data collector was started to capture and record the data from the sensors.

7) Tow

After the speed of track is up to the certain value and kept constant for seconds, the tow toggle was started up. At the same time, the electric magnetic release was activated and the model was released and departed from the bracket, which ditched into water and slide after free fall.

8) Stop the track

After the model ditched into water, track was decelerated and finally stopped.

9) Recycle the model and return to dock

After test, the model was picked up from the water by a steel hook and put to the hook on the hoist track. Once hoist track worked, model rise up and returned to the bottom of dock, while the towing track was restarted and back to dock finally.

10) Data collection and processing

After the model returned to dock, all data were collected by the micro data collecting system which was installed in the model, were transported and saved in the computer for further analysis.

9. Test result

The model experienced four stages totally during the test: acceleration, dropping, ditching and sliding. Photos at different moment were captured and shown in figure 4 as following.

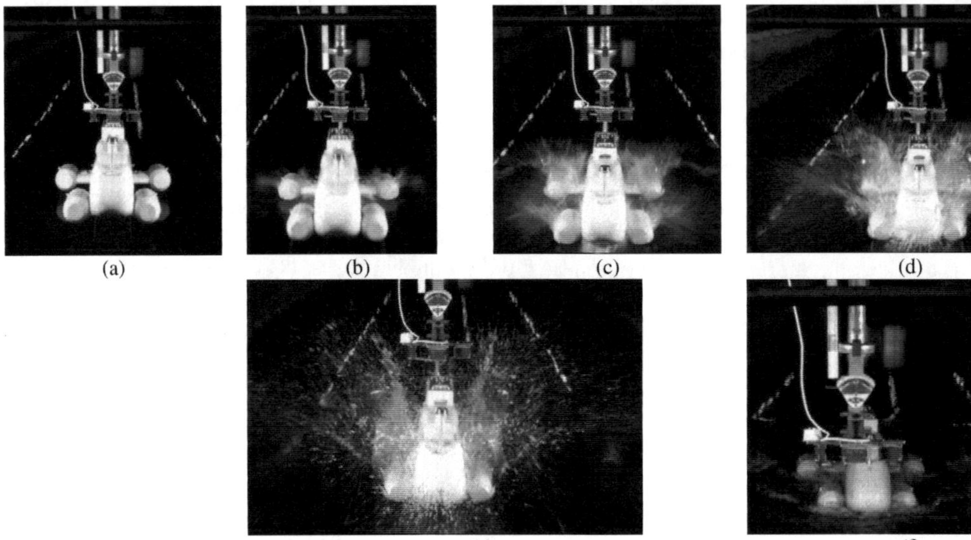
(a) (b) (c) (d) (e) (f)

Figure 4. Photos of the helicopter model at different moment in the experiment

After the ditching experiment, the data of each group was collected and filtered with frequency 50HZ and processed. Time history curves of pitch angle, roll angle, the vertical overload of C.G in each group were drawn and posted in figure 5 to figure 7 respectively.

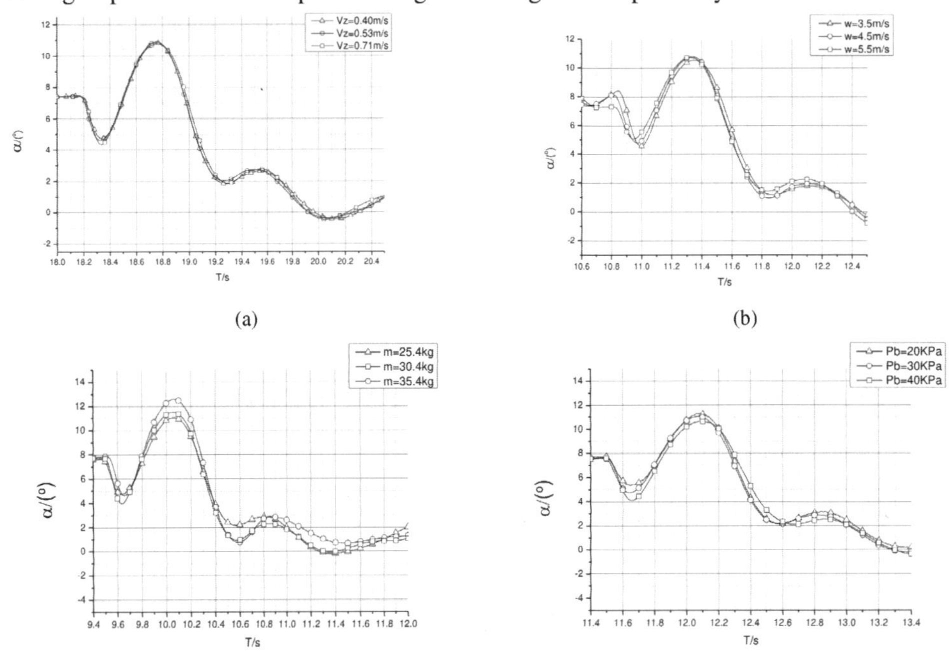

(a) (b) (c) (d)

Figure 5. Time history curves of Pitch angle in different groups

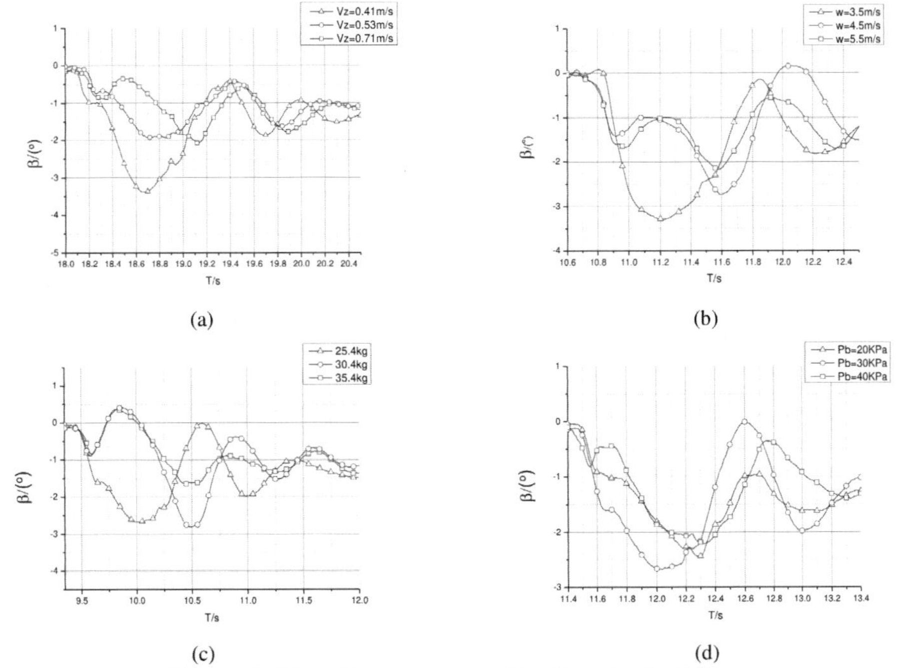

Figure 6 . Time history curves of Roll angles in different groups

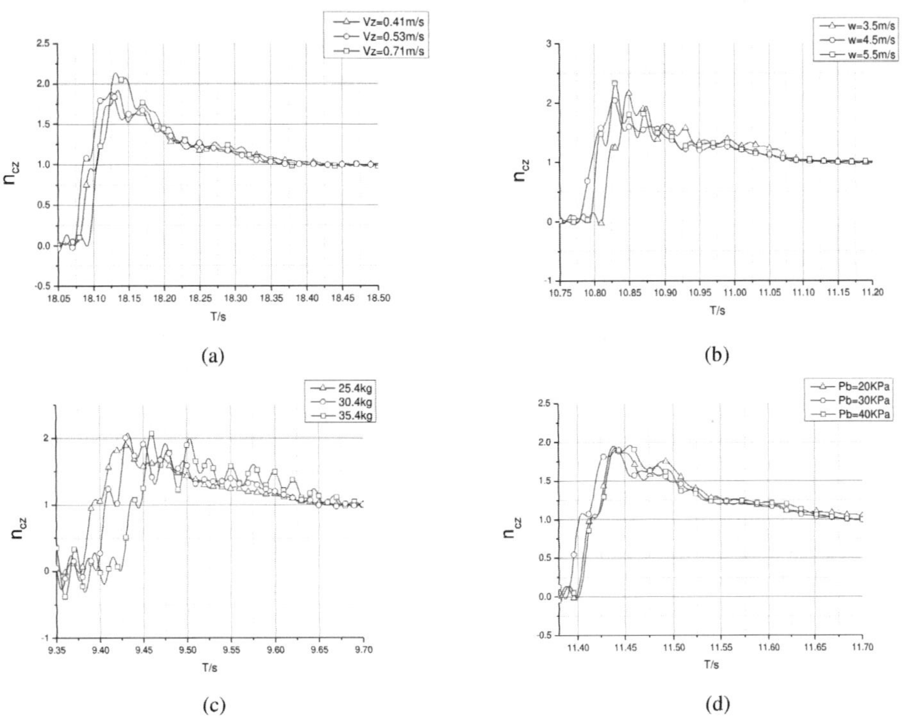

Figure 7. Time history curves of vertical overload of C.G in different groups

From the results, it can be concluded that:

a) As shown in figure 5 to figure7, the general variation tendency of the pitching and rolling angle as well as the vertical overload of C.G in each group is roughly the same. The pitching and rolling angle

enters a concussion at first after ditching and finally recovers to stable rapidly in approximately 2 seconds, which proves that the pitching and rolling movements of the helicopter is stable during the sliding period after the ditching.
b) For the pitching angle in figure 5, the first trough is obviously relative to the vertical velocity and the weight as well as the pressure of bladder. A higher velocity, greater weight and higher pressure of bladder are more likely to generate a lower trough. Nevertheless, because the time of shaking duration of curves are much close in each groups in figure 5, it is speculated that the dynamic stability of pitching is rarely effected by those parameters.
c) Other than pitching angle, the curve shapes of the roll angle are much different from each other in each group. Because of the inevitable influence by some factors, such as the initial roll angle of model installed, the posture at the moment when the model touched the water and the slight fluctuation of water surface, the time history of roll angle is full of randomness. However, the converge speed of roll angle is inclined to be faster in the groups of 35.4 kg and 20Kpa respectively, which shows that a larger weight and lower pressure of bladder are better for the lateral dynamic stability.
d) In the figure7, it is figured that a higher velocity, larger weight and higher bladder pressure bring out a heavier overload in Z direction of center of gravity. However, the vibration of curves is finally terminated after the same seconds in each group.

10. Summary

As the application of the helicopter increases rapidly, more and more attention is paid to the safety. Thus the survivability and crashworthiness has been an important assessment items for the helicopter design. In this paper, a ditching experiment with a model of helicopter scaled 1:8 and float bladders was designed and carried out, in which different parameters were set for comparison in groups. The dynamic response curves of loads and especially the posture were captured. Based on the results, the influence by different parameters was discussed. It is concluded that the model of helicopter was statically and dynamically stable in both lateral and longitudinal movements and some advices are therefore given for the design. Moreover the design and process of the experiment was introduced detailedly and proved effective and available, which probably provides the certification of helicopter ditching with references.

Acknowledgments

Supported by the project of Civil Helicopter pre-research (MJ-2014-F-15).

Reference

[1] Westland Helicopters Limited, “A review of UK Military and World Civil Helicopter Water Impacts over the Period 1971-1992”, Stress department report SDR146, November 1993.
[2] FAA. Airworthiness Standards: Transport Category Rotorcraft[S].Washington DC, 2011.
[3] Civil Aviation Administration of China. Chinese civil aviation regulations Parts 29 [S].CCAR29R1-88.Beijing:Civil Aviation Administration of China,2002:24-26.
[4] Alcedo, A.M., Design and Test of Float Landing Gear System for Helicopters,1979 Regional Lichten Award Paper Presented at the American Helicopter Society Southwest Regional Meeting, Fort Worth, TX,1979.
[5] M . J . Reilly,Lightweight Emergency Flotation System For The CH-46 Helicopter NADC-791-69-60, 1981.
[6] Wilson F T, Tucker R C S.Ditching and Floation Characteristics of the EH101 Helicopter[C].Thirteenth European Rotorcraft Forum,1987,Vol.2: 79.1-12.
[7] Li Mingqi. The subscale test and simulation on water impacting characteristic for emergency flotation system[D].Nanjing:Nanjing University of Aeronautics & Astronautics,2008.
[8] Wang Zhengzhong, Chen Lixia, Suo Qian,Ma Yujie.Test Research on Helicopter Ditching Load[J].Journal of Nanjing University of Aeronautics & Astronautics. 2017,49(2):258-263.
[9] Wang Zhengzhong, Ma Yujie, Jiang Ting. Computation and Test Validation of Lateral Stability for Helicopter on Water[J].Helicopter Technique,173(4):1-7.

IOP Conf. Series: Materials Science and Engineering **449** (2018) 012025 doi:10.1088/1757-899X/449/1/012025

Prevention of Hazardous Failure of the Turbine Rotor Due to Its Overspeed

Y.A. Nozhnitsky[1] and A.N. Servetnik[1]

[1]Central Institute of Aviation Motors (CIAM), 2, Aviamotornaya st., Moscow 111116, Russia

nozhnitsky@ciam.ru

Abstract. Causes of defects that can lead to the turbine rotor overspeed and subsequent hazardous consequences are considered. Methods of preventing unacceptable overspeed and confirmation of the sufficient strength of the turbine rotor are analyzed. Particular attention is paid to the developed method of calculating burst speed of the rotor. A comparison of the numerical and experimental data on basis of relation between radial displacements during spin tests, burst speed, high-speed image acquisition, fracture origin and fracture confirms possibility to use the developed method of calculations. The presented method allows to adjust the value of a disc rotation rate during certification spin tests to ensure that a disc with the worst mechanical properties will not burst in operation and to take into account differences in the design and loading conditions between the disc for spin tests and disc for operation.

If the identification of model's material parameters was performed based on disc's tensile tests and spin tests, then for a disc of another design from the same material only calculations can be used.

1. Introduction.

Fracture of the turbine rotor during overspeed is one of the most potentially hazardous defects of a gas turbine engine (GTE). Only in rare cases (for some auxiliary engines) it is possible to reliably exclude hazardous consequences (in particular, escape of fragments with high kinetic energy through the engine cases; uncontrolled fire; failure of engine mounts, leading to separation of the engine from the aircraft) when the turbomachine rotor (disc, blisk, bling, drum, centrifugal wheel, hereinafter referred to as a disc) fails.

This article examines some issues related to providing and confirming load-carrying capacity of a disc, taking into account possible overspeed. Turbomachine disc must have sufficient strength to withstand both normal operation conditions and conditions when maximum permissible operating speed is exceeded, including conditions after failure of the most critical component with respect to rotor overspeed or system or a combination of this failure with any failure of a component or system, which cannot be detected during a routine preflight check or during normal flight operation. It should be confirmed that even a disc made of a material with the combination of most adverse material properties and dimensional tolerances has sufficient strength [1-3].

2. Causes of a turbine rotor overspeed and measures to prevent unacceptable overspeed.

The turbomachine disc can experience overspeed, for example, due to a defect in engine control system, which results in an increased amount of fuel being supplied to combustion chamber; due to failure of a reduction gear connecting turbine rotor and power consumer; failure of fan's (compressor's) impeller, etc. However, in practice, the most difficult task is to confirm sufficient strength of the turbine disc after breaking kinematic connection with a power consumer as a result of failure, displacement or disconnection of shafts. In this case, the overspeed of the turbine disc before its failure can occur in a fraction of a second and reaction speed of a pilot does not allow to take any action in such a short time. At the same time, when the disc is fractured under the conditions of overspeed, the kinetic energy of the fragments is especially high. Therefore, in accordance with requirements of airworthiness standards [1-3], fracture, disconnection or displacement of shafts should not lead to hazardous failure [4-6]. Experience shows that failure of a shaft system can occur due to various reasons – failure of rotor's coupling parts, shaft's manufacturing flaws (including material defects); corrosion of the shaft's material (including stress corrosion); overheating of the shaft's material due to fire (for example, when fuel mixes with oil in a fuel-oil heat exchanger), due to a breakthrough of hot gas into the shaft cavity, due to loss of spline coupling lubrication, due to the shaft contact with adjacent parts (including, for example, a result of flexural deformations, a significant imbalance of the rotor after blade-off, a failure of a bearing or other part adjacent to the shaft), including event, when load reduction device has been activated to reduce the load after blade-off conditions; fatigue (as a result, for example, of flexural or torsional oscillations, particularly, when pressure in the fuel system of the engine fluctuates), etc.

Fracture of a high-pressure turbine (HPT) shaft usually immediately leads to surging; so HPT rotor overspeed due to the shaft failure is limited. Nevertheless, the absence of a hazardous failure should be confirmed in the case of HPT shaft failure. More dangerous are shaft failures in intermediate-pressure turbines (in three-shaft engines), low pressure turbines (LPT), free power turbines, and turbounits (turbostarters, turbo-generators, turbo-refrigerating units).

To prevent a hazardous effect due to shaft fracture, a set of measures is usually used, including prevention of a defect, which can lead to shaft failure; employment of diagnostic methods of engine technical condition, which allow identifying such defects at an early stage of their development; decrease in criticality of defect development. For example, to prevent failure of LPT shaft, a large set of measures is used, selected on the basis of FMECA analysis including provision of high quality shaft manufacturing (like for the main engine parts); substantiated choice of coatings and lubricants for shafts and adjacent parts (taking into account possible corrosion); provision of sufficient shaft strength under operating conditions, including confirmation of the shaft's durability taking into account possible defects and, possibly, limitation of the designated service life; elimination of the possibility of hazardous overheating of a shaft material due to friction of the shaft with other parts by installation of a bumper between high and low pressure rotor shafts, application of heat-resistant coating of LPT shaft; installation of a bearing on an intermediate part, and not directly on the shaft; selection of pressure distribution, which excludes possibility of hot gas breakthrough into the shaft cavity; elimination of the possibility of fuel and oil mixing; application of rotor vibration dampers; etc.

Experience shows that probability of turbomachine shaft failures exceeds extremely low probability, and causes of the shaft failures are very numerous. Therefore, the design of the engine should ensure that turbine rotor overspeed is limited, and that sufficient load-carrying capacity of the rotor, taking into account possible overspeed, must be reliably confirmed.

Design solutions that limit a turbine engine's rotor overspeed in an event of shaft failure, are shown in Table 1.

Table 1. Design solutions that limit turbine rotor overspeed in an event of a shaft failure

Solution	Advantages	Disadvantages and issues

Ranking of integrity in a "blade airfoil – blade root – disc" system, which ensures the initial fracture of the blades in their root section during rotor overspeed	Effective regardless of at which section shaft was fractured (to the left or to the right of a thrust bearing)	It is necessary to prevent escaping through engine case fragments, formed during fracture of multiple blades at overspeed conditions
		High cost of subsequent engine repairs
Fitting a rotor turbine onto stator after axial downstream displacement of the rotor as a result of shaft failure (for example, using banana-shaped nozzle blades to arrange guaranteed connection of nozzle vanes and rotor blades and subsequent (mass fracture "meshing") of the blades)	Relative simplicity	Effective only when shaft is fractured at the right side of thrust bearing
		Time required for rotor displacement in axial direction to contact a stator should be much less than time to rotor fracture due to overspeed (a fraction of a second).
		Guaranteed fitting of the rotor on the stator is necessary, which eliminates formation of a sliding bearing and provides dissipation of the rotor energy due to friction between rotor and stator, as well as fracture of blades and vanes
		Difficulty of determining possible rotation rate during overspeed (without special test)
		High cost of subsequent engine repair
Cutting turbine rotor blades at rotor displacement after shaft failure (usually used in single-stage turbines)	Relative simplicity	The same as with axial displacement of a rotor with meshing of rotor blades and nozzle vanes
		Possibility of nonlocalized rotor fragments escaping case (for example, outside of turbo-starter's radial armor protection)
Mechanical system for fuel cut-off after shaft failure	Possibility of ensuring required speed	Impossibility of monitoring system's workability at service conditions
	Relatively good condition of the engine after an event	
Electronic system for abrupt increase of fuel supply for instantaneous surging after shaft failure	Relatively good condition of the engine after an event	Effective only in some cases (to prevent the intermediate-pressure turbine rotor overspeed)

	Possibility of monitoring system's workability at service conditions	It is needed to test the electronic system's performance under external influences (EMC, etc.)
		It is needed to prevent the engine shut down on a false signal
Electronic system for fuel cut-off after shaft failure	Relatively good condition of the engine after an event	Reliability and efficiency depend on the choice of type, number, layout of sensors, choice of structure (number of channels) and algorithm of the system
	Possibility of monitoring system's workability at service conditions	It is needed to test the electronic system's performance under external influences (EMC, etc.)
		Difficulty of determining possible rotation rate during overspeed
		Required limit of a rotor's rotation rate (speed) is not always ensured

When using a system based on the meshing of rotor and stator blades of an engine turbine in an event of axial displacement of the turbine rotor after the shaft fracture, it should be shown that the shaft fracture to the left of the thrust bearing (in which case the rotor is not displaced in the axial direction) is an extremely unlikely event. In this case, all possible causes of shaft failure should be considered. Additionally, it is important that, in an event of axial displacement of the turbine rotor, contact between engine's rotor and stator occurs at airfoils, otherwise a sliding bearing may form and rotor's overspeed will continue.

The electronic system is usually realized as a separate unit from the main control system and is provided with the power supply capability from the backup source. To increase reliability of the engine's turbine's electronic overspeed protection system, a two-channel system with two sensors in each channel is used. The choice of sensors' location is also important for ensuring their high reliability. Two-channel solution allows the engine to continue to operate when one of the system's channels fails, which is important for ensuring reliability centered maintenance (RCM).

Sometimes protection against unacceptable overspeed of an engine's turbine disc should be carried out by combined use of various solutions – for example, by combined use of an electronic fuel cut-off system, and in conditions where the speed of this system may not be sufficient – a system based on integrity ranking of elements in the "blade-disc" system.

In some cases, to confirm the absence of a hazardous failure in an event of engine's shaft fracture, it is necessary to carry out specific expensive engine test with shaft cutting. However, more often confirmation of the sufficient strength of a turbine disc under conditions of its possible overspeed is carried out on the basis of a spin testing of the disc and (or) a calculation of disc burst speed.

An estimation of possible rotation rate of a disc during overspeed is an independent task, which is not considered here. The method for calculating burst disc speed is considered below.

3. Numerical simulation of a disc burst speed

3.1. Description of a model

The finite element method (FEM) is the most effective tool for calculating the stress-strain state of a disc under conditions of overspeed. In papers [7-8], possibilities of using FEM for the calculation of plastic deformations in turbine discs GTE using the theory of plasticity are shown and a satisfactory conformity between calculated and experimental values of a disc's burst speed is obtained (the difference was not more than 6%).

The basis of the calculation method is a step-by-step calculation of a disc using FEM with an increase of the rotation rate. We consider active (unidirectional) loading, in which hardening of the material is isotropic. We consider loading of the disc as a quasi-static process. Such assumptions are advantageous for reducing the time of the iterative solution. In this case, the parameter characterizing hardening of the material must be obtained at a deformation rate corresponding to the deformation rate of the disc during overspeed.

The process of transition of a disc from an elastic to an elastoplastic state performs step-by-step. First, the most stressed local zones (holes, fillets, slots, etc.) undergo plastic deformation, and then the plastic deformation extends further into the hub and web of the disc. Considering that the time it takes the disc to overspeed under operating conditions is a fraction of a second, we assume isothermal deformation process of the disc, i.e. neglected both by the internal release of heat due to plastic deformation, and by heating of the disc from external sources.

Taking into account the type of the disc's loading, we use isothermal theory of plasticity with isotropic hardening. Total deformation increments are split into elastic $\{\Delta\varepsilon^{el}\}$ and plastic deformation $\{\Delta\varepsilon^{pl}\}$:

$$\{\Delta\varepsilon\} = \{\Delta\varepsilon^{el}\} + \{\Delta\varepsilon^{pl}\}. \tag{1}$$

The increments of the elastic strain components are related to the increments of the stress components $\{\Delta\sigma\}$ by Hooke's law through the matrix of elastic moduli [D]:

$$\{\Delta\varepsilon^{el}\} = [D]^{-1}\{\Delta\sigma\}. \tag{2}$$

Plastic deformations are determined by introducing three characteristics of the material's nonlinear behavior: the yield function f, which characterizes conditions of plastic flow during complex stress state, the flow rule determining the direction of development of plastic deformation, and the hardening function $\bar{\sigma}$, which determines how f varies during plastic flow.

During plastic flow stresses must be on the loading surface, i.e. two conditions must be fulfilled simultaneously:

$$f(\{\sigma\},\bar{\sigma}) = 0, \tag{3}$$

$$df = 0. \tag{4}$$

Increments of the plastic deformation components with the use of associative plasticity are connected through the derivatives of the yield function f with respect to the corresponding stress components:

$$\{\Delta\varepsilon^{pl}\} = \lambda\left\{\frac{\partial f}{\partial\sigma}\right\}, \tag{5}$$

Where

λ – non-negative parameter that determines the value of the increment of the plastic flow.

In the practice of disc numerical simulation, the yield function with a Mises yield condition is widespread:

$$f = \left(\frac{1}{2}[(\sigma_1-\sigma_2)^2 + (\sigma_2-\sigma_3)^2 + (\sigma_3-\sigma_1)^2]\right)^{1/2} - \bar{\sigma}(\bar{\varepsilon}^{pl}, T). \tag{6}$$

Where

σ_1, σ_2, σ_3 – principal stresses;

$\bar{\sigma}(\bar{\varepsilon}^{pl}, T)$ – function of material hardening, which depends on the accumulated plastic deformation ε^{pl} and temperature T.

Assuming that this function is independent of the type of stress state, we use tensile curves at different temperatures. Usually $\bar{\sigma}(\varepsilon^{pl}, T)$ is given in the form of piecewise linear functions, and the intermediate values are determined by linear interpolation.

The justification for using (6) is the multiple validations performed for various classes of metallic materials. However, calibration of this model during the spin rig testing of an aircraft's GTE turbine disc made from Udimet 720 at normal temperature [9] have shown that the best correspondence of the

IOP Conf. Series: Materials Science and Engineering **449** (2018) 012025 doi:10.1088/1757-899X/449/1/012025

numerical and experimental data is achieved when using Tresca yield condition, which can be represented in yield function as:

$$f = \frac{1}{2}[(\sigma_1 - \sigma_2) + (\sigma_2 - \sigma_3) + (\sigma_3 - \sigma_1)] - \bar{\sigma}(\bar{\varepsilon}^{pl}, T). \quad (7)$$

To study the influence of the yield condition on the burst speed, we used the non-quadratic yield function with Hosford yield condition [10]:

$$f = \left(\frac{1}{2}[(\sigma_1 - \sigma_2)^a + (\sigma_2 - \sigma_3)^a + (\sigma_3 - \sigma_1)^a]\right)^{1/a} - \bar{\sigma}(\bar{\varepsilon}^{pl}, T). \quad (8)$$

Where

a – parameter.

Tresca (a=1) and Mises (a=2 or a=4) yield conditions are special and limiting cases of the Hosford yield condition [10]. For a> 4, the yield condition lies between the limit curves and tends to Tresca yield condition for a→∞.

Calculation of plastic deformations within each iteration is performed using the return mapping algorithm [11]. The main problem of calculating plastic deformations under overspeed conditions is a reliable determination of material model's parameters, including yield condition, hardening function and failure criterion. These model's parameters can be identified from tensile testing of smooth specimens and spin testing of a model disc of the same material [12-13]. For the purpose of experimental adjustment of the method, two identical workpieces of a disc made of heat-resistant steel were manufactured. The disc for burst spin test was made from the first workpiece, and from the second - ten cylindrical specimens with a diameter d_0=6 мм, cut from different areas of the disc.

3.2. Identification of material model parameters

3.2.1. Identification from tensile test. The tensile test was carried out in accordance with ASTM E8M Standard at normal temperature at a rate of 0.3 mm/min with the installation of an extensometer on the body of the specimen.

Based on mean values of mechanical properties, the tensile curve is shown in Figure 1a (the dashed line shows the area after the beginning of the necking, which is not used in calculations). True stress vs. true plastic strain diagram (Figure 1b) was determined assuming a constant volume of material before and after plastic deformation, based on the following relations:

$$\sigma_{true} = \sigma(1 + \varepsilon), \quad (9)$$

$$\varepsilon_{true}^{pl} = \ln(1 + \varepsilon) - \sigma_{true}/E. \quad (10)$$

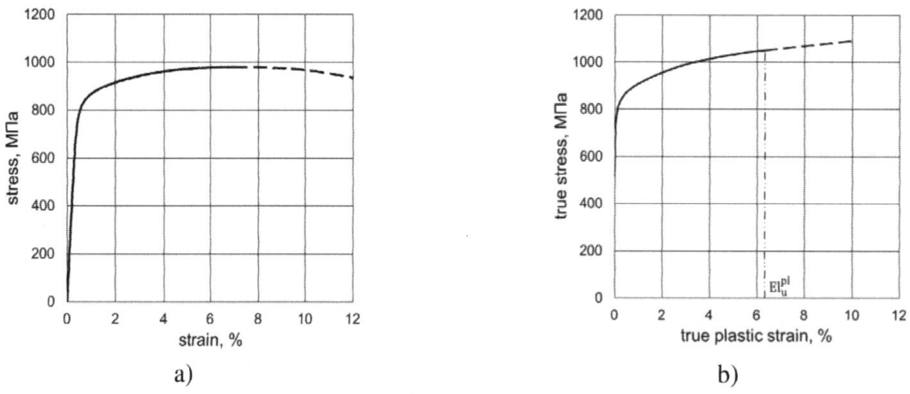

Figure 1. Average tensile curve (a) and true stress vs. true plastic strain diagram (b)

We used true stress vs. true plastic strain diagram in the calculations as a hardening function $\sigma(\varepsilon^{pl}) = \sigma_{true}(\varepsilon_{true}^{pl})$. The dashed 1 function to ensure the convergence of the solution. Uniform

IOP Conf. Series: Materials Science and Engineering **449** (2018) 012025 doi:10.1088/1757-899X/449/1/012025

plastic elongation El_u^{pl} (as a part of uniform elongation) was considered a critical strain, after which burst of the disc begins.ine in Figure 1b shows a linear extrapolation of the hardening

3.2.2. *Identification from burst spin test.* We carried out the burst spin test of the disc (Figure 2) at normal temperature with increasing rotation rate linearly [14]. Loading rates of the disc and samples were approximately equal. During the test, we recorded the rotation rate and elongation of the outer diameter of the disc by two linear eddy-current sensors (using two sensors located at diametrically opposite points).

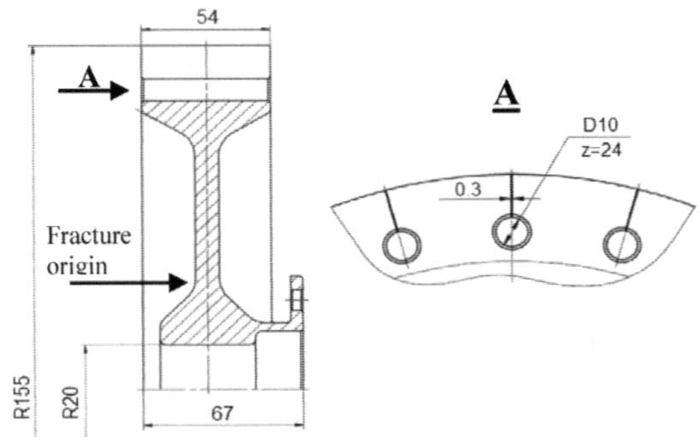

Figure 2. Drawing of the model disc (dimensions are in mm, z - number of orifices)

Disk burst at a rotation rate ω^{EXP} = 24282 rpm, while the radial displacement of the outer diameter of the disc at the time of burst was u^{EXP}=2.2 mm. Photography of disc after disk is shown in Figure 3. Fracture origin of the disc occurred along a cylindrical cross-section with the formation of a necking. Fracture of the disc along the meridian cross-sections passing through the holes is secondary.

We performed simulation of the disc overspeed via MSC. Marc on the disc sector with an angle α=15° (α=360/z). During formation of the finite element mesh, we took into account real geometric dimensions of the disc after manufacturing, determined by the results of 3D scanning. We divided the disk sector by first order hexahedral finite elements. On the boundaries of the sector, we set circumferential displacements to zero. Also, we fixed the model in one node in the axial direction to prevent it from moving as a rigid body. The values of the density and the Poisson's ratio are taken from the reference data.

The increase in the rotation rate was carried out with an even step: the first step corresponded to a rotation rate of 15,000 rpm; at the next steps rotation rate increased by 300 rpm.

A comparison of the results of the disc radial displacement calculations along the outer diameter under the condition of the Mises yield condition and the close to the Tresca yield condition (a=2 and a=100 in Eq. (8)), with experimental data is shown in Figure 4. In a range of rotation rates from 0 to 18000 rpm, the disc is deformed elastically, and correlations between calculated/experimental displacements of the disc coincide in this range. After reaching rotation rate of 18000 rpm, a deviation of the radial displacements from the elastic line is observed, which indicates elastoplastic deformation of the disc. Numerical displacements obtained with Mises yield condition and experimental displacements have a similar pattern over the entire range of rotation frequencies. For this material, Mises yield condition ensures better conformity of calculated and experimental data on radial displacements than Tresca yield condition. In this case, the accepted hypotheses about isotropic hardening and the associated plasticity can be considered acceptable.

Figure 3. Photography of disc after burst

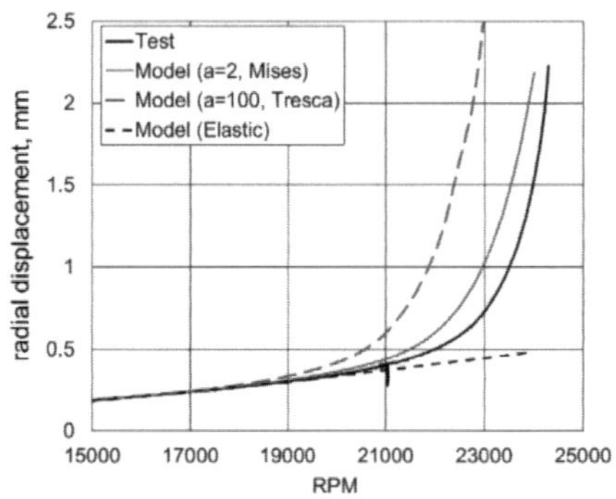

Figure 4. Experimental and numerical (a=2 and a=100) results of disc radial displacement

Distributions of the plastic strain intensity ε_i^{pl} of the disc at rotational rates of 23400, 23700 and 24000 rpm are shown in Figure 5. Plastic deformation reaches a maximum value in the disc web in the area, which corresponds to the fracture origin. At a rotation rate of 23400 rpm, the value of plastic deformation was 6.0% (see Figure 5a) and almost reached a critical strain El_u^{pl}= 6.3 %. Given the presence of a necking in the disc after the tests, we can assume that the fracture of the disc occurred with higher El_u^{pl}. At a rate of 24000 rpm (see Figure 5c), plastic deformations are already distributed over the entire section A-A.

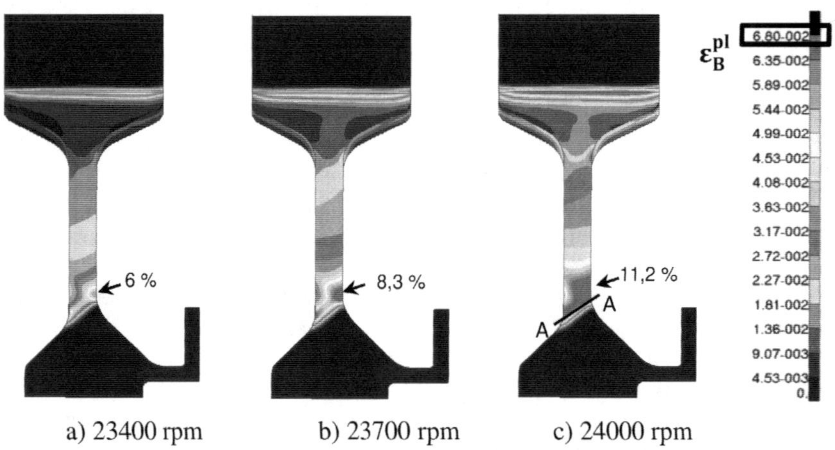

a) 23400 rpm b) 23700 rpm c) 24000 rpm

Figure 5. Distribution of the plastic strain intensity in a disc at different rotation rates

If we take into account that at the last loading step the calculated value of the radial displacement of the disc (u_r=2,18mm at 24000 rpm) is close to the experimentally determined value of the radial displacement of the disc at the moment of burst (u^{EXP}=2,2 mm at ω^{EXP} = 24282 rpm, see Figure 4), then the limiting state at which the disc burst, can be considered an attainment of plastic strain intensity equal to critical strain in an entire unsafe cross-section:

$$\varepsilon_{i,\ \text{cross-section A-A}}^{pl} = El_u^{pl}. \tag{11}$$

3.2.3. Discussion on the results

According to (11), the burst speed of the disc is $\omega^{NUM(1)}$ = 24000 rpm, and the difference between numerical and experimental burst speed – $\Delta^{(1)}$=1 %. The fracture criterion (11) is not local. An additional justification for this fact is the detection of a necking at a radius R=55 mm after burst of the disc.

The smooth specimen necking is determined by the reduction of diameter:

$$\psi_d = \frac{d_0 - d_k}{d_0} = 1 - \sqrt{1 - \psi}, \tag{12}$$

Where
d_k – diameter of the specimen in the neck area after burst;
ψ – reduction of area.

The web disk necking is determined by the reduction of web:

$$\psi_h = \frac{H_0 - H_k}{H_0} \tag{13}$$

The average value of reduction of diameter is ψ_d=0.26, and reduction of web – ψ_h=0.21. Thus, the change in the dimensions of the specimen and the disc in the plastic deformation localization zone turned out to be sufficiently close to each other.

In practice, local fracture criteria are used to calculate burst speed, which are presented in papers [6-9, 15, 16]. For example, if as the fracture criterion we consider the equality of the maximum plastic deformation and critical strain:

$$\varepsilon_{i,\ max}^{pl} = El_u^{pl}, \tag{14}$$

then the burst speed $\omega^{NUM(2)}$ = 23450 rpm and the difference between numerical and experimental burst speed – $\Delta^{(2)}$=3.5 %.

To estimate the value of burst speed, it is often used average radial stress criterion [17]: a disc will burst when average radial nominal stress equals the tensile strength of the material. As well as for the average hoop stress criterion [18], one of the main drawbacks of the method is the lack of possibility to take into account the plastic characteristics of the material. According to this criterion $\omega^{NUM(3)}$=22400 rpm and $\Delta^{(3)}$=8 %.

Thus, to calculate the burst speed of a disc made from steel, it is preferable to use the fracture criterion (11), where plastic deformation is determined based on the isothermal theory of plasticity with the Mises yield condition and isotropic hardening. The advantage of the method is also detection of the fracture origin and the critical disk section.

Using the method described above, for a number of model disks and full-size discs of various designs made from steels, titanium and nickel alloys tested at normal and operating temperatures, we performed a comparison of the numerical and experimental data on basis of relation between radial displacements during spin tests, burst speed, high-speed image acquisition, fracture origin and fracture mode analysis. Based on the results of the research, we gave recommendations for disc burst speed analysis. It should be noted that for the disc made from nickel based superalloys Udimet 720 and EP741NP, for which fracture of the specimens in the tensile test occurs without necking, it has been shown that the use of the Tresca yield condition [3,10] and the local fracture criterion (14) is advantageous.

When calculating the load-carrying capacity of a disc, it is necessary to take into account the most unfavorable operating conditions of the engine during overspeed, the unfavorable combination of dimensional tolerances and the statistically minimum values of the material's mechanical characteristics [1-3]. The presented method also allows to adjust the value of a disc rotation rate during certification spin tests to ensure that a disc with the worst mechanical properties will not burst in operation and to take into account differences in the design and loading conditions between the disc for spin tests and disc for operation.

If the identification of model's material parameters was performed based on disc's tensile tests and spin tests, then for a disc of another design from the same material only calculations can be used.

Acknowledgments
All activities have been carried in CIAM. The authors are grateful for permission to publish the paper. The authors are grateful to the team led by M.E.Volkov for testing samples in CIAM laboratory.

References
[1] Aviation regulations. Part 33. Airworthiness of aircraft engines 2012 IAC
[2] 14 CRF, Part 33. Airworthiness Standards. Aircraft Engines 2014 FAA
[3] Certification Specifications for Engines (CS-E) 2015 EASA
[4] Engine and Turbosupercharger Rotor Overspeed Requirements of 14 CFR §33.27 2011 FAA, AC33.27-1A
[5] Turbine Overspeed Resulting from Shaft Failure 2012 EASA, Certification Memorandum EASA CM-PIFS-003 Issue 01
[6] Nozhnitsky Y A and Kuevda V K 2018 Prevention of a dangerous failure of a gas turbine engine due to fracture, disconnection or displacement of the shafts *Proc. of the Scientific and Technical Congress of Engine-building* Moscow 2 pp.267-276
[7] Servetnik A N 2012 Energy-based method for gas turbine engine disc burst speed calculation, *28th Congress of the International Council of the Aeronautical Sciences* ICAS 2012 Brisbane Australia 23-28 September 2012 pp.2443-2448
[8] Nozhnitsky Y A, Karimbaev K D and Servetnik A N 2012 Numerical simulation of spin testing for turbo machine discs using energy-based fracture criteria *Proc. ASME Turbo Expo*, Copenhagen, Denmark 11-15 June 2012 V.7 pp.35-40
[9] Maziere M, Besson J, Forest S, Tanguy B, Chalons H and Vogel F 2008 Overspeed burst of elastoviscoplastic rotating discs Part II: Burst of a superalloy turbine disc *European Journal of Mechanics A/Solids*, V.28 pp.428-432
[10] Hosford W 2005 Mechanical behavior of materials Cambridge University Press p.425
[11] Simo J C and T J R Hughes 1998 Computational Inelasticity Springer-Verlag New York, P.392
[12] Kuzmin E P and Servetnik A N 2014 Yield surface investigation of alloys during model disc spin tests *Science and Education: Scientific Publication* №5 pp.330-339
[13] Servetnik A N and Kuzmin E P 2014 Yield surface investigation of alloys during model disc spin tests *Proc. ASME Gas Turbine India Conf* GTINDIA2014 India Delhi 15-17 December 2014 p.3
[14] Nozhnitsky Y A, Fedina Y A, Shadrin D V, Servetnik A N, Baluev B A, Kanachkin A V, Lepeshkin A R, Tomashiov A A and Cherhyshev S A 2015 New possibilities of using spin rigs to provide gas turbine engine strength reliability *Vestnik of the Samara State Aerospace University* V.14 №3 part 1 pp.71-87
[15] Karimbaev K D and Servetnik A N 2008 Numerical modeling of turbo machine disc burst speed test *Herald of Aeroenginebuilding* №3 pp.130-134
[16] Servetnik A N 2012 Load-carrying capability simulation of aviation gas turbine engine disc *Handbook. An Engineering Journal* V.187 №10 pp.44-49
[17] Birger I A, Shorr B F, Dem'yanushko I V, Dulnev R A and Sizova R N 1975 Thermal strength of machine parts Mashinostroenie Moscow Russia p.454
[18] Stodola A 1927 Steam and Gas Turbine V.2 McGrow-Hill Book Co. Inc. New York N.Y. p.1080

IOP Conf. Series: Materials Science and Engineering **449** (2018) 012026 doi:10.1088/1757-899X/449/1/012026

Design of Universal PCIe Interface Module Based on Vs

Tingting Du[1] and Qingzhong Jia[1]

[1]Key Ministry Education Laboratory of Dynamics and Control of Flight Vehicle, Beijing Institute of Technology, BIT, Beijing, China

bitdtt6979@163.com

Abstract. Due to wide variety of interfaces and complex bus protocols, the general detection system has problems such as high bus usage threshold and poor portability of the host computer software, which greatly reduces the user's development efficiency of the detection system. In response to the issue, this paper designs a universal interface module for PCIe based on vs. The module provides unified functional interfaces, through which the data transmission of all bus devices can be easily and reliably controlled. In this paper, the design idea of the universal interface module is explained. Finally, the module is illustrated and the result shows that the designed scheme is feasible.

1. Introduction

With the development of computer technology, the detection system consisting of the main control computer and various expansion modules is widely used in various fields[1]. Due to variety of interfaces of the measured object, the detection system is required to be able to integrate multiple interfaces, and these interfaces need to be able to change as the measured object changes. At present, the interconnected buses with high performance and wide application are mainly Gigabit Ethernet, PCIe, RapidIO, etc[2]. Among them, PCIe (peripheral component interconnect express), a serial computer expansion bus standard, is widely used in the field of embedded detection because of its high bandwidth, optional link number and high transmission speed[3].

For specific application of bus interfaces, each module manufacturer provides mature, complete and powerful APIs to meet various possible needs of transmitting data. However, in most detection systems, only a small part functions of bus interfaces are used, the complicated API parameters and the interfaces that cannot be transplanted due to different application protocols impose high requirements on developers of the detection system. At the same time, the interface matches the respective PC software, so the diversification of interfaces in the detection system results in poor software portability, a large amount of repetitive work for the development of the PC software and high development cost of the detection system. Therefore, how to improve program portability[4], reduce system development costs, and generalize interface applications have become some of the focuses of current research for computer detection systems.

By using the design pattern idea[5] and the technology of virtual function overloading, this paper establishes a unified user-oriented interface model and studies the generalized design of bus interface. The purpose is to make bus interfaces of the detection system intuitive and easy to use, improve the scalability and portability of system software.

2. Universal interface model establishment

Considering the generality requirement of interfaces, a universal interface model is established as shown in figure 1.

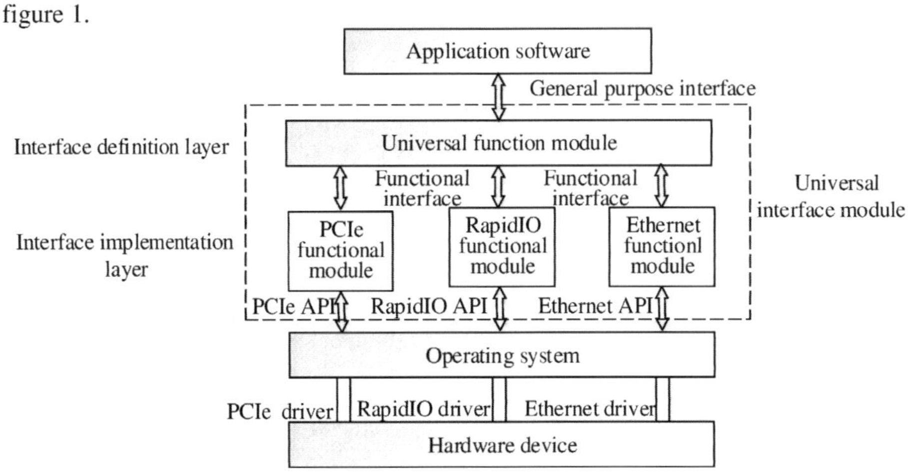

Figure 1. Universal interface model.

The whole model consists of four parts, in which the application software is responsible for human-computer interaction, and the hardware devices follow the bus protocols to achieve their respective functions. These two parts are not described in detail in this paper. As a link between the hardware device and the operating system, driver enables the host computer to operate the hardware device indirectly by calling APIs (Application Program Interface) provided by the operating system. Referring to the seven-layer architecture of the Open System Interconnect (OSI)[6], the universal interface module designed in this paper is equivalent to the top layer of OSI, the application layer. Its function is to provide the service interface to the users directly and complete the connection between the application and the network operating system.

As a focus of this design, the universal interface module can be divided into two layers. The upper interface definition layer abstracts and defines various operations and provides general purpose interfaces to users. In addition to basic operation interfaces for data transmission, it also includes additional function interfaces and user-setting interfaces. Through this layer, the application can realize the operation of interfaces of various buses without differentiation, and users can control the data transmission process intuitively, simply and reliably.

The lower interface implementation layer distinguishes the specific application, and further designs and encapsulates the basic APIs provided by the operating system into functional interfaces for the upper layer to call, thereby completing the concrete implementation of the general purpose interfaces defined by the upper layer. This layer further designs and improves the original reading and writing functions of the respective bus. Taking PCIe bus as an example, this layer implements functions such as periodic transmission, packet transmission, task scheduling and fault tolerance mechanism setting to ensure reliable data transmission. The interface definition layer and the interface implementation layer together form a bridge between users and different bus devices.

3. Universal interface module design

The universal interface module is divided into an interface definition layer and an interface implementation layer. Taking PCIe, RapidIO and Ethernet protocols as examples, the interface definition layer introduces generality and functionality. The design and packaging ideas of this layer are illustrated through the class diagram and important codes. For the interface implementation layer, it takes PCIe as an example to describe how to further encapsulate the operating system APIs into functional interfaces through working timing diagrams, thereby realizing the concrete implementation of general purpose interfaces reliably.

3.1. Interface definition layer

Analogous to the black box, this layer shields diversity of different protocols, abstracts APIs under different protocols into general purpose interfaces and provides them to the users, realizing that the application does not differentiate the operations of peripheral no matter which protocol is followed. General purpose interfaces defined by this layer are shown in figure 2.

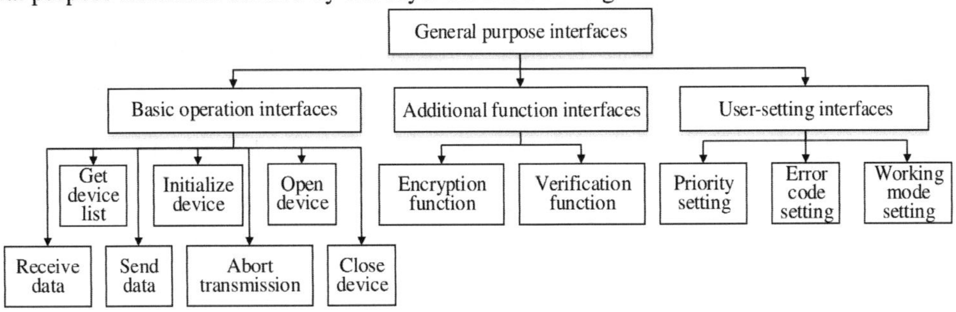

Figure 2. General purpose interfaces.

Through the overloading of above general purpose interfaces, the specific problems solved by this layer include: establishment of unified user-oriented interface model, parameterized design of interface configuration, introduction of general additional functions, design of task scheduling and working pattern. These key issues will be explained in the next section.

3.1.1. Unified user-oriented interface model. According to the defined general purpose interfaces, the main class diagram of the unified user-oriented interface model is shown in figure 3.

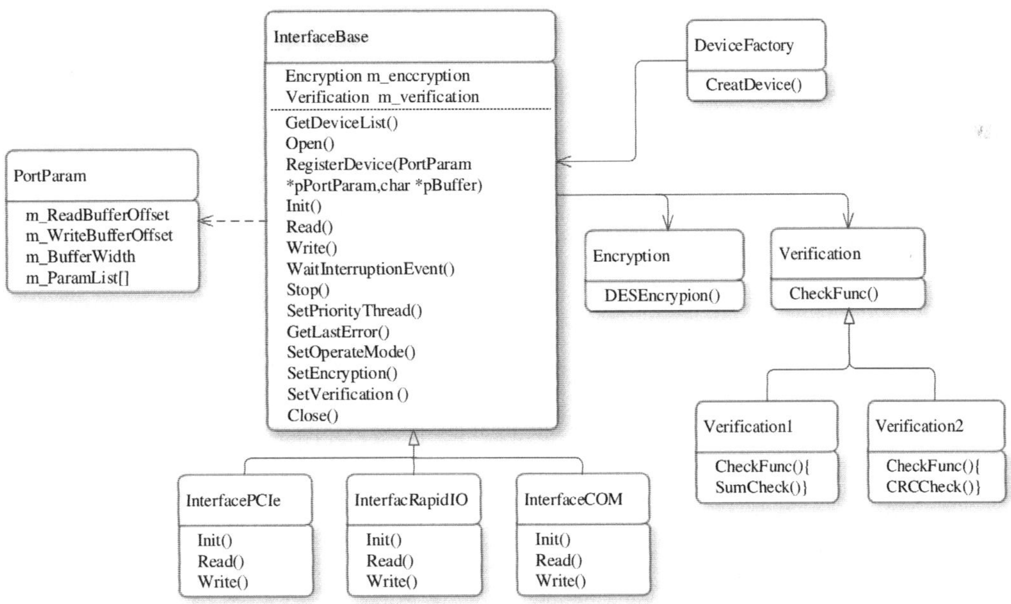

Figure 3. Class diagram of the interface definition layer.

The InterfaceBase class in the class diagram is the base one for all classes and it encapsulates general purpose interfaces defined for users in the data transmission process. Under the premise of satisfying the versatility, the class simplifies interfaces as much as possible, reduces the number of interfaces, lowers the threshold of use and facilitates users to operate the interfaces intuitively as well as control the entire data transmission process. These interfaces include obtaining the current device list, opening the device, initializing, reading, writing, aborting the transmission, setting the task priority, obtaining the error code, setting the working mode, encrypting, verifying, and turning off the device, which embody the "commonness" of interfaces. The "personality" embodied by different devices is implemented by the factory mode, that is, the DeviceFactory class in the figure 3.

After executing GetDeviceList() in the base class, we can obtain the flag word and the specific number device_num from zero without repetition corresponding to each device. According to the flag word (PCIe device is "PCIe"), the CreateDevice() method in the DeviceFactory class instantiates the InterfacePCIe object, thereby completing the overloading of virtual function under the PCIe protocol. The factory mode makes it unnecessary to change functional functions of each bus class inherited by the base class every time the device is replaced, just instantiating the corresponding object, thereby reducing the amount of code modification and repetition, and achieving the purpose of separating "realization" and "change". Because functional modules of different buses have structural similarities, the PCIe is used for description. The specific implementation code takes c++ as an example:

```
    class  DeviceFactory   //The DeviceFactory class
{
Public:
Void CreateDevice(string DeviceName)
        {
         InterfaceBace  *p_inter_base;
         case(DeviceName)
              {
               PCIe://Taking PCIe as an example
               InterfacePCIe  m_inter_PCIe;
               p_inter_base=&m_inter_PCIe;
               // Instantiate other bus objects
               }
        }
}
```

By instantiating different interface objects, virtual functions in the base class are overloaded in the corresponding interface subclass, and functions under the respective bus protocol are completed in the overloaded functions, thereby realizing the polymorphism required by the system. In other words, the unified user-oriented interface model completes the conversion of dedicated interfaces to general purpose interfaces.

3.1.2. Interface initialization. The types and amounts of configuration information required by the interfaces of different buses vary. To parameterize the initialization work, the PortParam class in figure 3 is designed as follows:

```
    class PortParam    //The PortParam class
{
public:
long m_ReadBufferOffset;  // Reading buffer offset address
long m_WriteBufferOffset;  // Writing buffer offset address
long m_BufferWidth;  // Buffer size
char  m_ParamList[16];
}
```

There is no method in this class, only the property set to configure the interface information. The reading/writing buffer offset address and the buffer size represent the configuration information shared by all buses. In addition, the 16-byte array m_ParamList[] is declared to store different configuration information required by each bus. For example, CAN requires baud rate and ID number. Ethernet requires IP address and port number. The volume of this array can be adjusted.

All searched bus devices should be registered uniformly before the interfaces are initialized, that is, RegisterDevice(PortParam *pPortParam, char *pBuffer) in the base class. All configuration information obtained from users stores in the pBuffer array. According to the size of attributes in the PortParam class, 0-39 bytes in the array represent the configuration information of interface 0, 40-79 bytes represent of interface 1, and so on. In this function, the same number of *pPortParam objects are instantiated to the number of devices, then the configuration information in the pBuffer array is assigned to the corresponding object. The other array DeviceArray[] with the same capacity as the number of devices is created. The assigned *pPortParam pointers are stored in this array, that is, each element in the array represents the configuration information of one device, and the number of each array element represents the number of device, which corresponds to the device_num. In this way, regardless of the addition and deletion of device, it will not affect the registration and configuration of other devices.

In the subclass, the initialization virtual function is overloaded, that is, Init(char *pDeviceArray, device_num). According to the device_num, we can find the corresponding *pPortParam and then parse the configuration information in it, thereby completing the initialization of the interface. At this point, the generalization processing of the universal interface module is also completed.

3.1.3. General additional function. In order to ensure the reliability and security in data transmission process, the module adds an encryption interface and a verification interface. This design uses the strategic mode to implement the introduction of multiple encryption algorithms and verification algorithms. Through this mode, whether encryption/verification or not and which encryption/verification algorithms are used will not affect the client program, which will achieve high cohesion and low coupling of code, and improve the flexibility of the application.

This mode is scalable and can continue to add verification and encryption algorithm without changing the rest of code. Users can expand the module according to this mode if the system needs to support other data processing functions.

3.1.4. Task scheduling. Because the system has multiple bus interfaces to share the tasks together, in order to achieve smooth and reliable work during the data transmission process, task scheduling among interfaces is essential. In the case where users do not set the priority of interfaces, the tasks enter the queue by default in the order of arrival, and are executed in order. In order to meet the user's own requirements for the bus scheduling order, the InterfaceBase class in figure 2 reserves the SetPriorityThread (PKTHREAD Thread, KPRIORITY Priority) interface, which is convenient for the users to customize the bus priority. The priority of the transmission task in the program should correspond to the user-defined priority. If there is a task command with higher priority during the task transmission process, the corresponding thread should be created to ensure that the data transmission process conforms to the bus priority set by users.

3.1.5. Working pattern. In order to make users more intuitive and more convenient to control the data transmission process of the bus device, general purpose interfaces also include GetLastError() and SetOperateMode(). Between them, GetLastError() is used to obtain the error code. Users can customize the error codes according to the possible errors during transmission process and visually monitor the status of the data transmission through the interface.

SetOperateMode(int type, unsigned int port_period, unsigned int interrupt_period) allows users to set the working mode of interfaces manually. Taking PCIe as an example, the first formal parameter represents the custom mode word, including periodic transmission mode: 1, big data transmission mode: 2, and small data transmission mode:3. For the aperiodic transmission, the second and the third parameters default, that is, the timing transmission period and the interruption period to stop periodic transmission. The program determines the required working mode automatically according to the data size when there is no working mode setting. Once the mode is set, it is forced to execute according to the user's command so that user has full control over interfaces.

The universal interface module is designed and packaged through the above modes, with clear hierarchy, well-organized code and high maintainability. And the designed interfaces meet the generalized and functional requirements on the basis of ensuring the controllable bus.

3.2. Interface implementation layer

In order to improve the reliability of interfaces, in the interface implementation layer, the inherent APIs of each bus are redesigned and encapsulated into functional interfaces, which can be invoked by the interface definition layer to realize generalization of interfaces. Compared with the inherent API, the functional interface makes the data transmission process more reliable. Next, taking PCIe as an example for detailed explanation.

Typical API functions and corresponding callback routines in the kernel driver of the PCIe bus are shown in table 1.

Table 1. Typical APIs and callback routines of PCIe.

API function	KMDF callback routine
CreateFile	EvtDeviceCreate
ReadFile	EvtIoRead
WriteFile	EvtIoWrite
DeviceIoControl	EvtDeviceIoControl
CloseHandle	EvtFileCleanup EvtFileClose

By calling API functions and executing the corresponding routines, the basic communication functions between the host computer and the hardware device following PCIe protocol can be realized[7]. For the data transmission process, this layer firstly solves the selection of bus working mode. Taking PCIe as an example, the design of working mode judgement is shown in figure 4.

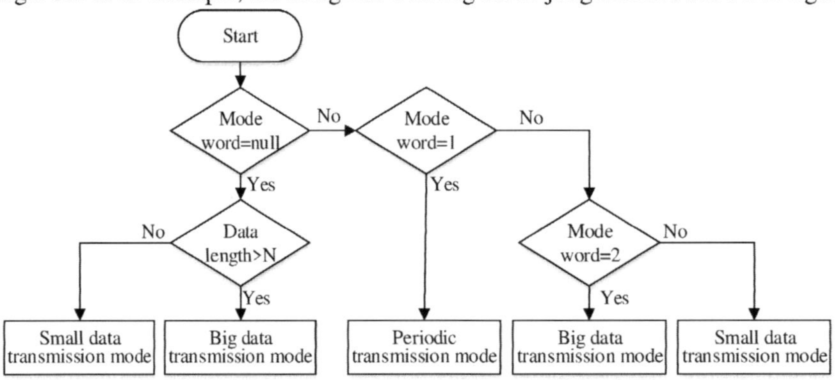

Figure 4. selection of the data transmission mode.

Next, according to the user's settings, two types of working modes, periodic transmission and single transmission, will be introduced in detail. In the periodic transmission process, the operation timing diagram is used to explain the judgment and processing of the periodic transmission routine. In the single transmission, PCIe packet transmission mechanism, priority mechanism and fault tolerance mechanism are introduced in detail through timing diagrams of reading and writing processing.

3.2.1. Periodic transmission. Because the periodic transmission can be regarded as resending data multiple times in a fixed cycle. it is necessary to judge whether the data is periodically sent before processing the data transmission tasks. According to the user's settings, the working timing of the periodic transmission is as shown in figure 5.

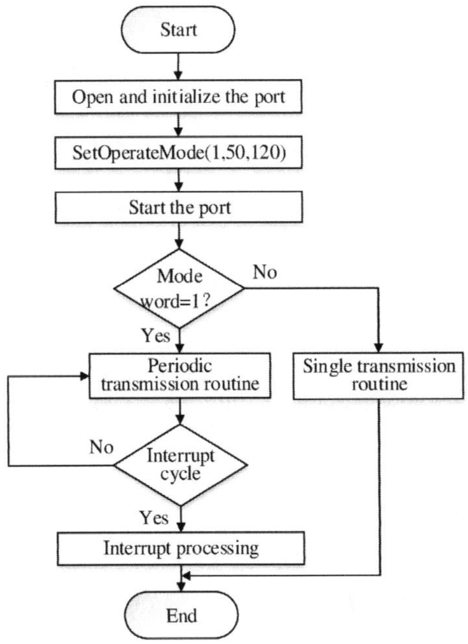

Figure 5. Working timing of periodic transmission.

According to the user's setting SetOperateMode (1, 5, 120), the flag word is 1, which means the data is sent periodically. In the periodic transmission routine, one timer is initialized according to the interface period set by the user for 5s, and the data is transmitted every 5 seconds. The interrupt timer is initialized simultaneously in the routine. After 120s, the transmission is aborted and the corresponding routine of interrupt processing is executed.

3.2.2. Single transmission. A round of single data transmission includes the most basic reading and writing functions required by the detection system. According to the user's demand for the amount of the transmitted data, this design adopts two working modes: DMA transmission for big data[8] and register transmission for small data.

In order to avoid long-term occupation of the bus caused by data transmission, a packet transmission mechanism is introduced, which is convenient for other operations in the packet transmission gap. In order to ensure the controllability and correctness in the data transmission process, the custom transmission protocol is designed to perform fixed encapsulation processing on the transmitted data in the form of a structure. The data packet protocol is shown in table 2.

Table 2. Data packet protocol.

Type	Byte number	Remark
Frame head	2	Fixed value
Target, source	1	The upper four bits are the target, the lower four bits are the source; 1 is the PC, and 2 is the hardware device.
Way ID	1	1:Event 2:Packet head 3:Sub-packet transmission 4:Packet cancel 5:Sub-packet retransmission 6:Small data transmission

Type	Byte number	Remark
Event ID	1	User-defined event content
Frame ID	2	Total packet number in the header packet transmission;Sub-packet number in the sub-packet transmission
Data total length	4	Special for packet head
Data section	10240	
Checksum	4	

As shown in the data packet protocol, the data portion occupies 10 KB of space, which is intended to select the data transmission mode automatically according to the data size. When there is no working mode setting, if the data size is less than 10KB, IODeviceControl direct transmission mode is adopted, and if it is larger than 10KB, DMA transmission mode is adopted. The DMA packetization method uses a header packet plus sub-packets. Each of packets has fixed length, and the last one is filled with 0 if it is not full. The sender is the process of unpacking and sending data, while the receiver is the process of receiving and combining packages. Following the data packet protocol, the writing timing is shown in figure 6.

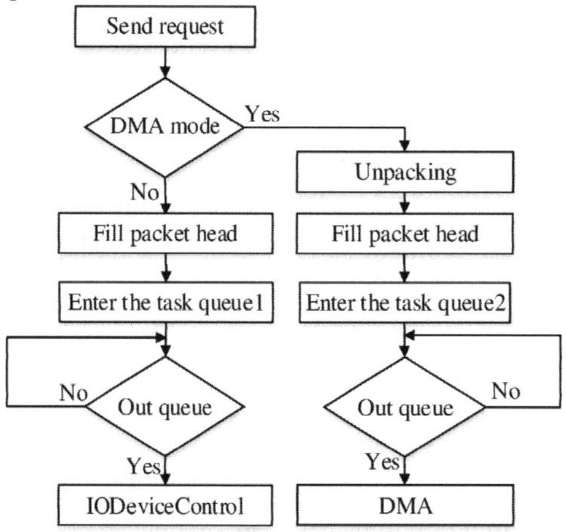

Figure 6. Working timing of writing process.

Unlike sending data, receiving data has no periodic concept and it is a passive behavior for the host computer. When the hardware device sets the interruption signal of transmitting data, the corresponding event is triggered in the kernel program, that is, WaitInterruptionEvent() in the base class. This interface listens to the event triggers of all buses. Once the event trigger is detected, it will enter the message response function of specific bus according to the trigger message, that is, the overloaded Read() in the corresponding subclass, thereby realizing the generalization of reading process. The reading timing under this design is shown in figure 7.

It can be seen from the reading timing that the fault tolerance mechanism needs to be set for receiving data. In this design, if the verification fails, data will be retransmitted automatically. If the number of retransmissions is more than three, the cause of error will be recorded for later investigation and the receiving process will be cancelled. To continue the data transmission process, sender must resend actively.

In addition, in order to ensure the real-time performance in the data transmission process, considering the dual influencing factors of waiting time and task time[9], this design sets the task priority mechanism - double queue mechanism. The data packets of big data and small data are respectively stored in the two queues, the corresponding number of data packets in the two queues are alternately popped in a fixed number ratio (e.g. large: small = 5:1). After one queue pops up packets, it switches to the other queue. If the number of packets in the queue is less than the specified number, all packets in it are popped up. If the number is zero, the other queue is switched. This kind of priority mechanism, which is mainly based on transmission of big data packets and supplemented by small data packets, has a simple algorithm and can also solve the problem that the time of big data transmission is so long that the system cannot respond in time. At the same time, it can avoid the situation that the big data transmission waits too long due to the accumulation of small data packets.

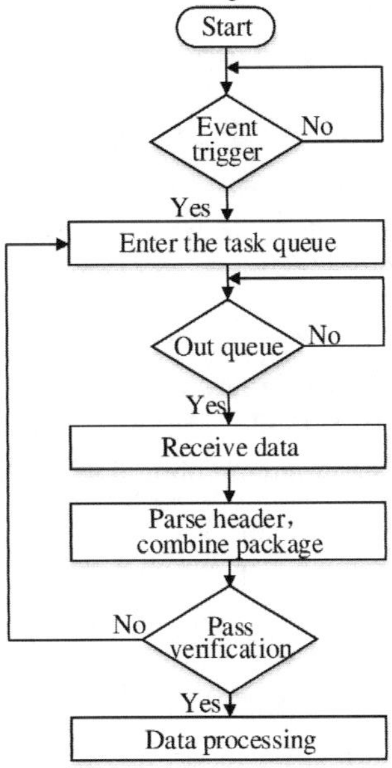

Figure 7. Working timing of reading process.

By introducing the design of periodic transmission and single transmission, this section describes design ideas and implementation methods of the interface implementation layer completely. The functional interfaces designed in this layer are secondarily encapsulated by the interface definition layer to implement general purpose interfaces that meet the design requirements.

4. Analysis of experimental result

In order to realize the communication bridge with reliability, functionality and versatility, this design proposes a universal interface model. To verify the feasibility of the scheme, this paper uses the PCIe interfaces applying to this model as an example to test. The test contents include reading and writing under the register transmission mode and the DMA transmission mode, and periodic transmission between the host computer software and DSP. The experimental results show that the error rate of data transmission process is low and the communication function of PCIe bus is implemented reliably.

Through multiple tests on the previous designed interfaces, this module is well adapted to the communication needs of different data sizes and multiple transmission modes. The universal module works stably and has simplified, intuitive and unified user-oriented interfaces. Its feasibility is demonstrated by the illustration of PCIe.

5. Conclusion

The universal interface model designed in this paper is intended to provide users with a simple, intuitive and easy-to-use universal interface module. By modelling, designing and packaging the interfaces, the program is deeply decoupled, which improves its portability and scalability.

Although it is inevitable to complicate the design and sacrifice certain performance while obtaining flexibility and scalability of the interfaces, it is acceptable for most application environment of the detection systems. What's more, the generalized and modular design of the detection system, as well as the scalability and portability of the application software, are greatly improved.

Reference

[1] Song LZ, Zhou HC and XU YP, The design and development of the test system for numerical control equipment interfaces, Modular Machine Tool & Automatic Manufacturing Technique, pp 26-9, 20 Dec. 2012.

[2] Wood.B, Backplane 101: RapidIO, PCIe, Ethernet, *World of Computer Automation*, pp 61-9, 2010.

[3] Yu W, Zhang Y, Xu H, Huang J, Gan C and Lu W, High Performance PCIe Interface for the TPCM Based on Linux Platform,*8th Int. Symp. on Computational Intelligence and Design* (ISCID 2015),11 May. 2016, Vol.2, pp 422-425.

[4] Lam H, Kirchgessner R, George A D, Reconfigurable Computing Middleware for Application, Portability and Productivity,*24th Int. Conf. on Application-Specific Systems, Architectures and Processors*, 5-7 June.2013, Washington, DC, USA, pp 211-18.

[5] Yan H and Zhang YR, Application of design pattern to communication interface design, *Computer Systems & Applications*, vol.5, pp 172-5,2012.

[6] Orzen S N, Interaction Understanding in the OSI Model Functionality of Networks with Case Studies,*9th Int. Symp. on Applied Computational Intelligence and Informatics* (SACI),15-17 May.2014, Timisoara, Romania, pp 327-30.

[7] Meng S and Lu J, Design of a PCIe Interface Card Control Software Based on WDF, *5th Int. Conf. on Instrumentation and Measurement, Computer, Communication and Control* (IMCCC 2015), 11 February.2016, pp 767-70.

[8] Hou HC, Wang YW and Li H, A high-speed DMA transmission system based on PCI express bus, *Microelectronics*, vol.43, pp 383-6, 20 June.2013.

[9] Zhao CY, Yan CX and Yu P, Real time scheduling of messages on 1553B bus, *Optics & Precision Engineering*, vol.18, pp 732-40, March.2010.

IOP Conf. Series: Materials Science and Engineering **449** (2018) 012027 doi:10.1088/1757-899X/449/1/012027

A PSK Signal Symbol Rate Estimation Algorithm based on the Combination of Stochastic Resonance and Wavelet Transform

Zheng Zhang, Jinquan Ma and Xuecheng Wan

Information system engineering institute, Information Engineering University, Zhengzhou, Henan, China.

zz2018zheng@gmail.com

Abstract. In non-cooperative communications, the deterioration of the channel makes the SNR of the received PSK signals at a low level, resulting in the failure to estimate the symbol rate. Stochastic resonance can use noise energy to strengthen the weak signals, and wavelet transform can effectively detect the transient phase. This paper proposes to use stochastic resonance combined with wavelet transform to estimate the symbol rate of the PSK signals. It not only overcomes the drawback that stochastic resonance disperses easily as a nonlinear system, but also reduces the influence of the optimal size of the wavelet. The simulation results show that this method can improve the output peak to a certain extent and reduce the threshold of SNR, which is of great significance for the symbol rate estimation for weak signals.

1. Introduction

In digital communication system, the symbol rate is related to a series of signal processing steps, such as demodulation, recovery of baseband sequence, estimation of time lapse, and so on [1]. In non-cooperative communications, it is often necessary to estimate the symbol rate blindly. However, if the signal-to-noise ratio (SNR) is extremely low, it is impossible to estimate the symbol rate accurately, and thus it is difficult to demodulate the signal, which may cause the loss of important information. Therefore, studying the symbol rate estimation problem under low SNR is of great value.

Stochastic resonance is a kind of nonlinear system in which periodic output is enhanced by the cooperation with periodic signals and random interference [2]. With the deepening of the research, scholars have found that stochastic resonance can also transfer the energy of noise to non-periodic signals with certain bandwidth, which enables stochastic resonance to be applied in many fields including signal processing. Meanwhile, if the stochastic resonance is only used for pre-processing the signals [3], the result is not stable due to the non-linear feature of the system, and it is usually not ideal. The wavelet transform can effectively detect and extract the signal phase when estimating the symbol rate [4], but the optimal scale is difficult to grasp, and its effect will deteriorate when the noise is strong.

To address the intermediate frequency PSK signals in low signal-to-noise ratio, a method of symbol rate estimation based on the combination of stochastic resonance and wavelet transform is proposed. First, the receiving signal is processed by the bistable stochastic resonance system, transferring some of the out-of-band noise energy to the in-band target signal. Then, the transient information can be detected by wavelet transform after energy enhancement. Finally, symbol rate is determined by calculating the spectrum of the squared envelope signal. This method can combine and strengthen the advantages of stochastic resonance and wavelet transform. After noise is greatly reduced and signal energy is enhanced, transient characteristics can be captured and extracted in time.

2. Classical PSK symbol rate estimation algorithm

2.1. Spectrum of the squared envelope signal

PSK can be written as

$$S(t)=\sum_{n}(\sqrt{S_n})g(t-nT_B-t_0)\exp[j(2\pi f_c t+\phi_n+\theta_c)]+\Gamma(t),\ \ 0\le t\le nT \tag{1}$$

where A denotes the amplitude, φ_n the phase with $\varphi_n\in\{(2m-1)\pi/M\mid m=1,2,...,M\}$, T_B the symbol duration, $R_B=1/T_B$ the symbol rate, t_0 a channel delay, θ_c the initial phase of the carrier, $g(t)$ a baseband pulse made of square-root raised cosine roll-off filter, and $\Gamma(t)$ the additive complex white Gaussian noise.

The classic method of estimating R_B depends on the spectrum of the squared envelope signal, because of the fact that $g(t)$ is band-limited, therefore discrete component can appear in the form of the single line spectrum which stands for the symbol rate after calculating the spectrum of the squared envelope signal [5]. $\hat{R}_B$ is estimated as follows:

$$\hat{R}_B=\arg\max_f|FFT[|s(t)|^2]| \tag{2}$$

The formula indicates that the symbol rate is the frequency point corresponding to the maximum spectral line in the field of frequency. This algorithm is not affected by the carrier and is easy to operate. Since it does not pre-process the signal, it cannot detect weak signal effectively, thus, only applicable to the signals with high SNR.

2.2. Wavelet transform

Wavelet transform is a kind of signal processing tool with great local characteristics, and it can detect and extract the phase of signals. Therefore, it can be applied to the detection of PSK symbol rate of signals. In addition, *Haar* wavelet function is widely used because of its simple form and prominent effect.

The discrete *Haar* wavelet is defined as

$$\frac{1}{\sqrt{k}}\psi(\frac{n}{k})=\begin{cases}1 & n=-\frac{k}{2},-\frac{k}{2}+1,...,-1\\ -1 & n=0,1,...,\frac{k}{2}-1\\ 0 & otherwise\end{cases} \tag{3}$$

where k is the scale.

For PSK signals, when the wavelet transform is in a symbol period, its value is fixed, and when the phase changes suddenly, the wavelet transform value will produce a corresponding sharp change [6].

In order to ensure the validity of wavelet transform, the scale k must be chosen under the condition of $k<f_s/R_B$ [7], in which f_s denotes the sampling rate. Since the sampling rate in stochastic resonance must be sufficiently large, the condition can easily be satisfied.

3. Combination method

3.1. Principle of stochastic resonance

The rectangular-shaped PSK signal can be considered as a periodic signal within one symbol interval with the effect of the carrier, and if the shaping filter is considered, it can be regarded as a similar periodic signal.

Bistable system is a typical nonlinear system, which can be represented by the Langevin equation

$$\frac{dx}{dt} = ax - bx^3 + A\cos\left(2\pi f_0 t + \varphi\right) + \Gamma\left(t\right) \tag{4}$$

where a and b are parameters of the system, $s(t) = A\cos\left(2\pi f_0 t + \varphi\right)$ the detected signal, A the amplitude, f_0 the frequency, and $\Gamma(t)$ the additive complex white Gaussian noise with mean value of 0 and the intensity of D.

The output's power spectrum of the bistable system is expressed as

$$S_t(D,f) = S_{f_0}(D,f) + S_\eta(D,f) \tag{5}$$

which consists of two parts: signal and noise, both of which are the function of the noise intensity D and frequency f. The output signal power spectrum is equation (6), and it is an impulse function for the single frequency signal. $r_k = \frac{a}{\sqrt{2}\pi}\exp(-\frac{a^2}{4bD})$, denotes the transition rate of the bistable system between potential wells. And the output noise power spectrum is equation (7), whose energy is distributed averagely in full band.

$$S_{f_0}(D,f) = \frac{2\pi A^2 a^2 r_k^{\,2}}{D^2(4r_k^{\,2} + 4\pi^2 f_0)}\delta(f - f_0) \tag{6}$$

$$S_\eta(D,f) = [1 - \frac{2A^2 a r_k^{\,2}}{D^2(4r_k^{\,2} + 4\pi^2 f_0^{\,2})}]\frac{4ar_k}{4r_k^{\,2} + 4\pi^2 f^2} \tag{7}$$

Considering narrow-band signal processing, the sampling rate determines the frequency band width. The input energy of the noise starts form 0 to f_s, so the signal to noise ratio of the output is

$$SNR_{out} = \frac{\frac{1}{\pi}\int_0^{+\infty} S_{f_c}(D,f)df}{\frac{1}{\pi}\int_0^{+\infty} S_\eta(D,f)df} = \frac{a^2 A^2 \pi \exp(-\frac{a^2}{4bD})}{\sqrt{2}D^2(\frac{a^2}{2\pi^2}\exp(-\frac{a^2}{2bD}) + \pi^2 f_c^2)} \bullet \frac{\arctan\frac{\pi f_s}{\frac{a}{\sqrt{2}\pi}\exp(-\frac{a^2}{4bD})}}{1 - \frac{A^2 a r_k^{\,2}}{D^2(\frac{a^2}{\pi^2}\exp(-\frac{a^2}{2bD}) + 2\pi^2 f_c^{\,2})}} \tag{8}$$

Stochastic resonance can converge the whole band energy and then transfer it to low frequency band [3], to enhance the specific signal energy accordingly. From the equation (8), we can conclude that the higher the sampling rate, the greater the output SNR gain, but the larger the computational complexity. Therefore, all these factors should be considered in the application.

Stochastic resonance not only has good performance which can greatly reduce the noise when extracting spectral lines, but also has some deficiencies: firstly, the target signal must be sampled at high sampling rate, which is usually greater than the bandwidth of more than 50 times; secondly, the system parameters are difficult to adjust[8][9], and when transferring the noise energy of the whole bandwidth to the low-frequency signal energy, the noise background at the bottom of the signal will be raised accordingly.

3.2. Combination with wavelet transform

Because processing the signal by stochastic resonance has certain limitations, the innovation in this paper focuses on the combination of stochastic resonance and wavelet transform, absorbing the advantages of them both and making up for their inadequacy, as long as the parameters of stochastic resonance be in a certain value range (in the same order of magnitude with carrier), and it is unnecessary to find the optimal parameters on purpose, which simplifies the steps. In the meanwhile, it also makes up the shortcoming of the wavelet transform whose optimal scale is difficult to confirm, because the ascension of the sampling rate makes it easy to meet the condition. Furthermore, the combination with the classic mode avoids moving the signal to baseband which require the precise estimation of carrier's frequency.

The system composition is given below, as shown in figure 1.

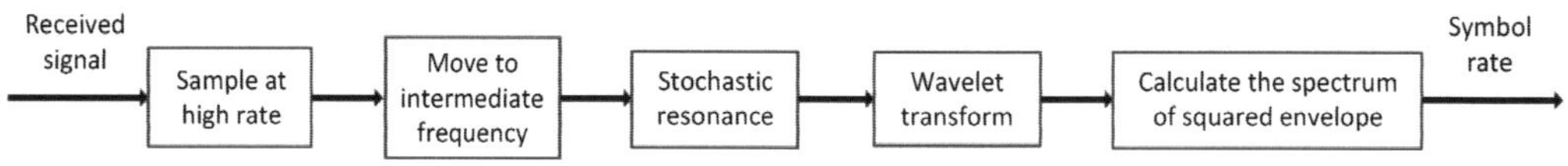

Figure 1. System block diagram of combining stochastic resonance with wavelet transform to estimate the symbol rate.

The algorithm is implemented as follows: Sample the PSK signals sent by message channel at high rate in order to satisfy the requirement of the stochastic resonance system; Move signals to the appropriate intermediate frequency to facilitate the determination of the system parameters of stochastic resonance; Pass the stochastic resonance system using iterative calculation; Use *Haar* wavelet function to transform the output of the last procedure with the minimum size of the wavelet scale as 2; Calculate the spectrum of squared envelope processed signal and observe the peak position, thus the symbol rate being obtained.

4. Simulation experiment and performance analysis

4.1. Effectiveness validation

Set the parameters of simulated signal: the symbol number N is 5000 and modulation way is QPSK, using ascending cosine shaping filter with roll-off factor of 0.6. The carrier frequency f_c is 10 Hz, the symbol rate R_B is 8 Hz, and the sampling rate f_s is 2000 Hz. Compare these three ways: the squared envelope method, the wavelet transform method, the stochastic resonance method with the new method: combining stochastic resonance with wavelet transform in the conditions which $Es/N0$ is 6dB. In addition, the scale k of the wavelet is 60, and the parameters of stochastic resonance system is $a = b = 10$. The results are shown in the figures below:

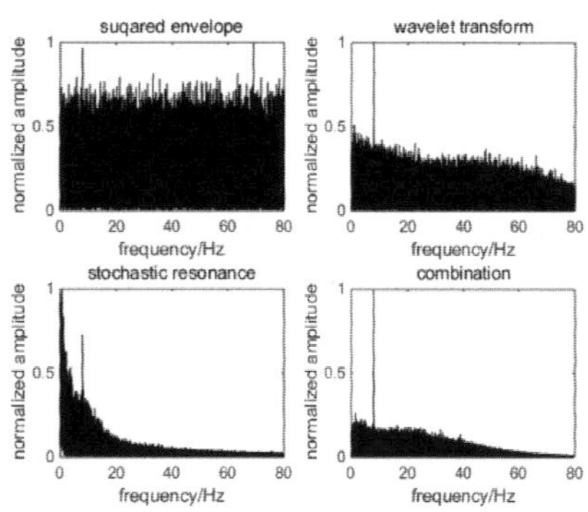

Figure 2. Output comparison diagram of the four methods.

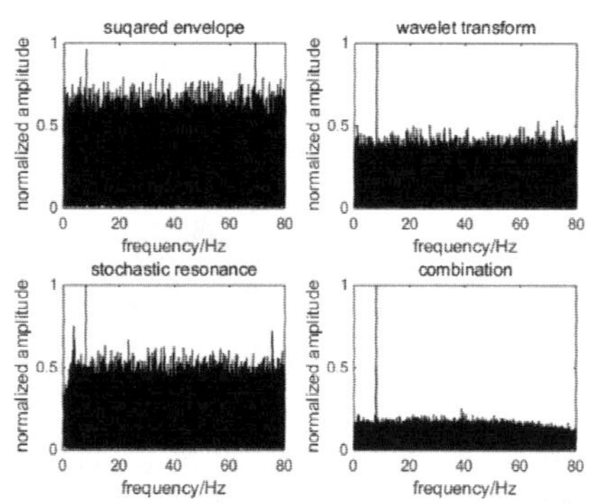

Figure 3. Output comparison diagram of the four methods after suppressing the background colour noise.

Figure 2 shows the output graphs of the four methods which have distinct differences. So as to automatically search for the spectral peak, a nonlinear filter is used to suppress the background colour noise [10], as shown in figure 3. It can be seen that, in the case of the failure of classical algorithm, the method of wavelet transform or stochastic resonance can be used to improve the performance, but at the same time, the output noise is strong either. The method of stochastic resonance combined with wavelet transform can greatly improve the spectral line and also remarkably reduce the background noise.

4.2. Analysis of the relative error

Make $Es/N0$ increase from -5 to 10, with the remaining parameters unchanged, and use the Monte Carlo simulation method in which the simulation number is 100. Because the symbol rate can be coarsely estimated according to the bandwidth, the peak value can be found within the range of $[0.5R_B, 1.5R_B]$. The relative error is defined as $\frac{|\hat{R}_B - R_B|}{M \times R_B}$ where M is the times of simulation. The graph of the relative error of the symbol rate estimation with the symbol signal-to-noise ratio is shown in figure 4.

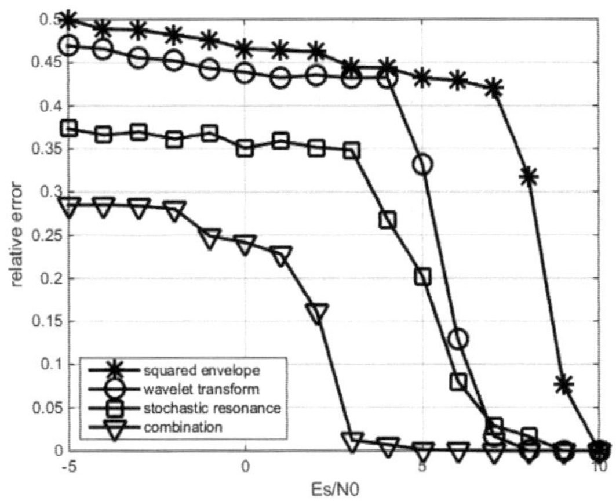

Figure 4. Relative error of QPSK signal varies with $Es/N0$.

It can be seen from the graph that with the increase of the $Es/N0$,the relative error of four methods are decreasing gradually, and there is an obvious steep descending process in each curve. The method of combining stochastic resonance with wavelet transform appears steep fall firstly, with the relative error lowering than other methods under the same noise condition.

4.3. The experimental summary

Through the simulation, compared the classic squared envelope way, the combination method, it can reduce the noise threshold at least by 6 dB for PSK (M = 2,4,8) signals. The anti-noise performance also has a higher degree of ascension than only using wavelet transform and stochastic resonance method. Table 1 shows the noise threshold of the four methods (the relative error is less than 0.05).

Table 1. Noise threshold of the four methods ($Es/N0$)

	BPSK	QPSK	8PSK
squared envelope	9	10	10
wavelet transform	6	7	7
stochastic resonance	6	7	7
combination	3	3	4

It is necessary to note that, from equation (9)(l is the oversampling rate), when $Es/N0$ is 5, SNR is about -7 dB under this experimental condition, which is due to the high sampling rate in the simulation, so our main concern is the degree of reduction of threshold under the same noise condition.

$$E_s / N_0(dB) = 10\log_{10}(l) + SNR(dB) \tag{9}$$

In addition, during the experiment, when signal-to-noise ratio is slightly lower, the stochastic resonance exists some failure situation, that is to say, when the energy is conversed from the high frequency to the low frequency, most of it moves to nearly zero frequency, which is caused by the nonlinearity of the system. The above problem is inevitable during the promotion and application of the stochastic resonance, and if only use the stochastic resonance method, it is unable to estimate the symbol rate (these singular values have been removed in the statistical noise threshold). But after the wavelet transform extracting the transient information, we can estimate the symbol rate accurately in the same resonance conditions, which shows that wavelet transform can excavate the transient information from the unideal results of the stochastic resonance.

5. Conclusion

In this paper, considering the characteristics of wavelet transform and stochastic resonance, a new method which combines them both to estimate the symbol rate is proposed and discussed, reducing the harsh conditions of using stochastic resonance method with the simple *Haar* wavelet. After the experimental verification, the combination method is still valid in a certain range of signal-to-noise ratio even though those two methods are invalid separately. Wavelet transform can improve the unideal results of the stochastic resonance, while stochastic resonance can make effective enhance pre-treatment for wavelet transform. The requirement of high sampling rate of stochastic resonance can also make the wavelet scale easy to meet the condition. In a word, this method is a new attempt to the symbol rate estimation, and it can reduce the nose threshold by about 6 dB compared with the squared envelope, wavelet transform and stochastic resonance those methods. The combination method is suitable for low SNR signals and can expand the train of thought for weak signal analysis, processing and parameter estimation.

References

[1] S E Luo, X H Yuan and L Y Luo 2011 Sub-band processing based symbol rate estimation method for MPSK signals *Journal of Signal Processing* (Beijing) vol 27 pp 304-308

[2] N Q Hu 2012 Stochastic resonance weak signal detection theory and method (Beijing: National Defence Industry Press)

[3] D Y Kong, P Hua and J Q Ma 2017 Application of stochastic resonance in narrow-band PSK signal processing *Journal of Signal Processing* (Beijing) vol 33 pp 721-731

[4] Y T Chan, J W Plews and K C Ho 1997 Symbol rate estimation by the wavelet transform *proc. of 1997 IEEE Int. Symp. on Circuits and Systems:* (Hong Kong) pp 177-180

[5] B S Koh and H S Lee 1993 Detection of symbol rate of unknown digital communication signals *Electronics Letters* vol 29 pp 278-279

[6] K C Ho, W Prokopi and Y T Chan 2000 Modulation identification of digital signals by the wavelet transform *proc. Radar, Sonar and Navigation* vol 147 pp 169-176

[7] W J Diao and J L Zhang 2006 and L Y Luo 2011 A New Kind of Arithmetic for Baud-Rate Estimation with Combination of Wavelet Transform and FFT *communication countermeasure* (Beijing) vol 2 p 18

[8] D Y Kong, P Hua and J Q Ma 2017 Adaptive stochastic resonance method based on artificial-fish swarm optimization *Chinese Journal of Electronics* (Beijing) vol 8 pp 1864-1870

[9] Y Feng 2016 Detection of cycle signal and non-periodic signal based on the principle of stochastic resonance *Computing Technology and Automation* (Hunan: Hunan University) vol 35 pp 86-88

[10] S P Liu, X Wen and L Jin 2007 A new nonlinear filtering algorithm for coloured background self-noise suppressing of symbol rate estimation *Chinese Journal of Electronics* (Beijing) vol 35 pp 95-99

IOP Conf. Series: Materials Science and Engineering **449** (2018) 012028 doi:10.1088/1757-899X/449/1/012028

Short Term Load Forecasting by Adaptive Neural Network

Hong Li

Computer Systems Technology, New York City College of Technology of the City University of New York, Brooklyn, New York 11201 USA

hli@citytech.cuny.edu

Abstract. The Generation and load balance is required in economic scheduling of the generating units and in electricity market trades. Energy forecasting became very important to mitigate some of the challenges that arise from the uncertainty in the resource. The paper presents a structure of artificial neural network with an adaptive learning algorithm used for a short-term forecasting of hourly electric power load. Historical data are sourced from Global Energy Forecasting Competition 2017 (GEFCom2017) including forecasting in the domains of electric load, weather, wind power, solar power, and electricity prices. An adaptive learning algorithm is derived from analysis of system stability to ensure convergence of training process. A simplified condition of learning factor is driven for use of computer simulation. An upper bound of learning factors is derived from the theory of convergence. At iteration of network training, a learning factor is defined to satisfy the convergence condition. The simulations with different initial state of network structure demonstrate that training error steadily decrease with an adaptive learning factor starting at different initial values whereas errors behave volatile with constant learning factors. The comparison demonstrated that a learning factor arbitrarily chosen out of the predefined stability domain leads to an unstable identification of the considered system; however, an adaptive learning factor satisfying the conditions chosen for this study ensures the stability of the identification system.

1. Introduction

Electricity energy cannot be stored as it should be generated as soon as it is demanded. Since there is no "inventory" from generation to end users (customers), ideally, power systems need to be built to meet the maximum demand, the so called peak load, to ensure that sufficient power can be delivered to the customers whenever they need it. Therefore, Electric Power Load Forecasting (EPLF) is a vital process in the planning of the electricity industry and the operation of electric power systems. Accurate forecasts lead to substantial savings in operating and maintenance costs, increased reliability of power supply and delivery system, and correct decisions for future development. However, forecasting, by nature, is a stochastic problem rather than deterministic. Since the forecasters are dealing with randomness, the output of a forecasting process is supposed to be in a probabilistic form, such as a forecast within error range under such value. Many researches have been focusing on load forecasting [1]. In work [2], a short-term load forecasting was presented using multi parameter regression. In work [3], a scholastic method is investigated mainly based on decomposition and fragmentation of time series. A review of short term load forecasting using artificial neural network (ANN) is given in [4]. It concluded that the artificial intelligence-based forecasting algorithms prove to be potential techniques for this challenging job of nonlinear time series prediction. This research uses an adaptive learning algorithm which was proven to guarantee the convergence of the training process by updating the learning factor at iteration. The data are sourced from the Global Energy Forecasting Competition 2017 (GEFCOM2017) which also including forecasting in the domains of electric load, weather, wind power, solar power, and electricity prices. In the following section, the artificial neural network and an adaptive algorithm are described and simulation results are presented.

2. Statistical forecasting models

2.1. Non-learning approach models

The Non-learning models describe the connection between solar irradiance from numerical weather predictions and electric power production directly by statistical analysis of time series from historical data without considering the physics of the system. This connection can be used for forecasts in the future plant outcomes. Many regression models are already implemented as time-series forecasting models, some of which include autoregressive integrated moving averages, and multiple linear regression analysis model.

2.2. Artificial neural network (ANN) model

The Artificial intelligence (AI) methods are used to learn the relation between prior electric load along with predicted weather condition and the power output. AI methods use algorithms that are able to implicitly describe nonlinear and highly complex relationship between input data and output power. High quality time series data consisting of weather predication and power output of the past are very important. One of the most common statistical learning models is the artificial neural network.

The ANN is a set of processing elements (neurons or perceptrons) with a specific topology of weighted interconnections between these elements and a learning law for updating the weights of interconnection between two neurons. In work [5], the Lyapunov function [6, 7] approach was used to provide stability analysis of Backpropagation training algorithm of such network. However, the training process can be very sensitive to initial condition such as number of neurons, number of layers, and value of weights, and learning factors which are often chosen by trial and error. The Backpropagation algorithm is used for learning – that is, weight adjusting. In the Backpropagation concept, information flows in one direction between the neurons (nodes) and the errors backpropagate in the opposite direction, changing the strength (weights) of the synapses (links) between the nodes while attempting to minimize the errors by using appropriate optimization technique such as the gradient descent method. After sufficient training iterations with known input data, the weights between the nodes are adjusted until the give correct response. Then, the ANN will give predicted output to the (unknown) input data. More sophisticated algorithms are introduced for training ANNs with different optimization methods to improve the performance [7].

The Least Square error function is defined and verified satisfying the Lyapunov condition so that it guarantees the stability of the system. In the work [5], the analysis carries out a method that defines a range of value of learning factors at iteration which ensure the condition for stability are satisfied. In simulation, instead of selecting a learning factor by trial and error, author defines an adaptive learning factor which satisfies the convergence condition and adjust the connection weight accordingly.

The ANN model problem can be outlined as follows: a set of data is collected from the system including input data and corresponding output data observed or measured as the target output of the ANN model. The set is often called “training set”. An ANN model with parameters, called weights, is designed to simulate the system. When the output from neural network is calculated, an error representing the difference between target output and calculated output from the system is generated.

The learning process of neural network is to modify the network, the weights, to minimize the error.

Consider a system with N inputs $X = \{X_1, \dots, X_N\}$ and M output units Y $=\{Y_1, \dots , Y_M\}$. A recurrent network combines number of neurons, called nodes, feed forward to next layer of nodes. A system of a single layer with M outputs can be expressed in form of

$$Y_{jp} = f(Z) = f\left(\sum_{i=1}^{N} w_{ij} X_{ip} + \sum_{i=1}^{D} v_{ij} Y_{j(p-i)}\right) \quad (1)$$

where w_{ij} is called connection weight from input X_i to output Y_j; v_{jj} is called connection weight of local feedback at *j*th node with *i*th delay; $f: R \to (-1, 1)$ is a nonlinear sigmoid function

$$f(Z) = \frac{1 - e^{-\theta Z}}{1 + e^{-\theta Z}} \tag{2}$$

with constant coefficient θ, called slope; $p = 1, \ldots, T$, T is number of patterns, D is number of delay used in local feedback.

The Backpropagation algorithm has become a common algorithm used for training feed-forward multilayer perceptron. It is a generalized Least Mean Square algorithm that minimizes the mean square error between the target output and the network output with respect to the weights. The algorithm looks for the minimum of the error function in weight space using the method of gradient descent. The combination of weights which minimizes the error function is considered to be a solution of the learning problem. A proof of the Backpropagation algorithm was presented in [7] based on a graphical approach in which the algorithm reduces to a graph labeling problem.

The total error E of the network over all training set is defined as

$$E = \frac{1}{T} \sum_{k=1}^{N_L} \sum_{p=1}^{T} e_k^2(p) \tag{3}$$

where $e_k^2(p)$ is the error associated with *pth* pattern at the *kth* node of output layer,

$$e_k^2(p) = (\, d_k(p) - Y_k^L(p))^2 \tag{4}$$

where $d_k(p)$ is the target at kth node and $Y_k^L(p)$ is the output of network at the *k*th node. The learning rule was chosen following gradient descent method to update the network connection weights iteratively,

$$\Delta W_j = -\mu \frac{\partial E}{\partial W_j}; j = 1, \ldots, M \tag{5}$$

$$\Delta v_j = -\mu \frac{\partial E}{\partial v_j}; j = 1, \ldots, D \tag{6}$$

where $W_j = (w_{1j}, \ldots, w_{Nj}) and\ v_j = (v_{j1}, \ldots, v_{Dj})$ are weight vectors in *j*th node; μ is a constant called learning factor.

From work [5], an extended and simplified condition was derived such that the system defined in (1) – (2) converges if the learning factor in (5) – (6) satisfies the following conditions:

$$2 - \theta\, |v^o| > 0\,, \tag{7}$$

$$\mu < \frac{\left(2-\theta\left|v_j^o\right|\right)^2}{4\theta(2+\theta(|w^o|-\left|v_j^o\right|} \tag{8}$$

3. Short term load forecasting

The changing energy landscape requires rigorous analysis to support robust investment and policy decisions. Power systems are complex; hence researchers and analysts often rely on large numerical computer models for a variety of purposes, ranging from price projections to policy advice and system planning. Such models include unit commitment, dispatch, and generation expansion models [8, 9]. These models require a large amount of input data, such as information about existing power stations, interconnector capacity, and yearly electricity consumption, and ancillary service requirements, but also (hourly) time series of load, wind and solar power generation, and heat demand. Fortunately, most of these data are publicly available, from sources such as transmission system operators, regulators, or industry associations.

3.1. Data source

The Global Energy Forecasting Competition 2017 (GEFCOM2017) provides historical data, including electric load, weather, and electricity prices, etc. The hourly data file is downloaded from the ISO New

England website which is publicly available to researchers. The object is to determine hourly electric load. Historical hourly load and temperature from year 2016 are used in this research.

Data from 2016 are used to train the ANN model. In Figure 1, hourly real time electric load of 60 days from January to February are in a scatter plot.

The scatter plot in Figure 2 demonstrated the relation of electric load that increases when the temperature is low.

The target is the hourly load. For every hour of electric load as output, the 13 inputs are defined as follows:

1 – 8: Electric load of every three hour from the previous 24 hours

9 – 16: Temperature of every three hour from the previous 24 hours

Data are used to setup a set of training data of 1400 patterns. Inputs and outputs are normalized to range from -1 to 1. After the ANN model is trained, the 10 hours forecast of electric load is calculated from the model and demoralized and then compared with the real time electric load.

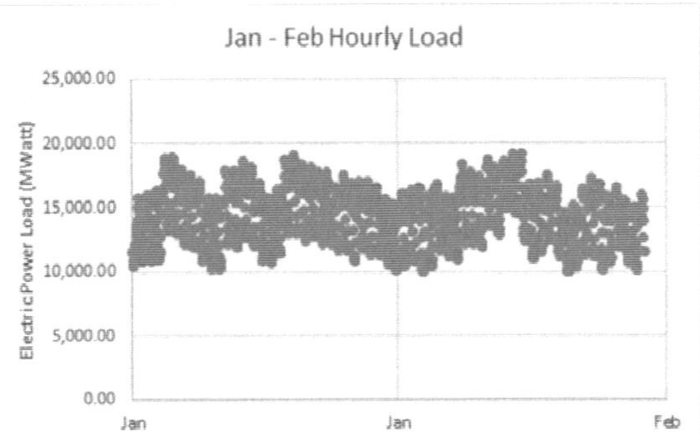

Figure 1 Scatter Plot of Load Data from Jan. and Feb.

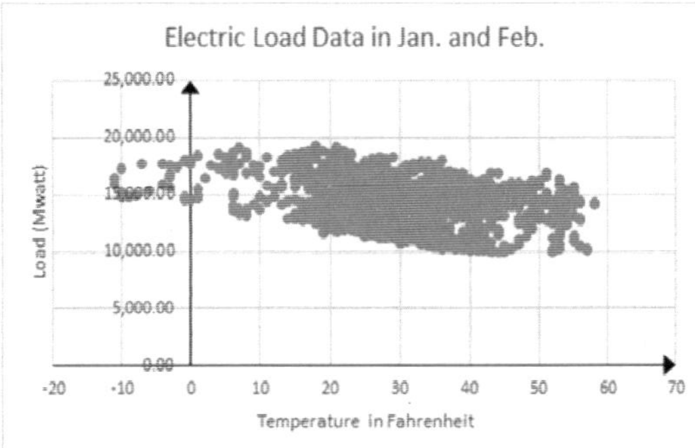

Figure 2 Scatter Plot of Load with Temperature Data in Jan. and Feb.

3.2. Adaptive learning factor

In neural network training using Backpropagation algorithm, the initial weights are randomly selected and the learning factor is preselected. The performance of the learning can sometimes be very volatile due to the selection of the learning factor. To find the optimal fit, the trial and error is common practice that runs the simulation with different values of learning factors. In this research, an upper bound of learning factors (8) is derived from the theory of convergence. At iteration of network training, the norm of weights is calculated and a learning factor is defined to satisfy the convergence condition (8).

3.3. Simulation results

A three layers neural network structure is selected with 16 inputs, 8 and 4 nodes in the hidden layer and one output. The connection weights are randomly generated between 0 and 1 as initial values. With learning factor, slope, and momentum term set as 0.8, 0.75 and 0.2 respectively, the training with the fixed learning factor stays at local minimum error 0.6964 shown in Figure 3 whereas with adaptive learning factor, the training reached to absolute error 0.031 shown in Figure 4. The Figure 5 and Table 1 demonstrate forecasting of electric load of 10 hours from the ANN model compared with the real time electric load.

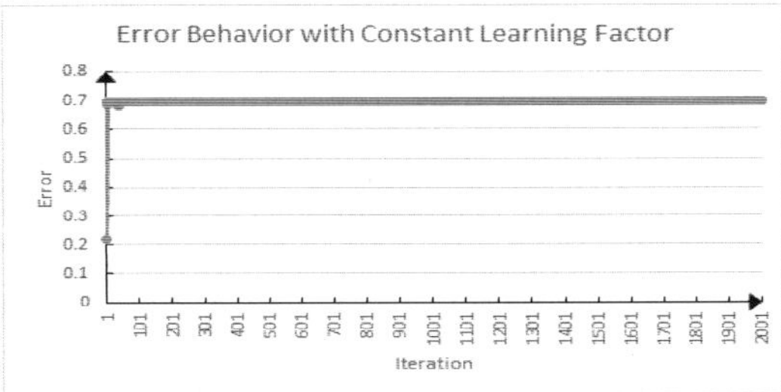

Figure 3 Error Behavior of ANN Training with Constant Learning Factor

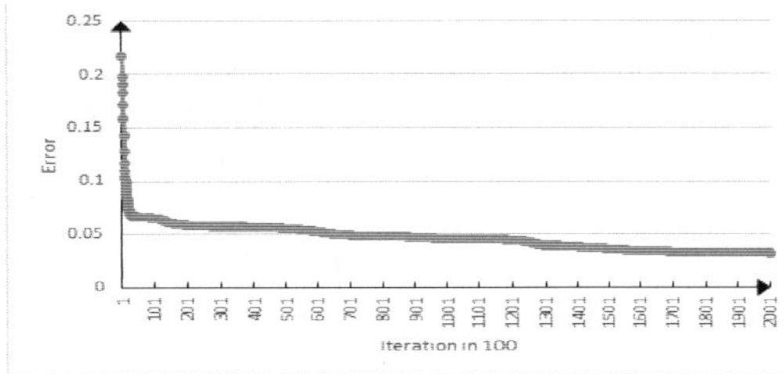

Figure 4 Error Behaviour of ANN Training with Adaptive Learning Factor

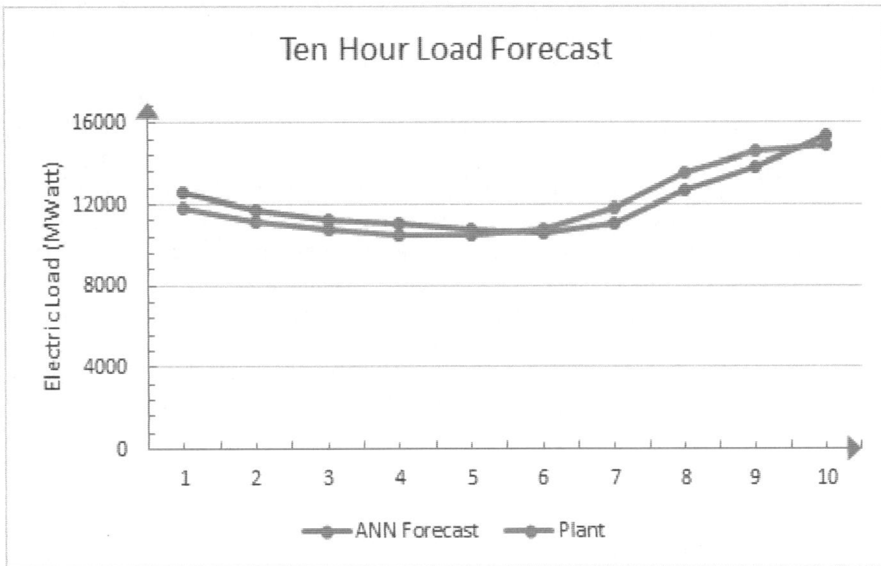

Figure 5 Ten Hour Load Forecasting

Table 1 10 Hours Forecasting of Electric Alload

hour	Electric Load (MWatt)	
	ANN Forecasting	*Real Time Load*
1	11696.8484	11120.818
2	11267.3753	10712.213
3	11054.60513	10485.787
4	10777.6042	10471.03
5	10537.2718	10778.012
6	11083.74692	11774.682
7	12692.034	13509.232

hour	Electric Load (MWatt)	
	ANN Forecasting	*Real Time Load*
8	13863.0863	14560.125
9	15355.1231	14851.957
10	15137.3306	14886.081

4. Conclusion

An Artificial Neural Network model is trained to give short term forecasting of electric load. The historical data of time-series electric load and weather temperature are used for training the model which successfully provide 10 hours forecasting. To ensure the convergence of training and avoid unstable phenomena, an adaptive learning factor is calculated at iteration of training following the analysis of convergence theory satisfying the convergence condition. The analysis results in a condition which provides an upper boundary of the learning factor. Instead of selecting a constant learning factor by trial and error, an adaptive learning factor is calculated at iteration satisfying the convergence condition. Furthermore, a more simplified condition was used to provide a feasible implementation of the adaptive learning factor. The simulations demonstrate for training with an adaptive learning factor as well as with a selected constant learning factor. The comparison demonstrated that a learning factor arbitrarily chosen out of the predefined stability domain leads to an unstable identification of the considered system; however, an adaptive learning factor satisfying the conditions chosen for this study ensures the stability of the identification system. The ANN network is trained with 60 days of data, including electric load and weather temperature and 10 hours forecasting is presented.

References

[1] J.P. Rothe, A.K. Wadhwani and S. Wadhwani, Short term load forecasting using multi parameter regression, International Journal of Computer Science and Information Security, Vol. 6, No. 2, 2009.
[2] H. Tao, D. Dickey, Electrical load forecasting: Fundamentals and Best Practices, Online open access eBook https://www.otexts.org/book/elf
[3] E. Almeshaiei and H. Soltan, A methodology for Electric Power Load Forecasting, Alexandria Engineering Journal (2011) 50, 137–144 Elsevier.
[4] A. Baliyan, K. Gaurav, and S.K. Mishra, A Review of Short , Term Load Forecasting using Artificial Neural Network Models, International Conference on Intelligent Computing, Communication & Convergence, 2015
[5] Hong Li (2014). Adaptive Learning Factor of Backpropagation in Feed Forward Neural Networks, International Journal of Modern Engineering, Vol. 14, No 2, Spring/Summer 2014, pp 47 – 53.
[6] R. Rojas, Neural Networks, Springer-Verlag, 1996
[7] Sontag, E.D., Mathematical Control Theory, Springer-Verlag, New York
[8] Massachusetts Institute of Technology Energy Initiative. The Future of the Electric Grid Chapter 3: Integration of Variable Energy Resources. Cambridge, MA: MIT, 2011. http://web.mit.edu/mitei/research/studies/documents/electric-grid 2011/Electric_Grid_3_Integration_of_Variable_Energy_Resources.pdf
[9] J. Kleissl, Solar Energy Forecasting and Resource Assessment. Elsevier, 2013.

IOP Conf. Series: Materials Science and Engineering **449** (2018) 012029 doi:10.1088/1757-899X/449/1/012029

Fault Tolerant Three Level ANPC Inverter Circuit with Finite Set Model Predictive Control

Li Xue-feng, Lei Xiao-ben, Han Jan-ding, Wang Chuan-qi and Zeng Jia-qi

Aeronautics Engineering Institute, Air Force Engineering University, Xi'an 710038, China

519132089@qq.com

Abstract. An improved active neutral point clamped (ANPC) inverter circuit topology is proposed. Through the fault-tolerant leg ,the fault-tolerant performance of the inverter circuit has been greatly improved. The fault-tolerant leg works as an actuating device which control the neutral point to remain stable, when there is no fault with the inverter. When some power tubes are broken the leg works as the redundant device to replace the power tubes. The finite set model predictive control method is used for this inverter circuit. This control method is quite suitable for the electronic circuit .This method is employed to lower THD and to advance its dynamic, so that the error tolerance and high power output of the aviation inverter are satisfied. The feasibility and effectiveness are proved by simulations.

1.Introduction

The power supply needed in the aircraft is often inconsistent with the main power supply, and there is no special AC power supply in some aircraft's DC power supply systems, but its electronic devices such as radar, gyroscope and radio navigation equipment require the alternating current of 36V or 115V of the frequency 400Hz. At the same time, in order to satisfy the emergency power supply for the flight instrument and the related equipment, the inverter circuit must be connected with the battery as the emergency power supply to convert the DC power into alternating current when the main power supply is broken. In order to satisfy the reliability of the aircraft power supply, the inverter circuit needs a higher fault tolerance. The design of its fault tolerant topology not only takes into account the increase of the failure probability caused by the increase of power devices, the multiplied increase in the complexity of the control methods, and the increase of the cost of the circuit.

A four-leg fault-tolerant NPC inverter was introduced in [1], and in some document the output of the fault phase is directly connected with the DC bus neutral point through an additional device or a the fourth leg. In document [2] the fourth leg is added to the inverter ,but this topology can not afford the fault situation that simultaneous breakage of two phase . Considering these methods and the fault situation of aeronautical inverter, we put forward a new topology based on the document.

The development of control theory has gone through the two stages of classical control theory and modern control theory. With the continuous development of material science and semiconductor technology, the topology of inverter circuit is becoming more and more complex. The types of control methods based on inverter circuit are more and more, and the characteristics of the mathematical model and control method can be obtained. It is divided into linear control and nonlinear control. the linear control is PID, and the nonlinear control conclude repetitive control [3], hysteresis Control[4], deadbeat control, Finite Control Set Model Predictive Control(FCS—MPC)[5-7].

The FCS-MPC can make full use of the discrete properties of power electronic circuits, through the FCS-MPC each possible combination of switch states are considered, and the switch state, which minimized objective function was used as circuit control, the complex mixed integer quadratic programming problem can be avoided.

2.The topology analysis of the fault tolerant ANPC inverter

In this paper, the traditional ANPC inverter circuit is improved by redundancy thought, and its topology is shown in Figure 1.

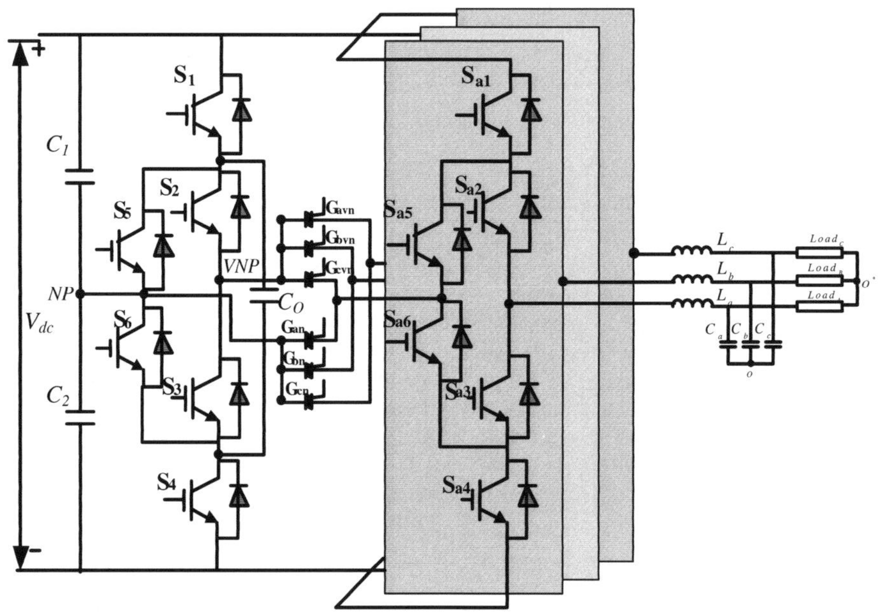

Figure 1. fault tolerant ANPC inverter topology

From the topology of figure 1, it can be seen that the fault tolerant ANPC inverter circuit structure has a redundant leg comparing to the original ANPC inverter . On the right side of the topology, it can be seen that the three phase leg is the same as the original ANPC inverter circuit, and the middle point of each phase is connected with the redundant leg. The point NP on the left side of the redundant leg is the neutral point of the all circuit and the VNP is the virtual point. Through the transformation between the NP and the VNP corresponding control signal and the fault isolation means, the circuit can realize the fault-tolerant function.

When the circuit works normally, the redundant leg acts as a device for adjusting the location in the circuit. The capacitor C_o is charged to Vdc/2 before the operation of the circuit. At this time, the voltage is maintained by the virtual middle point VNP. The ΔU and ΔI are defined as (1).

$$\Delta U = V_{dc} / 2 - U_{C_o} ; \Delta I = i_{s_A} + i_{s_B} + i_{s_C} \quad (1)$$

In formula (1), $i_{s_A}, i_{s_B}, i_{s_C}$ are the current which flow from the virtual neutral point VNP to the midpoint of two clamp power pipes in the three-phase bridge arm of A, B and C, and the current from point VNP into the direction of the three phase bridge arm is positive. When the circuit works normally, the switch tube S_5 and S_6 have been turned off.

In the situation of $\Delta U < 0, \Delta I < 0, U_{C_o} > V_{dc} / 2$, the switch S_1 and S_3 are engaged, the capacitor is in the discharge state, and the capacitor voltage drops. When $U_{VNP} = V_{dc} - U_{C_o}$ The voltage of the neutral point will gradually rise to Vdc/2.

In the situation of $\Delta U < 0, \Delta I > 0, U_{C_o} > V_{dc}/2$, the switch S_2 and S_4 are engaged, the capacitor is in the discharge state, and the capacitor voltage drops. When $U_{VNP} = V_{dc} - U_{C_o}$ The voltage of the neutral point will gradually drop to Vdc/2.

In the situation of $\Delta U > 0, \Delta I < 0, U_{C_o} < V_{dc}/2$, the switch S_2 and S_4 are engaged, the capacitor is in the charge state, and the capacitor voltage rises. When $U_{VNP} = V_{dc} - U_{C_o}$ The voltage of the neutral point will gradually drop to Vdc/2.

In the situation of $\Delta U > 0, \Delta I > 0, U_{C_o} < V_{dc}/2$, the switch S_1 and S_3 are engaged, the capacitor is in the charge state, and the capacitor voltage rises. When $U_{VNP} = V_{dc} - U_{C_o}$ The voltage of the neutral point will gradually rise to Vdc/2.

By comparing the magnitude of current and capacitor voltage, the switches in redundant leg are switched back and forth between S1, S3 switch state and S2 and S4 switch state. The advantage of this method is that it does not need to balance the relationship between the output performance and the offset of the loci by calculating the control amount of the algorithm, and it is easy to realize, but the disadvantage is obvious. This method is used to control the middle loci, which requires high switching frequency and large switching loss.

When the circuit works normally, the right three phase APNC output circuit is the same as the traditional ANPC operation mode. The switch states of one phrase of inverter is shown in table 1, similarly to the rest two phrases, when "1" represents "on", "0" represents "off".

Table 1. switch states of one phrase

Switch state	S_{a1}	S_{a2}	S_{a3}	S_{a4}	S_{a5}	S_{a6}	Output voltage
P	1	1	0	0	0	1	$V_{dc}/2$
OU1	0	1	0	0	1	0	0
OU2	0	1	0	1	1	0	0
OL1	0	0	1	0	0	1	0
OL2	1	0	1	0	0	1	0
N	0	0	1	1	1	0	$-V_{dc}/2$

When the power transistor fails, the topology of the circuit topology is reconstructed, taking the A phase fault as an example. When the A phase leg power tube S_{a1} break down, the A phase can not output P state and OL2 state in the circuit output state shown in table 1. Because the OL2 state output level is 0, the other redundant 0 level state can be supplemented, but the P state output voltage is Vdc/2, so the circuit must be fault-tolerant to guarantee the circuit work normally. The fault-tolerant scheme is shown in figure 2, it shows the current path in different states. The phrase A is linked to VNP by G_{avn} The phrase B, C are linked to NP by G_{bn} and G_{cn}.

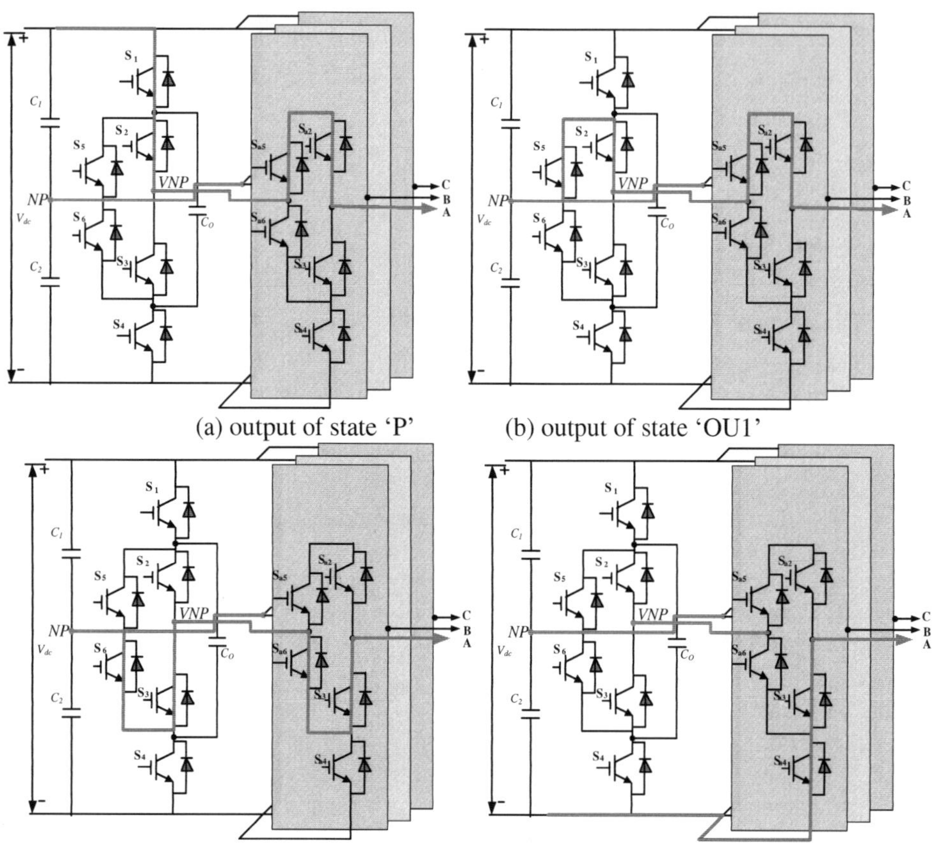

(a) output of state 'P' (b) output of state 'OU1'

(c) output of state 'OL1' (d) output of state 'N'

Figure 2. The fault-tolerant scheme of S_{a1}break down

when S_{a2} break down, phrase A can not output P state and OU1 and OU2 state. Its circuit topology reconfiguration scheme is shown in figure 3.

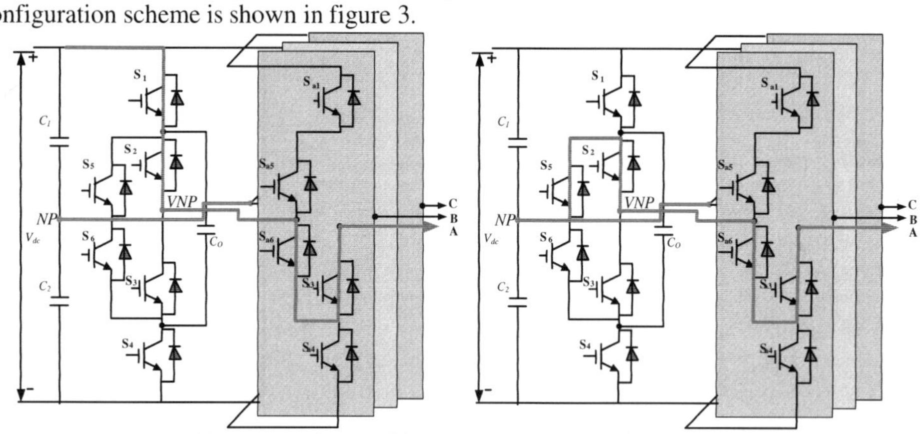

(a) output of state 'P' (b) output of state 'OU1'

IOP Conf. Series: Materials Science and Engineering **449** (2018) 012029 doi:10.1088/1757-899X/449/1/012029

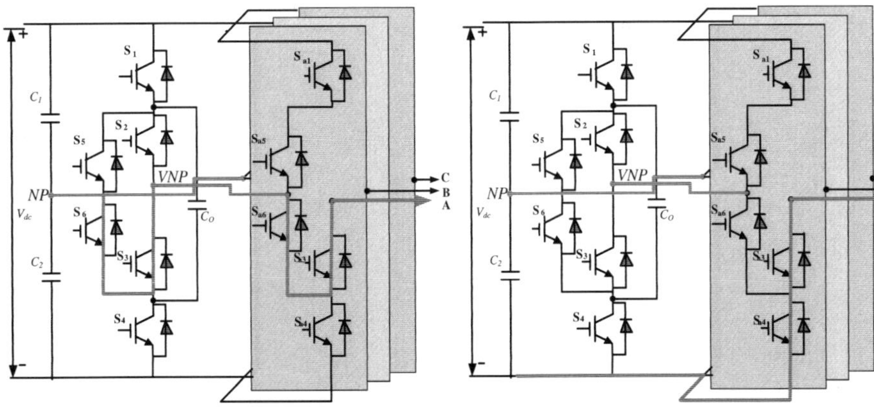

(c) output of state 'OL1' (d) output of state 'N'

Figure 3. The fault-tolerant scheme of S_{a2}break down

The situation of S_{a3}, S_{a4} break down are similar to the situation of S_{a1}, S_{a2} break down, so there is no longer to give unnecessary details.

When the same side power pipe fault between the different phase leg. For example, S_{a2} in A and S_{b2} in B break down at the same time, the failure can not output P state and OU1, OU2 state, in which OU1 and OU2 can be replaced by other redundant 0 level state. We will mainly introduces how to output P state. The topology is shown in figure 4.

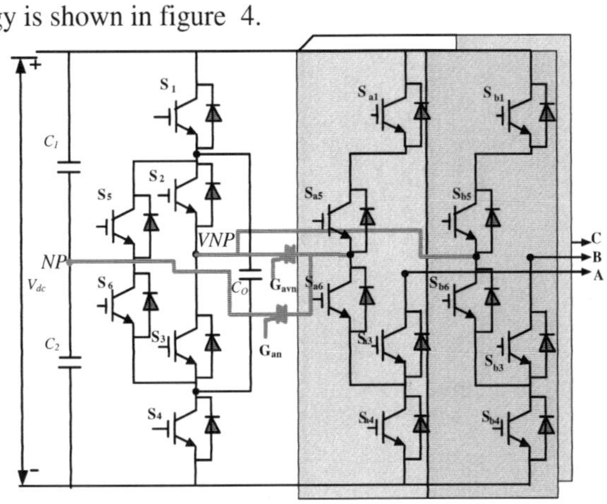

Figure 4. The fault-tolerant scheme of S_{a2}, S_{a2}break down

The neutral point of the B phase clamp power pipe is connected directly with the virtual medium site VNP, and the middle point of the A phase clamp power pipe is connected to the VNP and NP respectively through the G_{an} and G_{avn}.. When A and B two phase needs to output P state simultaneously, the A and B two-phase power pipes are in the fault tolerant scheme according to the fault of the single power pipe, connect the G_{avn}, disconnect G_{an}; when only one phase bridge arm needs to output P state, by controlling G_{avn}, and G_{an}, the bridge arm that needs to output P state is connected to VNP, disconnected with NP, and controlled according to the fault tolerance scheme of the phase single tube.

3.The FCS-MPC of ANPC inverter

Define the switch variables S_a, S_b and S_c to represent the output state of each phase leg. Take the A phase as an example, S_a=1 represents the A phrase output P state; Sa=0 represents the O state of the A phrase(OU1,OU2,OL1,OL2 are all belong to O state); Sa=-1 represents the output of P state, Thus there comes to 27 state form table 1 and figure 1 ,and the vector shows in figure 5. The expression of

IOP Conf. Series: Materials Science and Engineering **449** (2018) 012029 doi:10.1088/1757-899X/449/1/012029

voltage between the output of each leg and the virtual neutral point U_{AN} can be obtained as the expression(2)

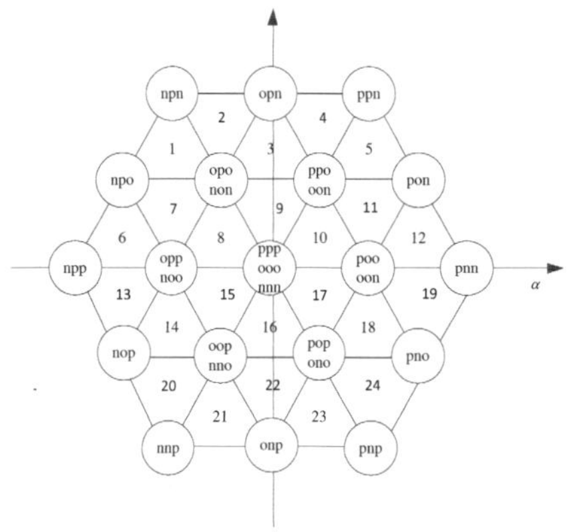

Figure 5. Output voltage vector of three level ANPC inverter circuit

$$u_{AN} = \frac{V_{dc}}{2} \times S_a \tag{2}$$

The voltage between the A, B and C of the three-phase and the point 'O' of the capacitance in the capacitor can be known from the topology are expression(3)

$$u_{AO} = u_{AN} - u_{ON}; u_{BO} = u_{BN} - u_{ON}; u_{CO} = u_{CN} - u_{ON}; u_{ON} = \frac{1}{3}(u_{AN} + u_{BN} + u_{CN}) \tag{3}$$

The u_{ON} is the voltage between O and VNP. Combining (2) and (3),there comes out expression(4)

$$\begin{bmatrix} u_{AO} \\ u_{BO} \\ u_{CO} \end{bmatrix} = \frac{V_{dc}}{6} \begin{bmatrix} 2 & -1 & -1 \\ -1 & 2 & -1 \\ -1 & -1 & 2 \end{bmatrix} \begin{bmatrix} S_a \\ S_b \\ S_c \end{bmatrix} \tag{4}$$

Defining space variables $\alpha = e^{j(2\pi/3)}$,there comes to u_o

$$u_O = \frac{2}{3}(u_{AO} + \alpha u_{BO} + \alpha^2 u_{CO}) \tag{5}$$

Similarly, the three-phase inductor current, capacitor voltage, output current, can be transformed into space vector form (6)

$$i_f = \frac{2}{3}(i_{af} + \alpha i_{bf} + \alpha^2 i_{cf}); u_C = \frac{2}{3}(u_{C_a} + \alpha u_{C_b} + \alpha^2 u_{C_c}); i_{O^*} = \frac{2}{3}(i_{aO^*} + \alpha i_{bO^*} + \alpha^2 i_{cO^*}) \tag{6}$$

According to Kirchhoff's voltage and current law and the working principle of inductance and capacitance, the formula (7) is available.

$$\begin{cases} \frac{\mathrm{d}x}{\mathrm{d}t} = Ax + B_1 u_O + B_2 i_{O^*} \\ y = Cx \end{cases} \tag{7}$$

Where $x = \begin{bmatrix} i_f \\ u_C \end{bmatrix}$; $A = \begin{bmatrix} 0 & -\frac{1}{L} \\ u_C & 0 \end{bmatrix}$; $B_1 = \begin{bmatrix} \frac{1}{L} \\ 0 \end{bmatrix}$; $B_2 = \begin{bmatrix} 0 \\ -\frac{1}{C} \end{bmatrix}$; $C = \begin{bmatrix} 0 & 1 \end{bmatrix}$

The mathematical model shown in formula (7) is a continuous mathematical model. In the finite set model predictive control, the continuous mathematical model needs discretization like (8).

$$\begin{cases} x(k+1) = A^* x(k) + B_1^* u_O(k) + B_2^* i_{O^*}(k) \\ y(k) = Cx(k) \end{cases} \tag{8}$$

Where $A^* = e^{AT_s}$, $B_1^* = \int_0^{T_s} e^{At} B_1 \mathrm{d}t$, $B_2^* = \int_0^{T_s} e^{At} B_2 \mathrm{d}t$, T_s is the sample time.

The reference voltage at the k+1 time $u^{ref}(k+1) = 3u_c^{ref}(k-1) - 2u_c^{ref}(k-2)$

In this paper, taking the output voltage optimization as an example, the selection of its objective function is described in(9).

$$g = \left[u_{C\alpha}^{ref}(k+1) - u_{C\alpha}(k+1)\right]^2 + \left[u_{C\beta}^{ref}(k+1) - u_{C\beta}(k+1)\right]^2 \tag{9}$$

$u_{C\alpha}^{ref}(k+1)$ and $u_{C\alpha}(k+1)$ are the real axis components of reference output voltage and predicted voltage at k+1 time respectively. $u_{C\beta}^{ref}(k+1)$ and $u_{C\beta}(k+1)$ are the imaginary axis components, respectively. The algorithm flow chart is shown in Figure 6

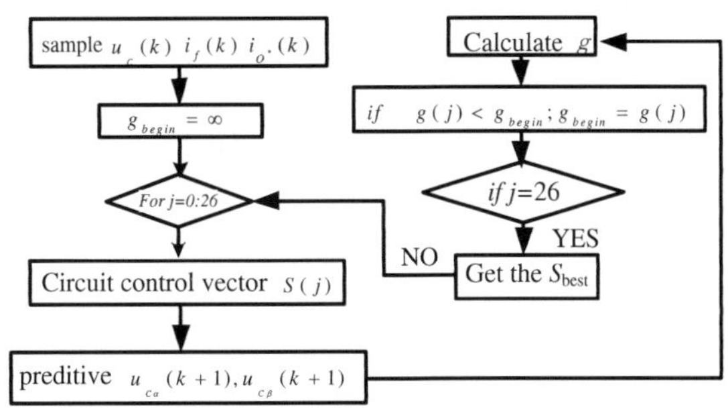

Figure 6 . Algorithm flow chart

According to the model of inverter circuit switching function constructed by formula (8), the predictive value $u_c(k+1)$ can be calculated by the controlled amount under the action of 27 possible vectors at the time of K which is made of by all the u_{AN}, u_{BN} and u_{CN}. Combining the $u_c(k+1)$, $u_c^{ref}(k+1)$ and formula (8), the best switch state can be got which goes to the g_{min}.

4.Simulation

According to the inverter topology shown in Figure 1, the strategy researched in this paper was proved by MATLAB/Simulink, main simulation parameters were provided as following. $V_{dc} = 270V$, filter inductance L=2mH, filter capacitance C= $40\mu F$, rated frequency is 400Hz, sampling period $T_s = 10\mu s$, The load is 80Ω,

The steady state experiment results of FCS-MPC based on the circuit were shown in Figure 7, THD=2.34%, which is easy for filter design. The transient experiment results are shown in Figure 8 The Figure 8 shows the change of load from empty to full load, and it shows that the FCS-MPC can meet the dynamic change requirements of the system. No mater the steady state or the transient state this method can make output has a good power quality, which meet the requirement of aeronautical inverter.

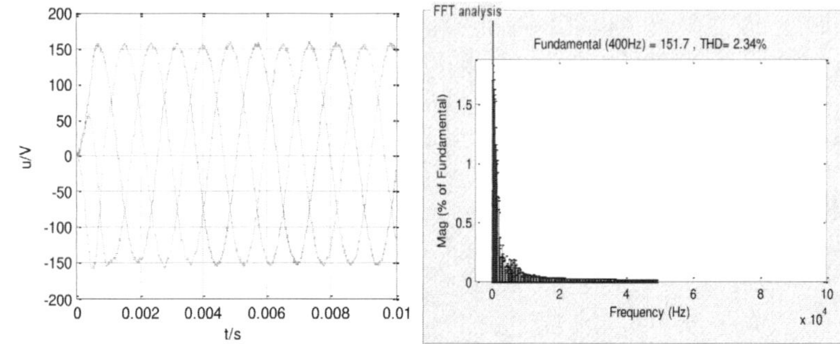

Figure 7.steady state experiment results

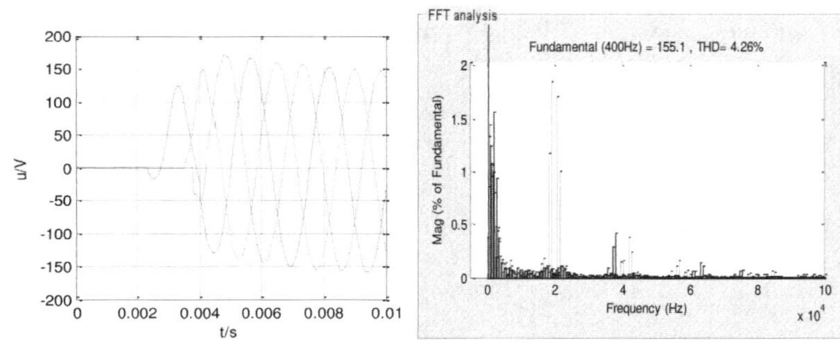

Figure 8. transient state experiment results

5.Conclusions

The tolerant three level ANPC inverter provide more fault tolerance function for the aeronautical inverter. The FSC-MPC make the low THD and fixed harmonic of output voltage are obtained, which was easy for filter design. Finally the steady state and transient properties of the strategy was proved by simulations.

References

[1] S. Ceballos, J. Pou, E. Robles, I. Gabiola, J. Zaragoza, J. L. Villate, D. Boroyevich, 2008 J IEEE Transactions on Industrial Electronics ,Three-level converter topologies with switch break down fault-tolerance capability, 55 (2008) pp:982–995.

[2] R Katebi , J He , N Weise 2017 *J IEEE Transactions on Power Electronics ,* Advanced Three Level Active Neutral Point Converter with Fault Tolerant Capabilities , 99 (2017)PP :125-1132

[3] G. Escobar, A. A. Valdez, J. Leyva-Ramos, 2007 *J IEEE Transactions on Industrial*, Electronics Repetitive-Based Controller for a UPS Inverter to Compensate Unbalance and Harmonic Distortion, 54(2008)pp:504-510.

[4] Uffe Borup, Prasad N. Enjeti, Frede Blaabjerg, 2001 *J IEEE transactions on industry applications* A New Space-Vector-Based Control Method for UPS Systems Powering Nonlinear and Unbalanced Loads ,37(6):1864-1870.

[5] Stefan Almer, Sebastien Mariethoz, Manfred Morari. 2012 *J. IEEE TRANSACTIONS ON CONTROL SYSTEMS TECHNOLOGY.* Sampled Data Model Predictive Control of a Voltage Source Inverter for Reduced Harmonic Distortion, VOL.24, NO.14,pp:89-95.

[6] S. Aurtenechea, M. A. Rodriguez, E. Oyarbide. 2007 *J. IEEE Trans. Ind. Electron,* Predictive control strategy for dc/ac converter based on direct power control. VOL.54, NO.12, pp:1261-1271,.

[7] Li Ning, Li Ying-hui, Lei Hong-li. 2013 *J. Power Electronics*, A Multi-fault Tolerant Inverter Topology. VOL.47, NO.3, pp:106-108.

Investigations on a Three-dimensional Model of the Moving-coil Linear Motor Applied in the Space-borne Micro Compressor

Y B Zhao[1,2], D Ding[1,2] and X C Mao[1,2]

[1]Shanghai Institute of Measurement and Testing Technology, Shanghai, 201203, China
[2]Shanghai Key Laboratory of Online Test and Control Technology, Shanghai, 201203, China

zhaoyb@simt.com.cn

Abstract. In order to optimize the magnetic circuits of the linear motor applied in the space-borne pulse tube cryocooler, a three-dimensional magnetostatic field model of the moving-coil linear motor in the micro compressor is established. The effects of several key dimensions on the distributions of the magnetic flux density inside the air gap are simulated and discussed by finite element method. The simulated results show that the magnetic flux density decreases with the increase of both width and thickness of air gap. According to the analysis, it will greatly facilitate the design, fabrication and future practical applications of the micro moving-coil linear compressor.

1. Introduction

The core component of a linear compressor is the linear motor. According to different moving parts, it can be divided into three main types: moving-iron, moving-coil and moving-magnet types. In recent years, both the moving-coil and moving-magnet linear compressors have wide applications in a number of scientific and industrial fields. Especially, the moving-coil linear motor realizes the complete elimination of the radial force due to its structural characteristics, and it does not generate axial force and torque on the current-carrying coil when it is open, and thus having several advantages such as high efficiency, low noise and high reliability. Therefore, it has become the preferred power source for space-borne pulse tube cryocoolers in the past thirty years [1–3].

The micro moving-coil linear compressor is designed based on the modified scaling principle [4], which assumes that the magnetic induction intensity B at the electrified inductor should be fixed. However, the magnetic induction intensity inside the air gap may be weakened due to the reduction in the size of the linear motor. Therefore, its magnetic circuit should be optimized before component processing and experimental testing. In this paper, a three-dimensional magnetostatic field model of the moving-coil linear motor in the micro compressor is established. The influences of several key dimensions on the magnet circuits are simulated and discussed.

2. Model setup

Figure 1 shows a schematic of the typical Oxford-type moving-coil dual-opposed linear compressor, in which the permanent magnet, the return iron and the moving coil form the magnetic structure of the linear motor. In order to study the characteristics of the magnetic field and then optimize the magnetic circuit configuration, a three-dimensional theoretical model of the micro magnetic structure is established, as shown in figure 2, which consists of four essential parts: the permanent magnet, the pole piece, the magnetic pole and the air gap. All of four parts are designed to be hollow cylinders for simplifying the calculation process.

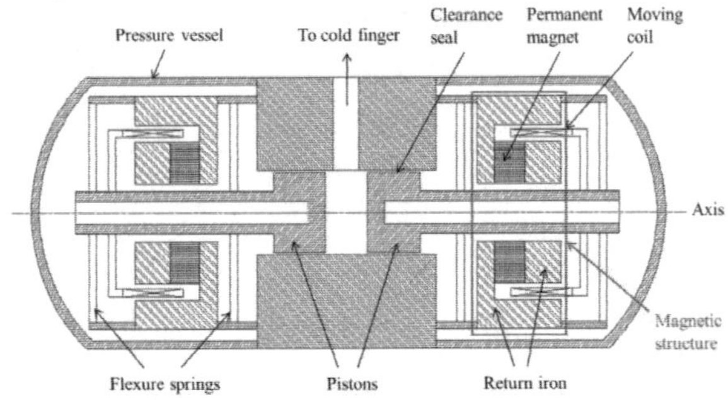

Figure 1. The typical Oxford-type moving-coil dual-opposed linear compressor

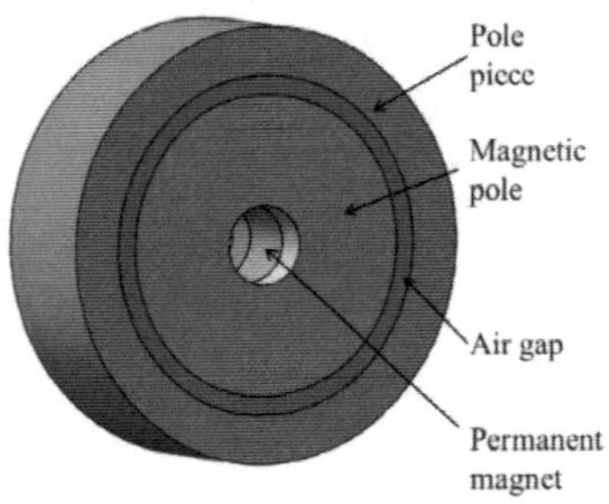

Figure 2. The 3-D model of the micro micro magnetic structure

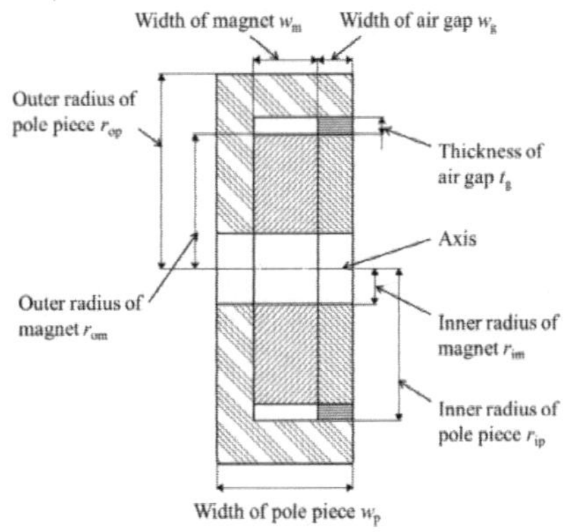

Figure 3. The cross-sectional drawing of the magnetic structure

Figure 3 shows the cross-sectional drawing of the 3-D micro magnetic structure model, in which several important dimensional parameters are marked, including the width of air gap w_g, the thickness of air gap t_g, the width of permanent magnet w_m, the width of pole piece w_p and the inner and outer radiuses of permanent magnet and pole piece r_{im}, r_{om}, r_{ip}, and r_{op}, respectively. In the whole simulation, both the inner and outer radiuses of the magnetic pole are equal to those of the permanent magnet, and the width of the magnet pole should keep the same with that of the air gap.

Table 1. Initial dimension parameters of the simulated magnetic structure

Dimension parameter	Value (mm)
Width of air gap w_g	7
Thickness of air gap t_g	2.4
Width of magnet w_m	9.3
Outer radius of magnet r_{om}	19
Inner radius of magnet r_{im}	5
Outer radius of pole piece r_{op}	27.5
Inner radius of pole piece r_{ip}	21.4
Width of pole piece w_p	21.6

Table 1 gives the initial dimension parameters of the simulated micro magnetic structure in the basic model, which is scaled down from mid-size moving-coil dual-opposed linear compressors developed in the authors' laboratory based on the scaling principles [5].

The material of the permanent magnet is selected as the NdFeB alloy. Its coercive force and maximum energy product are as high as 890 kA/m and 280 kJ/m^3, respectively, and its magnetization is in the axial direction. Both the pole piece and the magnetic pole are made of pure iron, and the air gap is filled with the working fluid of helium.

The governing equations for the 3-D micro magnetic structure model are based on Maxwell's equations, including the Faraday's law of electromagnetic induction, the Gauss's law, the flux continuity theorem and the Ampere's loop law, which are given by:

$$\nabla \times E = -\partial B/\partial \tau \quad (1)$$

$$\nabla \times H = J + \partial D/\partial \tau \quad (2)$$

$$\nabla \cdot B = 0 \quad (3)$$

$$\nabla \cdot D = \rho \quad (4)$$

where E, B, H, D, J, ρ and τ represent the electric field intensity, magnetic flux density, magnetic field intensity, electric displacement, conduction current density and volume charge density, respectively.

Each component in the model is divided into the tetrahedral mesh. The minimum length of the mesh in the air gap is set as 0.2 mm, while that in the permanent magnet, the pole piece and the magnetic pole is 0.5 mm. The total number of mesh elements is up to 140175 in the basic model and no noticeable difference about the simulated result is observed when the minimum length of the mesh in the air gap is decreased to 0.1 mm.

3. Simulated results and discussions

3.1. Width of air gap

The magnetic flux density to be investigated is inside the air gap, and therefore the size of air gap will have great impact on the distribution of magnetic field. Table 2 gives the dimensional parameters of the micro magnetic structure in Cases A1–A6, in which the width of air gap ranges from 1 mm to 11 mm, and the width of pole piece varies accordingly on the condition that other parameters remain constant. Case A4 is used as the basis of optimization, and its specific parameters are provided in table 1. The distributions of the magnetic flux density inside the air gap with different widths of air gap are calculated, and the effect of w_g on the specific thrust is also investigated.

Table 2. Parameter settings in Cases A1–A6

Case number	w_g (mm)	t_g (mm)	w_p (mm)	w_m (mm)	r_{om} (mm)	r_{op} (mm)	r_{im} (mm)	r_{ip} (mm)
A1	1	2.4	15.6	9.3	19	27.5	5	21.4
A2	3	2.4	17.6	9.3	19	27.5	5	21.4
A3	5	2.4	19.6	9.3	19	27.5	5	21.4
A4	7	2.4	21.6	9.3	19	27.5	5	21.4
A5	9	2.4	23.6	9.3	19	27.5	5	21.4
A6	11	2.4	25.6	9.3	19	27.5	5	21.4

Figures 4(a)–4(f) show the contours of the simulated magnetic flux densities inside the air gap in Cases A1–A6, respectively. It can be observed that, the magnetic flux density monotonously increases in radial direction from outside to inside in each case, except that there exist slightly uneven distributions in Cases A1 and A3. Therefore, the moving coil should be placed closer to the magnetic pole so as to get higher specific thrust. In addition, the magnetic flux density inside the air gap is between 1.02 T and 1.18 T in the basic case, and its mean value is about 1.10 T, which is basically the same with the tested value in the mid-size moving-coil linear compressor developed in the authors' laboratory. It proves the assumption in the scaling principles that the magnetic field intensity is kept constant.

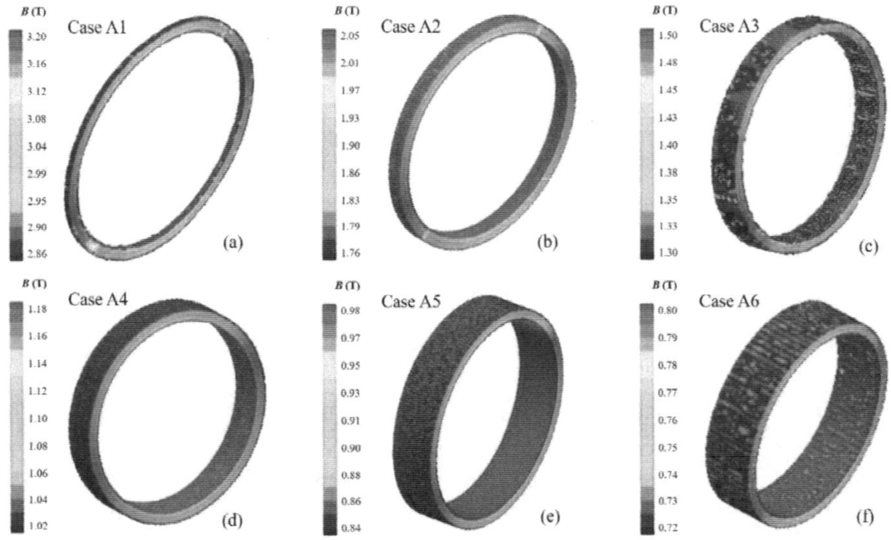

Figure 4. The contours of the simulated magnetic flux density in Cases A1–A6

It can also be found that the magnetic field intensity decreases with the increase of the width of air gap in figure 4. Figure 5 shows a detailed comparison of the radial distributions of the average magnetic flux densities in Cases A1–A6, where the radial location represents the distance between the test point and the inner boundary of air gap. The simulated values have been averaged in axial direction. The results show that the magnetic flux density drops sharply when w_g increases from 1 mm to 5 mm, and then decreases gradually when w_g increases from 5 mm to 11 mm. The reason is that the magnetic lines of force become more concentrated in the air gap with the smaller w_g. Moreover, it is worth noting that the difference between the maximum and minimum magnetic flux density also decreases with the increase of the width of air gap.

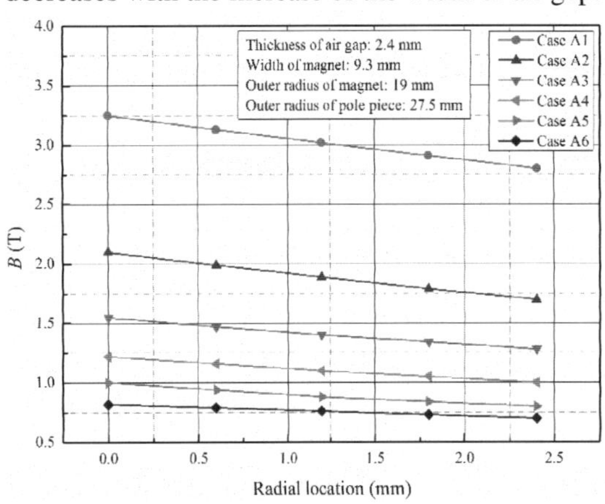

Figure 5. Radial distributions of the average magnetic flux densities in Cases A1–A6

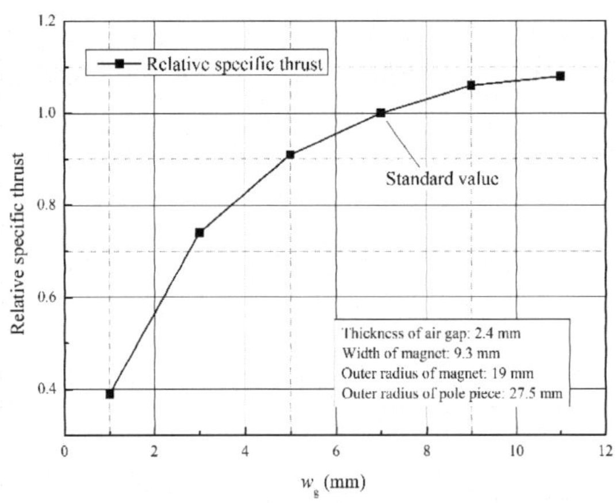

Figure 6. The relative specific thrust in Cases A1–A6

The specific thrust F_{LM} is given by:

$$F_{LM} = BL_w \propto Bw_g \tag{5}$$

where L_w represents the effective length of coil. The linear motor in the micro compressor employs the long coil design, and therefore L_w is proportional to w_g. As a result, the specific thrust is directly proportional to both the magnetic field intensity and the width of air gap. Figure 6 shows the relative specific thrust in Cases A1–A6, in which the result in Case A4 is set to the standard value of 1. With the increase of the air gap width, the magnetic flux density decreases as shown in figure 5, but the specific thrust shows an upward trend due to the greater increase of the effective coil length. Especially when the air gap width increases from 1 mm to 5 mm, the relative specific thrust rises by a wide margin from 0.39 to 0.91. However, the rising amplitude of the relative specific thrust drops sharply when the air gap width continues to increase. Therefore, based on the design principle to minimize the size of the linear motor under the sufficient motor force, Cases A3 and A4 are considered to be two ideal choices. However, as shown in figure 4(c), there exists slightly uneven distribution of the magnetic flux density in Case A3, and thus Case A4 is the best choice, and 7 mm is the optimum width of air gap, which is the same as the initial setup.

3.2. Thickness of air gap

The thickness of air gap t_g also needs to be optimized. Table 3 gives the dimensional parameters in Cases B1–B6. Only the air gap thickness is changed on the condition that other key parameters are constant. The working condition of B4 is exactly the same as A4.

Table 3. Parameter settings in Cases B1–B6

Case number	w_g (mm)	t_g (mm)	w_p (mm)	w_m (mm)	r_{om} (mm)	r_{op} (mm)	r_{im} (mm)	r_{ip} (mm)
B1	7	0.6	21.6	9.3	19	25.7	5	19.6
B2	7	1.2	21.6	9.3	19	26.3	5	20.2
B3	7	1.8	21.6	9.3	19	26.9	5	20.8
B4	7	2.4	21.6	9.3	19	27.5	5	21.4
B5	7	3.0	21.6	9.3	19	28.1	5	22.0
B6	7	3.6	21.6	9.3	19	28.7	5	22.6

Figure 7(a) –7(f) shows the distributions of the internal magnetic flux density inside the air gap in Cases B1–B6. The magnetic flux density also shows a trend of increasing outward from inside. However, it is worth noting that, for the two small thicknesses in Cases B1 and B2, the distribution of the flux density is seriously uneven, especially when the air gap thickness is only 0.6 mm. In addition, a small number of spots appear in Case B6. As a result, when the air gap is too large or too small, there may be uneven distributions of magnetic flux density.

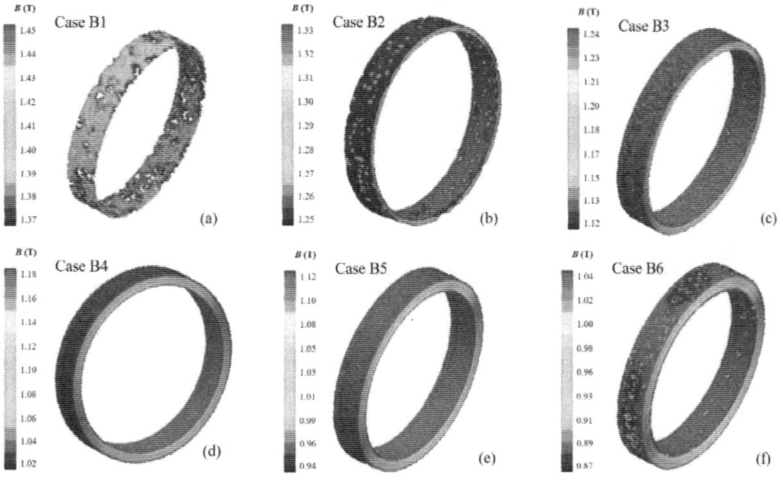

Figure 7. The contours of the simulated magnetic flux density in Cases B1–B6

Figure 8 further compares the distribution of flux density along the radial direction in Cases B2–B6. Due to the seriously uneven distribution in Case B1, it is not within the scope of discussion. As the air gap thickness increases, the magnetic flux density generally decreases uniformly. When the air gap thickness is greater than 2.4 mm, the average magnetic flux density in the air gap range is less than 1.1 T, so the premise of the modified scaling principle is not satisfied. Since the magnetic flux density distribution is uneven in Case B2, it is only necessary to further compare the specific thrust characteristics of Cases B3 and B4. According to the above design, the thickness of each layer of the coil is about 0.53 mm, and when the thickness of air gap is 2.4 mm, four layers of coils can be wound, and 1.8 mm is only enough to wind three layers. It is calculated that the ratio of the specific thrust in Cases B3 to B4 is about 1.16. Therefore, 2.4 mm can be considered as the optimal air gap thickness, which is still consistent with the initial design value.

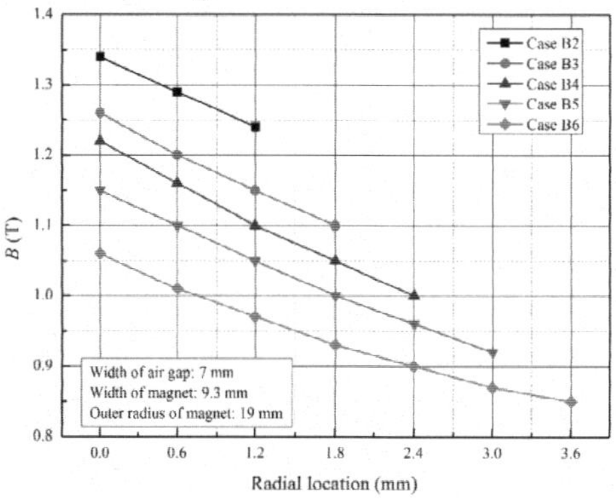

Figure 8. Radial distributions of the average magnetic flux densities in Cases B2–B6

4. Conclusions

In this paper, a three-dimensional magnetostatic field model of the moving-coil linear motor in the space-borne micro compressor is established to optimize magnetic circuits in the structure. The effects of some key dimensions such as air gap width on the distributions of the magnetic flux density inside the air gap are simulated and analysed.

The simulated results show that the magnetic flux density monotonously increases in radial direction from outside to inside. Therefore, the moving coil should be placed closer to the magnetic pole so as to get higher specific thrust. In addition, the magnetic flux density decreases with the increase of both width and thickness of air gap. Moreover, when the air gap is too large or too small, there may be uneven distributions of magnetic flux density.

According to the analysis, it proves that the modified scaling principle in designing the space-borne micro compressor is reasonable. The corresponding experiment will be conducted in the future to validate the simulation results.

References

[1] M. Petach, M. Waterman, G. Pruitt, et al., High frequency coaxial pulse tube microcooler. Cryocoolers 2009; 15: 97–103.

[2] D. Durand, E. Tward, G. Toma, et al., Efficient high capacity space microcooler. Cryocoolers 2014; 18: 59–64.

[3] J. R. Olson, P. Champagne, E. Roth, et al., Microcryocooler for tactical and space applications. Adv. Cryo. Eng. 2014; 59: 357–64.

[4] L. Zhang, H. Dang, J. Tan, et al., Theoretical and experimental investigations on the partial scaling method for the Oxford-type moving-coil linear compressor. Cryogenics 2015; 69: 26–35.

[5] Zhao YB, Dang HZ. CFD simulation of a miniature coaxial Stirling-type pulse tube cryocooler operating at 128 Hz. Cryogenics, Vol.73, pp.53–59, 2016.

Predesign of Nodes of Cable Net for Umbrella Reflectors Based on Bezier Surface

Xin Jin, Xiaofei Ma, Yesen Fan and Qiang Xue

China Academy of Space Technology(Xi'an), Xi'an 710000, China

1124589304@qq.com

Abstract. The umbrella antenna cable net is different from the ring truss antenna. Most umbrella antennas do not use triangular cable nets. In this case, a predesign method based on Bezier surface is proposed for a common umbrella antenna cable net, and the corresponding formula is derived.The relationship between the shape accuracy and the cable net shape is obtained. From the extremum condition of the function. the distribution form of the node with the highest accuracy is obtained. The new cable net structure is parallelogram and triangle alternation. The cable net can meet the requirements of umbrella antennas with different number of antenna ribs. Finally, the accuracy of the surface of the reflector obtained by this method is verified, and the design can meet the requirements of engineering accuracy. This method can quickly calculate the distribution of the reflector nodes meeting the engineering accuracy. This method is a useful attempt to design the reflector of the umbrella cable-net antenna, and it has some inspiration to the design of this kind of antenna.

1.Introduction

The large aperture umbrella satellite antenna consists mainly of the reflector and the expansion mechanism, in which the reflector is composed of several antenna ribs and the cable net attached to them, as shown in Figure 1 [1]. The cable net structure uses tension cable to divide the metal mesh into several units, and connects with the antenna ribs through the nodes. In the prestress state, the ideal paraboloid is fitted by some small planes or surfaces. In the antenna design stage, the fitting error of the cable net structure to the ideal paraboloid is the main factor affecting the accuracy of the antenna. For the umbrella antenna, the triangular cable net can not be used in many cases because of the characteristics of the structure, but the cable net structure shown in Figure 2 is used, which makes the unit surface not be a plane, so the empirical formula [2] that uses the maximum length of the triangular element surface will cause greater error.

Figure 1. Umbrella reflector

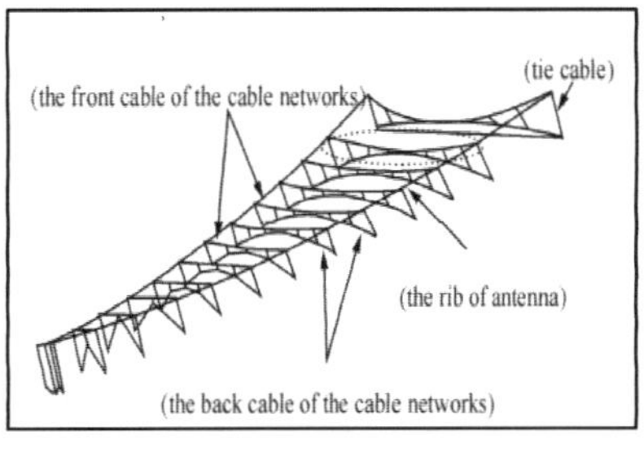

Figure 2. Cable net of an umbrella reflector

At present, the cable network structure predesign shown in Figure 2 is often reversed through experience or finite element method, and there is no convenient method to obtain the length of cable segment which is satifactory for requirement. In this paper, a method based on the double Bezier surface is proposed for this cable net structure, and an empirical formula is obtained, and a cable net form suitable for the umbrella antenna is obtained, which makes the pre design of the cable net possible.

2. Pre design method of cable net of umbrella antennas

The basic principle is to use double one Bezier surface to fit the unit surface, and get the relationship between the root mean square error and the cell size by integration.

2.1 Double one power Bezier surface

In three-dimensional space, there are $(n+1)\times(m+1)$ points $P_{ij}(i=0,1,\dots,n;\ j=0,1,\dots,m)$,we call the $n\times m$ power parametric surface

$$P(u,v)=\sum_{i=0}^{n}\sum_{j=0}^{m}P_{ij}B_{i,n}(u)B_{j,m}(v) \tag{1}$$

the $n\times m$ power Bezier surface.

in which:

$$B_{i,n}(u)=C_n^i u^i(1-u)^{n-i} \tag{2}$$

$$B_{j,m}(v)=C_m^j u^j(1-v)^{m-j} \tag{3}$$

The matrix form of Bezier surface is

$$P(u,v)=\left[B_{0,n}(u),B_{1,n}(u),\cdots,B_{n,n}(u)\right]\begin{bmatrix}P_{00} & P_{01} & \cdots & P_{0m}\\ P_{10} & P_{11} & \cdots & P_{1m}\\ \vdots & \vdots & \vdots & \vdots\\ P_{n0} & P_{n1} & \cdots & P_{nm}\end{bmatrix}\begin{bmatrix}B_{0,m}(v)\\ B_{1,m}(v)\\ \vdots\\ B_{m,m}(v)\end{bmatrix} \tag{4}$$

When $n=m=1$, it is a double one power Bezier surface. If there are $(n+1)\times(m+1)=2\times 2=4$ control points P_{00}, P_{01}, P_{10}, P_{11},that

$$\begin{aligned}P(u,v)&=[1-u\quad u]\begin{bmatrix}P_{00} & P_{01}\\ P_{10} & P_{11}\end{bmatrix}\begin{bmatrix}1-v\\ v\end{bmatrix}\\ &=(1-u)(1-v)P_{00}+u(1-v)P_{10}+(1-u)vP_{01}+uvP_{11}\end{aligned} \tag{5}$$

Its graphical representation is shown in Figure 3. We can prove that it is a piece of curved surface on a hyperbolic paraboloid.When *u*=0 and *u*=1, for them in (5), the two boundaries we get are straight lines; when *v*=0 and *v*=1, for them in (5), the two boundaries are straight lines too. The double one power Bezier surface is surrounded by four straight lines.

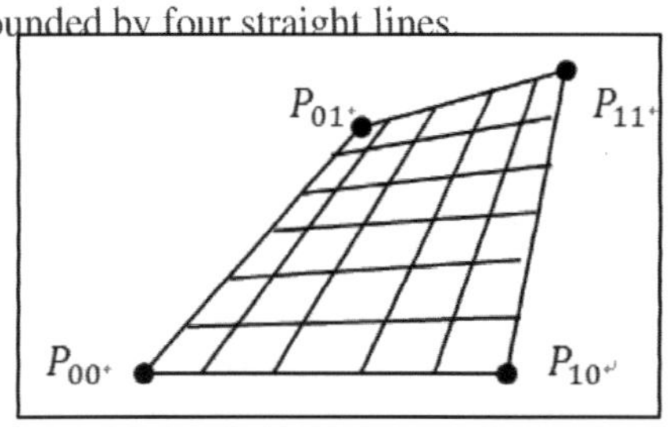

Figure 3. Double one power Bezier surface

The coordinate values of four vertices are substituted into the equation, and the parametric equation of the surface is obtained.

$$\begin{cases} x = w - uw \\ y = u - uw \\ z = uw \end{cases} \tag{6}$$

The surface equation can be obtained by eliminating the parameters.

2.2 RMS error analysis

Figure 2 shows that the four nodes of the cable net structure are taken as one unit, and the projection in the caliber plane is a trapezium. The surface on ideal paraboloid controlled by four node is a space quadrilateral, which satisfies the double Bezier surface. Establish a coordinate system on the projection plane of paraboloid, as shown in Figure 4.

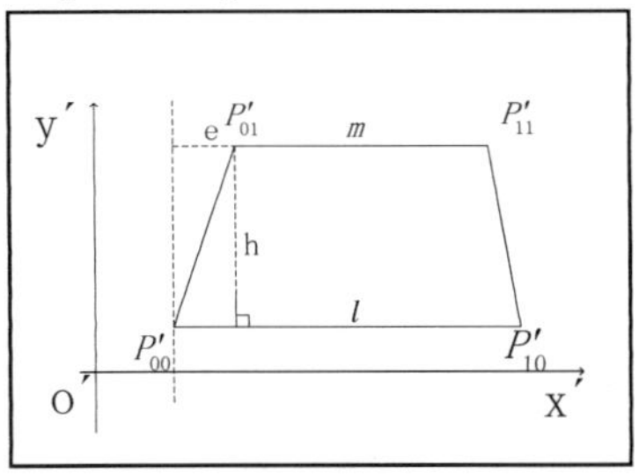

Figure 4. Cable net unit on $x'o'y'$ plane projection

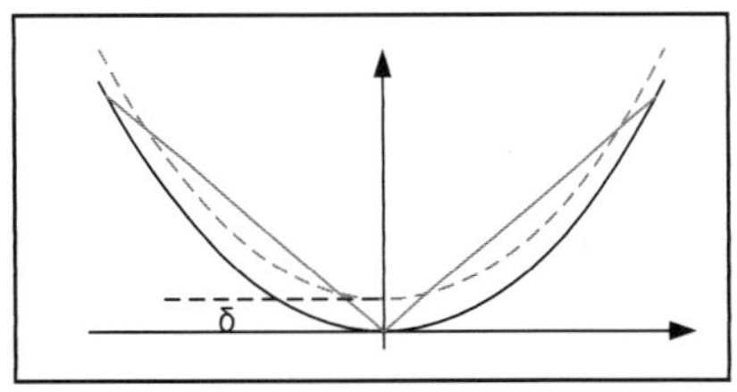

Figure 5. Paraboloid fitting

The projective quadrilateral is trapezoid, and the vertex coordinates are as follows:

$$\begin{aligned} &P'_{00}(x_0, y_0) \\ &P'_{01}(x_0 + e, y_0 + h) \\ &P'_{10}(x_0 + l, y_0) \\ &P'_{11}(x_0 + e + m, y_0 + h) \end{aligned} \tag{7}$$

The linear equation of the line $P'_{00}P'_{01}$ is $x = k_1 y + b_1$, and we substitute P'_{00}, P'_{01} into the straight line equation, get the following expression.

$$\begin{aligned} k_1 &= \frac{e}{h} \\ b_1 &= -\frac{ey_0 - hx_0}{h} \end{aligned} \tag{8}$$

The linear equation of the line $P'_{10}P'_{11}$ is $x = k_2 y + b_2$, and we substitute P'_{10}, P'_{11} into the straight line equation, get the following expression.

$$\begin{aligned} k_2 &= \frac{e - l + m}{h} \\ b_1 &= -\frac{ey_0 - hl - hx_0 - ly_0 + my_0}{h} \end{aligned} \tag{9}$$

Since the four vertices of the surface are on the paraboloid, the surface is on one side of the paraboloid. In order to further reduce the error, refer to the present method of triangle plane fitting[3], the surface is moved along the z axis as a whole, such as Figure 5, the displacement is δ. Therefore, the coordinates of P_{00}, P_{01}, P_{10}, P_{11} are:

$$\begin{aligned}&P_{00}(x_0, y_0, \frac{x_0^2+y_0^2}{4f}+\delta)\\&P_{01}(x_0+e, y_0+h, \frac{(x_0+e)^2+(y_0+h)^2}{4f}+\delta)\\&P_{10}(x_0+l, y_0, \frac{(x_0+l)^2+y_0^2}{4f}+\delta)\\&P_{11}(x_0+e+m, y_0+h, \frac{(x_0+e+m)^2+(y_0+h)^2}{4f}+\delta)\end{aligned} \tag{10}$$

We replace the four points above (5) and get the following:

$$(1-u)(1-v)\begin{bmatrix} x_0 \\ y_0 \\ \frac{x_0^2+y_0^2}{4f}+\delta \end{bmatrix}+u(1-v)\begin{bmatrix} x_0+l \\ y_0 \\ \frac{(x_0+l)^2+y_0^2}{4f}+\delta \end{bmatrix}$$
$$+(1-u)v\begin{bmatrix} x_0+e \\ y_0+h \\ \frac{(x_0+e)^2+(y_0+h)^2}{4f}+\delta \end{bmatrix}+uv\begin{bmatrix} x_0+e+m \\ y_0+h \\ \frac{(x_0+e+m)^2+(y_0+h)^2}{4f}+\delta \end{bmatrix}=\begin{bmatrix} x \\ y \\ z \end{bmatrix} \tag{11}$$

The parametric equation of the surface is obtained as following:

$$\begin{cases} x = x(u,v) \\ y = y(u,v) \\ z = z(u,v) \end{cases} \tag{12}$$

We eliminate the parameter to get the surface equation $z = z(x, y)$.

The root mean square error of the surface[4] relative to the ideal paraboloid is:

$$F(\delta, m, l, h, e) = \delta_{rms}^2 = \frac{2}{(m+l)h}\int_{y_0}^{y_0+h}\int_{k_1y+b_1}^{k_2y+b_2}\left(z-\frac{x^2+y^2}{4f}\right)^2 \tag{13}$$

We let $\frac{\partial F}{\partial \delta} = 0$,and get the following expression:

$$\delta = -\frac{e^2-el+em+h^2+l^2-lm+m^2}{24f} \tag{14}$$

Replace (13) into (12), and we get $F_{min1}(m, l, h, e)$. We let $\frac{\partial F_{min1}}{\partial e} = 0$,and get

$$\begin{cases} e_1 = \frac{l-m}{2} \\ e_2 = \frac{l-m}{2} + \frac{\sqrt{-4h^2-3l^2+6lm-3m^2}}{2} \\ e_3 = \frac{l-m}{2} - \frac{\sqrt{-4h^2-3l^2+6lm-3m^2}}{2} \end{cases} \tag{15}$$

When $e = \frac{l-m}{2}$, F_{min1} gets a minimum of F_{min2}. Through analysis, we can see that the quadrilateral $P_{00}'P_{01}'P_{11}'P_{10}'$ is an isosceles trapezium. In particular, when m=0, the isosceles trapezoidal degenerate into an isosceles triangle.

Considering the geometric form of the umbrella antenna, when $m = \frac{l}{2}$, the distribution of nodes is regular, it is easy to manufacture and install cable nets. At this time, the projection shape of $e = \frac{l}{4}$ cable net at the mouth is shown in Figure 6.

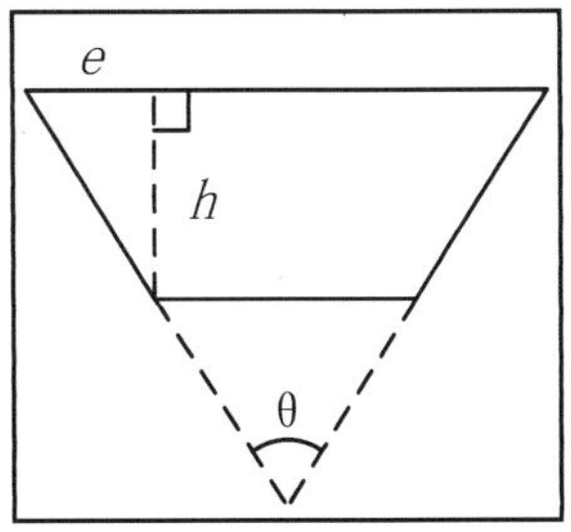

Figure 6. Projection shape

θ is the angle of projection of adjacent antenna ribs at the mouth surface. It is derived from the geometric relation:

$$\frac{e}{h} = tan\left(\frac{\theta}{2}\right) \tag{16}$$

That is

$$h = l/4tan\left(\frac{\theta}{2}\right) \tag{17}$$

The number of equable portions of the antenna ribs is

$$\mathrm{N} = \left[\frac{L}{h/cos\frac{\theta}{2}}\right] = \left[\frac{4Lsin\left(\frac{\theta}{2}\right)}{l}\right] \tag{18}$$

L is the length of the antenna ribs, and [*] is an integral function.

Substituting m, e, h into F_{min2}, we get the minimum mean square error at this time.

$$\delta_{rms}^2 = \frac{l^4(43\cos^2\theta - 84cos\theta + 42)}{184320f^2(cos\theta - 1)^2} \tag{19}$$

That

$$\delta_{rms} = \frac{l^2}{192\sqrt{5}f}\sqrt{42 + \frac{\cos^2\theta}{(\cos^2\theta - 1)^2}} = \frac{l^2}{192\sqrt{5}f}\sqrt{42 + \cot^2\theta} \tag{20}$$

At this time, the displacement is $\delta = -\frac{l^2}{384f}(11 + cot^2\theta)$.

According to this method, the cable net is predesigned. The cable net form is trapezoid and triangle alternating structure, and the distribution of nodes is regular.

3. An example of 2 meter aperture antenna

Suppose the reflector of an umbrella antenna is $D = 2m$, the focal length is $f = 4m$, the number of antenna ribs is 8, and the design error is $\delta_{rms} = 0.3mm$. This method is applied to pre design the cable net form and verify its accuracy.

After calculation, the cable net structure parameters are as follows: $N = 6$; $\delta = -0.61379mm$. The cable network structure is shown in figure 7 and figure 8

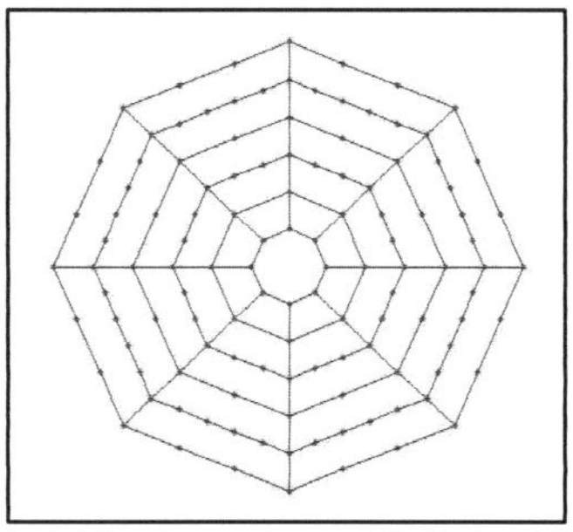

Figure 7. Planform of reflector

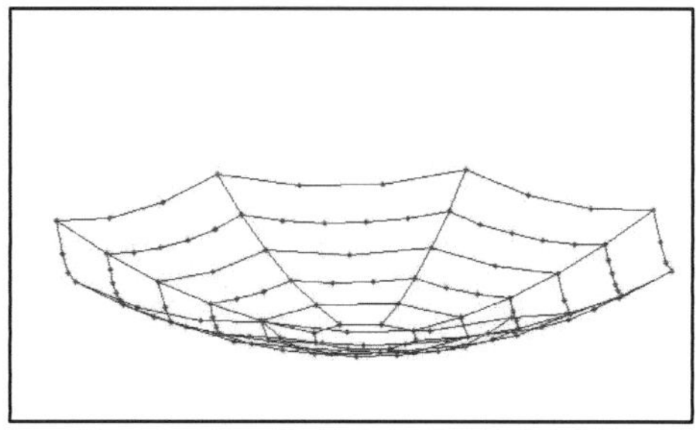

Figure 8. Lateral view of reflector

Due to the symmetry of the reflector, one of the sectors is selected as the research object to explore the accuracy of the reflector surface. First, according to the distribution of nodes, the reflector is divided into triangular and trapezoid shaped surface units. By interpolating, several sampling points are generated in each unit, and the shape surface precision of each unit is calculated approximately, and the shape surface precision of the whole sector is [5]. The concrete formula is as follows:

$$\delta_i = \sqrt{\sum_{j=1}^{m} (z'_{ij} - z_{ij})^2 \Big/ m} \tag{21}$$

$$\delta_{rms} = \sqrt{\sum_{i=1}^{n} A_i \delta_i^2 \Big/ A} \tag{22}$$

In which δ_i is the root mean square error of the i cable net unit, z'_{ij} is the ordinate of the j interpolation sampling point in the i cable net unit. z_{ij} is the ordinate of the corresponding points of z'_{ij} on the ideal paraboloid. m is the number of interpolated post sampling points for the i cable net unit. δ_{rms} is the root mean square error of the sector. A_i is the area of the projection plane of the i unit in the xoy plane. A is the area of the sector projection plane, and n is the number of cable net elements in the sector.

The calculated root mean square error of the reflector is $\delta_{rms} \approx 0.27132\text{mm}$, which meets the design requirement of 0.3mm. The error distribution is shown in Figure 9.

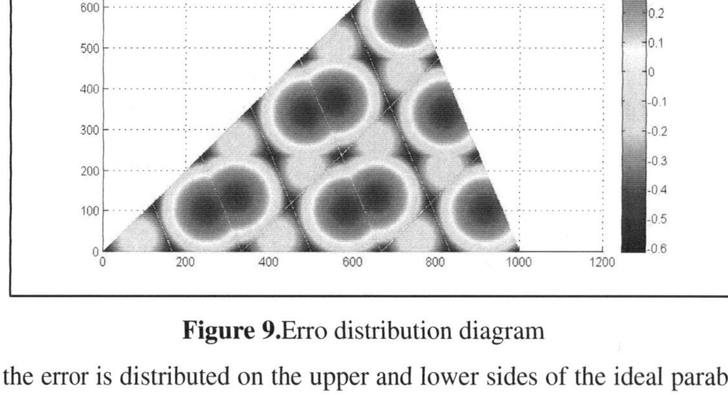

Figure 9. Erro distribution diagram

As we can see, the error is distributed on the upper and lower sides of the ideal parabola. Although the unilateral error is large, the error of the two sides counteract each other, and the overall error satisfies the requirement of the surface precision.

4. Conclusion

The effectiveness of the proposed method is verified by an example. Using this method, the design of cable net node can be completed quickly, and the precision requirement is satisfied. As long as the number of antenna ribs is adjusted, the node distribution that meets the requirement of precision can be distributed. Compared with the traditional triangular cable net structure, it can adapt to the reflector design of different number of antenna ribs, so that the number of nodes on each antenna rib is equal. At the same time, because the node has a certain displacement along the Z axis relative to the ideal paraboloid, it can increase the precision of the surface while the number of nodes is certain. When the precision of the surface is fixed, only less nodes are needed.

Reference

[1] Murphey T. Historical Perspectives on the Development of Deployable Reflectors[C]// Aiaa/asme/asce/ahs/asc Structures, Structural Dynamics, and Materials Conference. 2009.
[2] P.K. Agrawal, M.S. Anderson, M.F. Card, Preliminary design of large reflectors with flat facets, IEEE Trans. Antenn. Propag. 29 (4) (1981) 688–694.
[3] Dongwu Y.Spaceborne large deployable cable antenna structure design and profile adjustment[D].XIdian University,2010. DOI:10.7666/d.y1958605.
[4] Hill F S, Kelley S M. Computer Graphics Using OpenGL (3rd Edition)[M]. Prentice-Hall, Inc. 2007.
[5] YIJ-GN Sui-wei,HU Liang. Research On the Calculating Method for the Principle Error Of Parabolic Cable Antenna [J]. Techniques of Automation and Applications ,2017,36(05):11-14.

Radio Frequency Excited Plasma Discharge Simulation for Potential Helicon Plasma Thruster

Md Mahbubur Rahman[1] **and Ighor Uzhinsky**[2]

Center for Design and Materials Skolkovo Institute of Science & Technology Moscow, Russia

E-mail: mahbubur.rahman@skolkovotech.ru[1], I.Uzhinsky@skoltech.ru[2]

Abstract. The development of advanced propulsion is the key element in the implementation of a robust space exploration program. Advance thruster concepts such as the development of electrodeless Plasma thrusters with high-density helicon plasma sources expected to mitigate the existing problems of the finite lifetimes inherent in electric propulsion. Electrodeless plasma thrusters are potentially more durable than conventional thrusters that use electrodes such as a gridded ion, Hall thrusters, arcjets, and resistojets. The goal of the research is to simulate a compact high-power-density helicon plasma source operating, which is under examination for a potential electric propulsion application. Plasma modeling is performed in COMSOL Multiphysics plasma module. A Nagoya type III antenna is placed on the surround of a dielectric tube and electrically excited at 13.56 MHz. Plasma is formed in the ionization chamber, which contains argon gas at low pressure. The plasma is sustained utilizing electromagnetic induction.

1. Introduction

Electric propulsion is capable of achieving high specific impulse, and the ability to control thrust levels makes it suitable substitute to the conventional chemical thruster for in-space propulsion application. The recent improvement will enable the electric thrusters to be used as a primary propulsion systems for future spacecraft. Variable specific impulse thruster with multiple in-space refuelling capabilities will permit successful asteroid redirect mission [1].

Electric propulsion applies electrical energy to the propellant from an external power source. Electric thrusters divided into two broad categories: use the electricity to heat the propellant, which emerges as a neutral gas, and those use the electric or magnetic field to accelerate ions. Since the beginning of space era, chemical rockets dominated due to the high thrust to mass ratio. However, chemical rockets have limited lifetime.

Electric propulsion has a wide range of thruster based on the mechanism through which electric power is utilized to accelerate the flow. Characteristic specific impulse ranges of different space propulsion is shown in Table 1. Currently, electrothermal, electrostatic, or electromagnetic thrusters are used in space application. Electrothermal thruster's works on the principle of using electricity to heat the propellant at high temperature which creates the driving force by acceleration through the nozzle. Thermal energy of the propellant fluid is transformed to kinetic energy. Electrostatic thrusters make use of a static electric field to produce thrust by accelerating charged plasma. The electromagnetic field can be generated externally or self-induced [2].

Table 1. Specific impulses, thrust efficiency's and thrusts of electric thrusters, deployed and experimental.[3]

Thruster type	$I_{sp}(s)$	Efficiency (η_t)	Thrust
Plused Plasma	240-760	20%-30%	1×10^{-4} Ns
Gridded ion	7650	77%	0.43N
Hall	1600	50%-60%	$(2.5 - 12) \times 10^{-3}N$
Magneto-Plasma-dynamic	3670	38.8%	$\sim 250 \times 10^{-3}N$
Mini helicon	1000-4000	~ 20%	$10 \times 10^{-4}Ns$
Electron cyclotron resonance	429 (Xe)	3.5%	$0.86 \times 10^{-3}N$
Electrodeless Lorentz force (ELF)	1000-6000	50%	1.0 N(average)
Hybrid (VASIMR)	3400	56%	$3.6 \pm 0.2N$

Until now, all operating electric thruster uses electrodeless to ionize the propellant, for acceleration of the ions and electrons. The increasing of input power for ionization causes erosion, mainly by sputtering degrades lifetime limit on such devices. To address the point, the development of electrodeless thruster, such use electric or magnetic body forces to ionized gases to produce a propellant stream with directional velocity [3]. In the development of electrodeless plasma thruster, Helicon plasma sources are useful in various condition. They can create high density $\sim 10^{13}cm^{-3}$ helicon plasma with a spacious range of operating parameters. The experimental and theoretical approach of helicon plasma production and its acceleration schemes with described and characterized [4].

2. Helicon Plasma thruster

To realize a long lifetime of an electric propulsion system, we have been investigating various electrodeless electric propulsion concepts utilizing a helicon plasma source. Several research groups are currently working on the development of helicon plasma thruster (HPT). The study focuses on the implementation of the permanent magnet, optimizing magnets for required magnetic flux, operating power range, antenna optimization.

Mini helicon experiment (mHTX) at MIT, Helicon Plasma Hydrazine Combined Micro (HPHCOM) by ESA and Helicon Double Layer Thruster (HDLT) at Australian National University are potential HPT prototypes. The thruster performance of these various technology demonstration prototypes are presented in Table 2 and described below.

HPHCOM prototype operates at nominal power 700 W, magnetic field strength 200 G, novel resonant Antenna called S-helicon operates at frequency 13.25 MHz, managing power of the antenna is optimized to a very low range below 100 W. Permanent magnets were proposed with the different arrangement, the magnetic field strength range 400-1100 G.

HDTL uses electromagnets with a magnetic strength 100-200 G with operating range 200-800 W. The RF antenna has an operating frequency of 13.25 MHz. Experimental result detects a steady state current, expected thrust up to 6 mN and specific impulse 800 S using argon propellant.

Mini helicon thruster experiment (mHTX) projected a collimated plasma plume with a single magnet configuration, which is potential for electric propulsion application. Different measurement technique proved, mini helicon produces real thrust operating on atomic Ar and Molecular N_2. In comparing to HDTL its highly ionized propellants with fuel utilization efficiencies of 90%.Single magnet configuration is used, gas flows along the dielectric quartz tube with an antenna embracing it. The nominal operating power in the range 700-1000 W.

The antenna operating frequency is 13.25 MHz. Experimental results of mHTX presented 12% efficiency with 2000 S specific impulse and providing thrust up to 20 mN. Varying plasma beam velocity using different means of controls made the helicon most suitable for variable specific impulse [6].

Table 2.HPT main prototypes: summary of propulsive performances [5].

Prototype	Power	Thrust	$I_{sp}(s)$	$\eta(\%)$	$\eta_u(\%)$
HDLT	< 1.5kW	6mN	800	25-35	1-3
mHTX	700-1500W	20mN	2000	90	10-20
HPHCOM	50W	1.5mN	1200	90	13
PMEP	700W	3mN	500	< 50	1
HPHT	20-50kW	1-2N	> 2000	~ 100	55

The Variable Specific Impulse Magnetoplasma Rocket (VASIMR) is a hybrid thruster. Currently, under-development by Ad Astra Rocket Co. It is not exactly considered as a helicon plasma thruster, but it uses helicon plasma sources to produce plasma. Ion cyclotron resonance energizes the plasma and accelerates through the magnetic nozzle, in contrast, produces thrust. The nominal power of operation 200 kW, with a thrust efficiency, reaches near 50%, able to produce 3 N thrust. The RF operating powers estimated ~ 30 kW. Sizeable magnetic field >1T is necessary, which can only be achieved by the superconducting magnet [5]. Thermal behavior is also an important issue. High-temperature effects the optimal operations of the magnets, solar radiation effect of the different orbit is explained [7].

3. Helicon thruster regime of operation

Helicon plasma thruster prototypes are characterized by the RF power range and the magnetic field strength in which thruster operates. The design incorporates a wide range of controlled variables; Ionization Chamber length L, mass flow m, RF power P_{RF}, magnetic field strength B, RF frequency ω_{RF}. The prime design parameters define the performance of thrusters are ionization chamber dimension, input mass flow, input RF power and frequency. The antenna shape and location, the magnetic field strength. These design parameters interact in an insignificant way and influence outright processes in the thruster, from ionization to internal plasma losses and external expansion.

3.1. Ionization chamber

Efficient ionization from the neutral gas into plasma occurs in the ionization chamber. The neutral gas propellant flows along the Dielectric quartz tube; the RF antenna generates electromagnetic waves propagate and energizes the electrons, ionized the gas. During the ionization process, major loss mechanism at a low electron temperature below 5-7 eV is excitation, de-excitation or permanence as a meta-stable state.

To achieve effective ionization requires high electron temperature (>10 eV) and mean free path smaller than the chamber length $\lambda_{ion} << L$. The dimension of the chamber has to comply so that the electromagnetic wave mode can propagate in it.

Electron Larmor radius required less then quartz tube radius $l_e << R$ for better electron magnetization for low plasma losses to the wall. Dielectric quartz tube has high heat resistance, and low thermal expansion as the electrons and ions reaching the wall and constantly deposit energy and tube is exposed to constant heat flux. The strong magnetic field is required for the plasma confinement. The optimal position for magnet, antenna, and the tube is required for efficient thruster operation [8].

3.2. RF Antenna

The helicon antenna generates electromagnetic wave decomposite power to the plasma. Helicon waves are known as whistler waves, a low-frequency wave, where ω is below plasma frequency ω_p, in between lower hybrid frequency ω_{LH} and the electron cyclotron frequency ω_c, [9]. Helicon discharge has high ionization efficiency. With lower RF power 1-2 kW relatively fully ionized plasma $\eta \geq 10^{14}$ cm^{-3} can achieve[9].The applied current to the antenna and waves can be reflected by the plasma, partially be radiated to free space and re-absorbed by the antenna as reactive power. The high plasma resistance and power factor are the key factors for the optimal design of antenna [8].

3.3. Superconducting Magnet

Magnetic field generator is significant components of Helicon plasma thruster. The helicon plasma thruster design mainly incorporates radio frequency plasma discharge in the presence of a magnetic field. Optimal design of the electromagnets to create requisite magnetic field is one of the key factors that define the performance and efficiency of the thruster. Magnetic field allows the electromagnetic wave to propagate through the plasma, reducing the plasma losses to walls and creates a magnetic nozzle for the supersonic plasma expansion.

4. Regime of Operation

The schematic process of HPT thruster shown in figure 1.

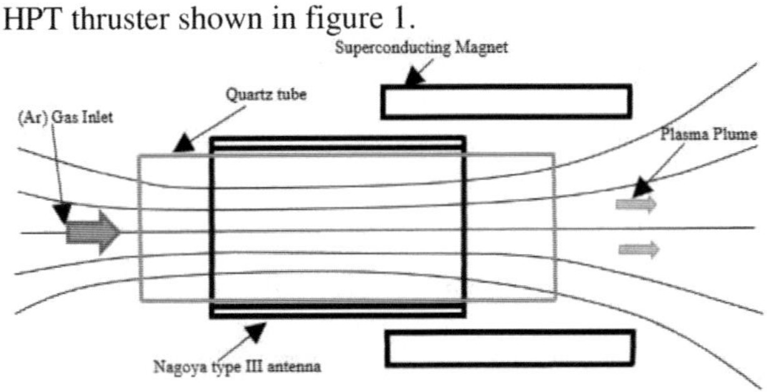

Figure 1. Sketch of mini-helicon thruster operation.

4.1. Design Constraints

Nominal values of plasma thruster for simulation is presented in Table 3.

Table 3. Nominal values of the plasma thruster prototype.

Ionization chamber dimensions [mm]	Length: 100; Inner radius: 23
Power input at RF generator [W]	1000
Magnetic field in chamber [G]	350
Antenna RF frequency [MHz]	13.56
Antenna geometry [mm]	Nagoya type III antenna, 50 mm long

5. Simulation COMSOL Multiphysics

The helicon plasma thruster design mainly incorporates radio frequency plasma discharge in the presence of a magnetic field. Radio Frequency plasma discharge was carried out using, COMSOL Multiphysics a cross-platform FEA multiphysics software. Plasma Module simulates low-temperature plasma sources, systems and discharges. Radio Frequency module is used to simulate antenna electromagnetic wave propagation, and the AC/DC module is used to simulate the magnetic field configuration. The simulation process is described below.

5.1. 2D Geometry

The 2D cross section of the thruster used in the simulation process is shown in figure 2.

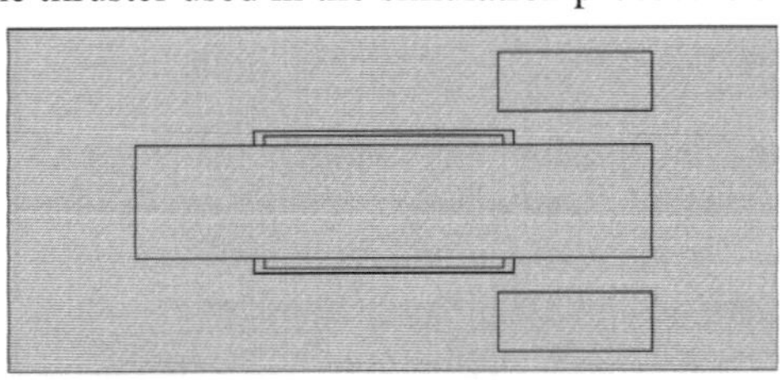

Figure 2. 2D Sketch of mini-helicon thruster simulation

5.2. *Simulation Steps*

5.2.1. Plasma Module: COMSOL solves for drift diffusion equation for the electron density and energy density. The Mathematical model of COMSOL plasma Simulation is explained in mathematical modeling section.

5.2.2. Magnetic field: The superconducting magnet configuration composed of Solenoids coil which surrounds the ionization chamber chosen for the required magnetic field strength. The solenoids coils are placed accordingly, so it limits interactions with the RF antenna.

5.2.3. Electromagnetic Wave Guide: A Nagoya type III antenna used for the simulation. The antenna is placed outside of the dielectric quartz tube. The length of the antenna is 100 mm and designed for the Argon Plasma. The RF antenna emits the 13.56 MHz frequency, and nominal power is about 1000 W.

5.3. *Mathematical modeling*

Mathematical model used in COMSOL Multiphysics, Frequency -transient study is described below [10].

Wave heated plasma discharge mostly higher number density. The electron density and mean electron density is computed by solving drift- diffusion equation.

Electron number density:

$$\frac{\partial(n_e)}{\partial t} + \nabla \cdot \Gamma e = R_e \tag{1}$$

$$\Gamma e = -(\mu_e \cdot E)n_e - D_e \cdot \nabla n_e \tag{2}$$

n_e Denotes the electron density, $(1/m^3)$, R_e is the electron rate expression $(1/(m^3 \cdot s))$, μ_e is the electron mobility ,($m^2/(V \cdot s)$), E is the electric field (V/m), and D_e is the electron diffusivity.

$$\frac{\partial(n_e)}{\partial t} + \nabla \cdot -n_e\,(\mu_e \cdot E) - D_e \cdot \nabla n_e = R_e \tag{3}$$

Electron energy density:

$$\frac{\partial(n_e)}{\partial t} + \nabla \cdot \Gamma_\varepsilon + E \cdot \Gamma_e = R_\varepsilon \tag{4}$$

$$\Gamma_\varepsilon = -(\mu_\varepsilon \cdot E)\, n_\varepsilon - D_\varepsilon \cdot \nabla n_\varepsilon \tag{5}$$

R_ε is the energy loss/gain because of inelastic collisions $(V/m^3 \cdot s)$, μ_ε is the electron energy mobility $(m^2/V \cdot s)$, E is the electric field (V/m), and D_ε is the electron energy diffusivity (m^2/s).

$$\frac{\partial(n_\varepsilon)}{\partial t} + \nabla \cdot -n_\varepsilon\,(\mu_\varepsilon \cdot E) - D_\varepsilon \cdot \nabla n_\varepsilon + E \cdot \Gamma_e = R_\varepsilon \tag{6}$$

Electrostatic field is computed,

$$-\nabla \cdot \varepsilon_o \varepsilon_r \nabla V = \rho \tag{7}$$

High frequency electric field is computed in the frequency domain,

$$\nabla \times (\mu_r^{-1} \nabla \times E) - k_o^2(\varepsilon_r - \frac{j\sigma}{\omega \varepsilon_0})\mathrm{E} = 0 \tag{8}$$

Where plasma current density and the electric field are more complicated,

$$\sigma^{-1} \cdot J = E \tag{9}$$

Plasma coupling between electromagnetic wave and magnetic field. σ is plasma conductivity tensor, function of electron density collision frequency and magnetic flux.

$$\alpha = \frac{q}{m_e(\vartheta_e + j\omega)}, \beta = n_e q \alpha \quad (10)$$

Where, q= Electron charge, m_e= Electron mass, n_e= Collision frequency and ω is angular frequency of electromagnetic field.

5.4. Cross Section Data Argon (Ar)

Plasma consists of three species ions, electrons and mostly neutral atoms and molecules. The density of electrons and ions are a small fraction of the neutral atoms and molecules, exceptions in earth-based plasma nuclear fusion where it's fully ionized. Plasma used in the application of industrial and commercial are not necessarily entirely ionized. Because of the high density of neutral atoms and molecules collisions between them and the electrons and ions become significant. The central parameter of the collisions process is the cross-section. Electrons collide with the gas particles or ions, several reactions are possible. The electron can ionize the gas, impart momentum to it, could excite the energy levels of the electrons around the molecules and ions, and could recombine with the ions and many more. In all of this stochastic process, electrons lose energy. The cross-section data for neutral gas Ar is presented in the Table 4, Cross section is denoted as the area available for electrons for the collision to happen. Cross section data permits the rate coefficient for a particular reaction to compute using Electron Energy Distribution Function (EEDF).

Table 4. Argon Collision and reactions modeled [10].

Reaction	Formula	Type	Δ(eV)
1	e+Ar=e+Ar	elastic	0
2	e+Ar=e+Ars	excitation	11.5
3	e+Ars=e+Ar	super-elastic	-11.5
4	e+Ar=2e+Ar+	ionization	15.8
5	e+Ar=2e+Ar+	ionization	4.24
6	Ars+Ars=e+Ar+Ar+	Penning ionization	-
7	Ars+Ar=Ar+Ar	metastable quenching	-
-	Ars=Ar	deexciation	-
9	e+Ar+=Ar	recombination	-

6. Result and discussion

COMSOL solves for drift diffusion equation for the electron density and energy density. Electron density distribution is shown in the figure 3 along the mid cross section towards the length of the ionization chamber.

IOP Conf. Series: Materials Science and Engineering **449** (2018) 012032 doi:10.1088/1757-899X/449/1/012032

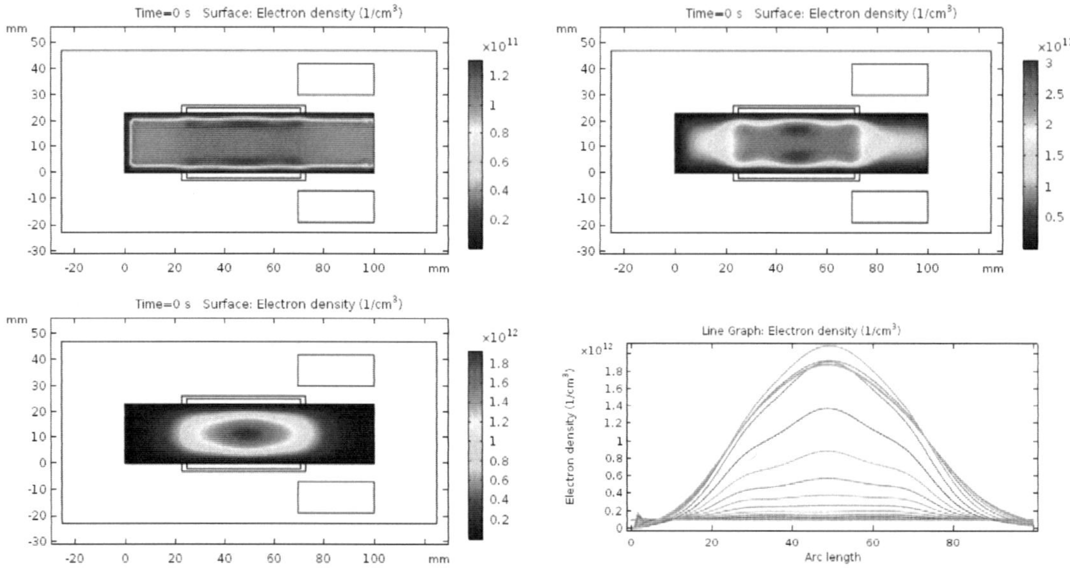

Figure 3. Electron energy density distribution along the length of ionization chamber.

The maximum electron density is in the plasma column and the value is about $1.8 \times 10^{12} cm^{-3}$. Nagoya III type helicon antenna is used and placed in the middle embracing quartz tube, ionization chamber. The critical factor of RF driven Plasma engine lies in the antenna, it energies the neutral flow of initial neutral gas into a magnetized plasma with respectively high ionization density. The plasma discharge initiates near to the antenna and expands towards the center position of the ionization chamber.

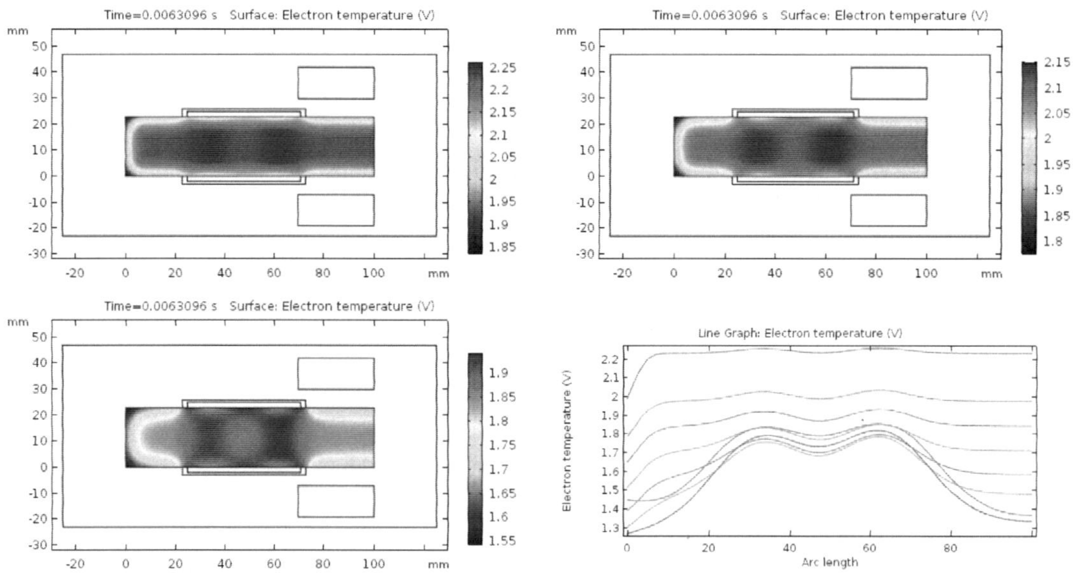

Figure 4. Electron Temperature distribution along the length of ionization chamber.

The electron temperature distribution is illustrated in the figure 4.Maximum electron temperature reaches 2.25 eV.In the presence of a magnetic field, electrons get magnetized and having cyclotron gyration collides with neutral atoms. In the process, electrons lose their energy because of atoms collisions and lose the ability to reach higher plasma temperature. Electron potential distribution presented in the figure 5.

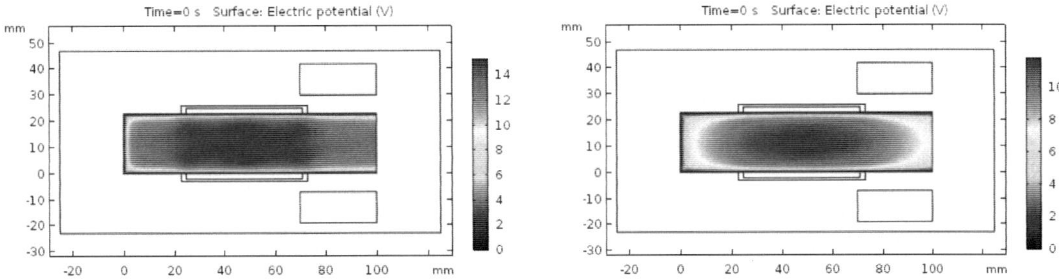

Figure 5. Electron potential distribution along the length of ionization chamber.

Nomenclature

λ_{ion}	Ion Mean free path
ω_{RF}	Radio frequency
B	Magnetic flux density vector
E	Electric field vector
EM	Electromagnetic wave
P_{RF}	Radio frequency power
HPT	Helicon plasma thruster
L	Ionization chamber length

References

[1] R. Gradinia, A. Golkar, I. Uzhinsky, and A. Ivanov, "Asteroid redirect mission concept using a variable isp plasma propulsion system," *68th International Astronautical Congress (IAC)*, 2017.

[2] M. Manente, M. Walker, J. Carlsson, C. Bramanti, S. Rocca, D. Currelia, Y. Guclu, and D. Pavarin, "Feasibility study of low power helicon thruster," pp. 1–14, 2008. [Online]. Available: http://www.esa.int/gsp/ACT/doc/PRO/ACT-RPR-PRO-2008-ISPCFeasibility-study-low-power-helicon-thruster.pdf

[3] S. Bathgate, M. Bilek, and D. Mckenzie, "Electrodeless plasma thrusters for spacecraft: a review," *Plasma Science and Technology*, vol. 19, no. 8, p. 083001, 2017.

[4] S. Shinohara, H. Nishida, T. Tanikawa, T. Hada, I. Funaki, and K. P. Shamrai, "Development of electrodeless plasma thrusters with high-density helicon plasma sources," *IEEE Transactions on Plasma Science*, vol. 42, no. 5, pp. 1245–1254, 2014.

[5] J. Navarro-Cavall, E. Ahedo, M. Merino, V. Gomez, and M. Ruiz, "Helicon Plasma Thrusters: prototypes and advances on modeling," *33rd International Electric Propulsion Conference, IEPC-2013-285*, pp. 1–11, 2013.

[6] O. V. Batishchev, "Minihelicon plasma thruster," *IEEE Transactions on plasma science*, vol. 37, no. 8, pp. 1563–1571, 2009.

[7] M. M. Rahman, D. Shankar, and S. Santra, "Analysis of radiation environment and its effect on spacecraft in different orbits," *68th International Astronautical Congress (IAC)*, 2017.

[8] M. Ruiz and G. Victor, "Design and development of a 1 kW-class helicon antenna thruster," pp. 1–12, 2015.

[9] F. F. Chen, "Physics of helicon discharges," *Physics of Plasmas*, vol. 3, no. 5, pp. 1783–1793, 1996. [Online]. Available: http://aip.scitation.org/doi/10.1063/1.871697

[10] C. Multiphysics, "Comsol Multiphysics Plasma simulation," pp. 1–12.

AUTHOR INDEX